An Army like No Other

An Army like No Other

How the Israel Defense Forces Made a Nation

Haim Bresheeth-Žabner

London • New York

First published by Verso 2020

1 3 5 7 9 10 8 6 4 2

Verso
UK: 6 Meard Street, London W1F 0EG
US: 20 Jay Street, Suite 1010, Brooklyn, NY 11201
versobooks.com

Verso is the imprint of New Left Books

ISBN-13: 978-1-78873-784-5
ISBN-13: 978-1-78873-785-2 (UK EBK)
ISBN-13: 978-1-78873-786-9 (US EBK)

British Library Cataloguing in Publication Data
A catalogue record for this book is available from the British Library

Library of Congress Cataloging-in-Publication Data

Names: Bresheeth-Zabner, Haim, 1946– author.
Title: An Army like no other : how the Israel Defense Forces made a nation / Haim Bresheeth-Zabner.
Description: London ; New York : Verso, 2020. | Includes bibliographical references and index. | Summary: "To understand Israel, you must first understand its army David Ben-Gurion, founder and first prime minister of the State of Israel, established that force in 1948 convinced that the 'whole nation is the army'. To his mind, the Israel Defense Forces (IDF) would be an army like no other—not merely an instrument of military might but the institution that would transform a diverse, mostly immigrant population into a nation. Haim Bresheeth-Zabner, a writer, academic and Israeli peace activist, charts the evolution of the IDF from the days of the Nakba, through conflicts with Egypt, Syria, Lebanon, Jordan, and Iraq, to its ongoing assaults on the vulnerable Palestinian population, culminating in the continuing attacks on Gaza. As he shows, the IDF has been the largest, richest and most influential institution in Israel's Jewish society and the Alma Mater of its social, economic, and political ruling class. Today, it is embedded in all aspects of daily life and identity in Israel, including the vast military-industrial complex that drives the national economy. An Army like No Other is an unflinching exploration of the war-driven origins of the Israeli state as well as the on-going story of Israeli militarism."—Provided by publisher.
Identifiers: LCCN 2019054518 (print) | LCCN 2019054519 (ebook) | ISBN 9781788737845 (hardback) | ISBN 9781788737869 (ebk) | ISBN 9781788737852 (ebk)
Subjects: LCSH: Israel. Tseva haganah le-Yiśra'el—History—20th Century.
Classification: LCC UA853.I8 B75 2020 (print) | LCC UA853.I8 (ebook) | DDC 355.0095694—dc23
LC record available at https://lccn.loc.gov/2019054518
LC ebook record available at https://lccn.loc.gov/2019054519

Typeset in Sabon by MJ & N Gavan, Truro, Cornwall
Printed and bound by CPI Group (UK) Ltd, Croydon CR0 4YY

Dedicated to the memory of my late parents,
Arie (Leib) Žabner and Machla Žabner-Zilberberg,
and to my beloved Yosefa Loshitzky

The most frustrating thing about history, however, is that so much in it escapes language, escapes attention and memory altogether.

Edward Said

Contents

Acknowledgements

This book would not have been possible without the tireless and inspiring work over decades of many researchers on a range of Middle Eastern subjects and related topics. I am especially indebted to the work of the following: Gilbert Achcar, Hannah Arendt, Nasser Arruri, Benjamin Beit-Hallahmi, Uri Ben-Eliezer, Sami Shalom Chetrit, Erskine Childers, Norman Finkelstein, Neve Gordon, Isabella Ginor, Adam Hanieh, Jeff Halper, Shir Hever, Ghada Karmi, Martin Kemp, Rashid Khalidi, Yizhak Laor, Ronit Lentin, Yosefa Loshitzky, Moshe Machover, Dina Matar, Nur Masalha, Benny Morris, Ilan Pappe, Uri Ram, Sara Roy, Maxime Rodinson, Yehuda Shenhav, Shlomo Sand, Tom Segev, Ella Shohat, Avi Shlaim, Shabtai Teveth, Zeev Sternhell, Jacobo Timerman, Patrick Wolfe, Nira Yuval-Davis, and Idit Zertal.

As this book was completed, I was able to see the new Hebrew volume by Uri Ben-Eliezer, dealing with many similar themes, even if from a different perspective. I regret that it was too late to refer to it in this work, though I make much use of his earlier texts.

I would especially like to thank to Avi Shlaim, who first suggested that I develop my ideas into a book after listening to a lecture I gave; to Nur Masalha, who helped me to conceive the structure of the argument; to Ronit Lentin, who has spared no effort in advising me; and most of all to my partner, Yosefa Loshitzky, who has had to live with this book and its arguments for as long as I have, and whose careful assistance, astute advice, and thoughtful support improved it and made it possible.

At Verso, I am in particular indebted to Tariq Ali, who appreciated the concept, and to Leo Hollis, whose careful work has helped make this book concise and reader-friendly; his sense of humor and great sensitivity helped to make the process enjoyable. The process of copyediting, managed by Mark Martin and the content editor Ida Audeh, has been crucial—Ida has used her Palestinian background to clarify and crystalize the text, and I am greatly indebted to her, for her knowledge and sensitivity.

The librarians at SOAS and Senate House have been most helpful as much of the work depended on Hebrew texts and I wish to thank them for their kind assistance.

Finally, it would be remiss of me if I failed to mention certain events that affected the production process of this volume. During the early part of the work with the Verso senior editor Leo Hollis, my desktop computer was professionally hacked, making all copies of the book files, both on the hard disc and on connected USB memory stick, unusable, as well as rendering the computer inoperable.

Fortunately—I thought—there was a single backup copy on my cloud server, and on superficial examination the text looked unaltered. Unbeknownst to me, a hostile and malicious modification was applied to this master file, with a view to discrediting the book after its publication. By sheer luck, during the final proofreading, I decided to check a range of footnotes, to be on the safe side. What started as a mere safety check became a full-scale examination and verification of *all* the footnotes. I have discovered that while I was working on the reconstitution of the text after the original hacking someone had affected a sophisticated modification of the file, without changing the main text; almost a hundred footnotes were tampered with. Wrong page numbers were implanted, and false book and journal titles, some invented, replaced the original ones. I had to check and correct more than 800 references.

This work, which took six weeks, could not have been completed without the professional support and devotion of Mark Martin and Leo Hollis at Verso, and the support of the SOAS librarians Bob Burns, who assiduously helped to identify some of the false titles and find the real ones, and Carl Newell, who assisted in finding the last of the missing sources—five books that were physically misplaced on the shelves or hidden behind other volumes, or even placed on the floor between shelves, making them inaccessible. To those four professionals I wish to extend my hearty thanks for enabling the timely completion of the book, which would be impossible otherwise.

A word to the wise—this is not the first instance of such hostile intervention in academic publishing, and it is not going to be the last, clearly. Extreme care is required of those publishing critical material on Israel. The informed reader shall have no difficulty in identifying the likely culprit—the one with the motive, ability and resources to effect such damage. On this occasion, the malicious efforts have failed.

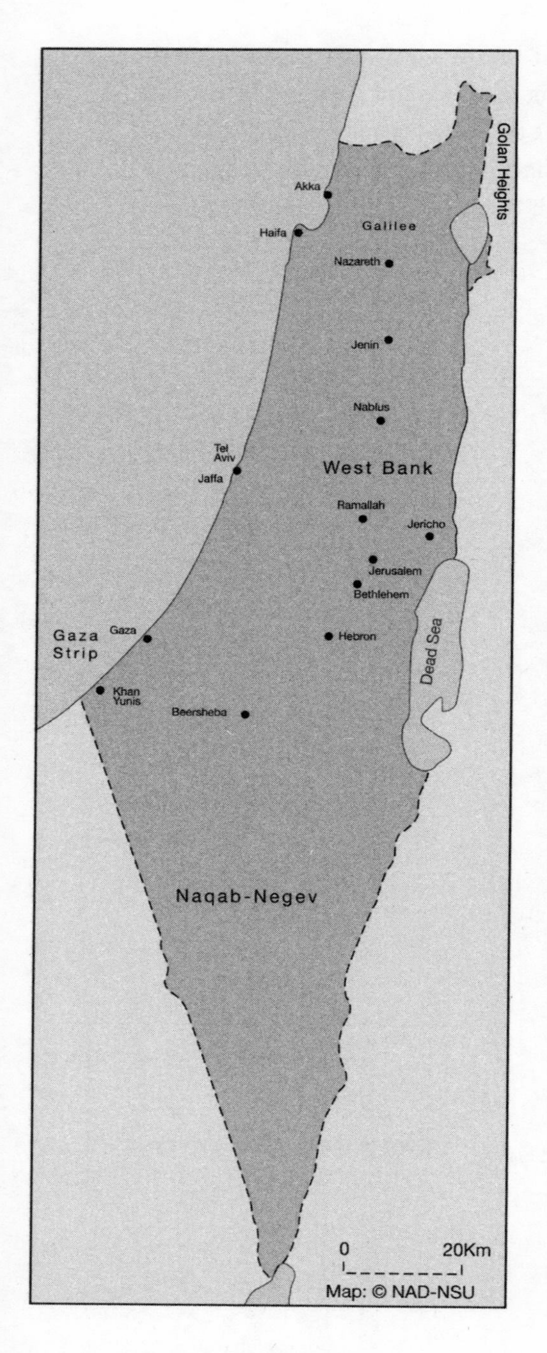

Mandatory Palestine

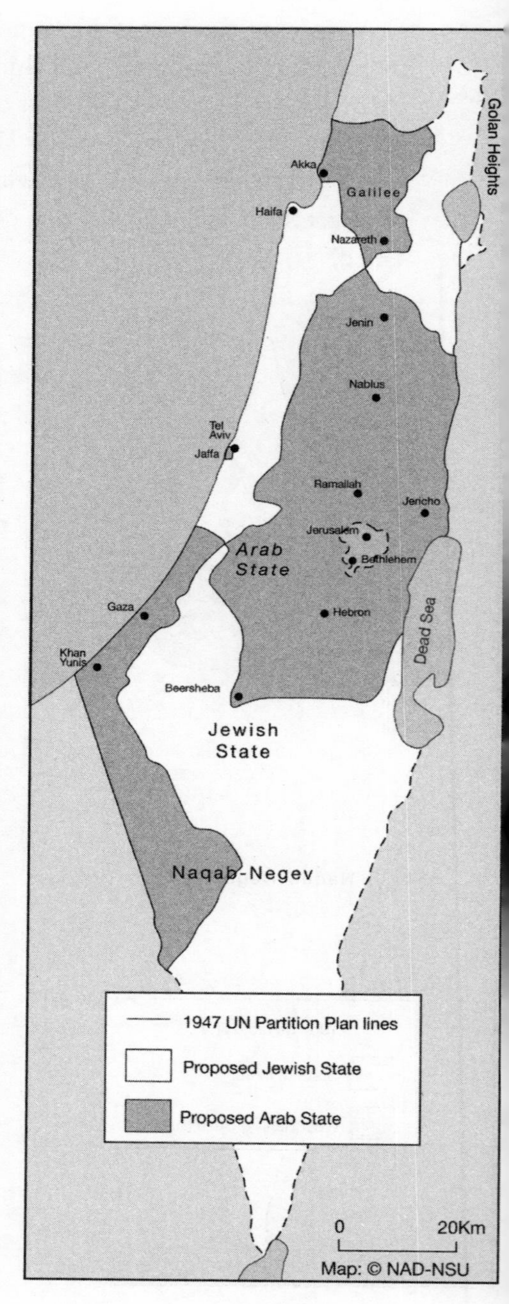

The 1947 UN Partition Plan

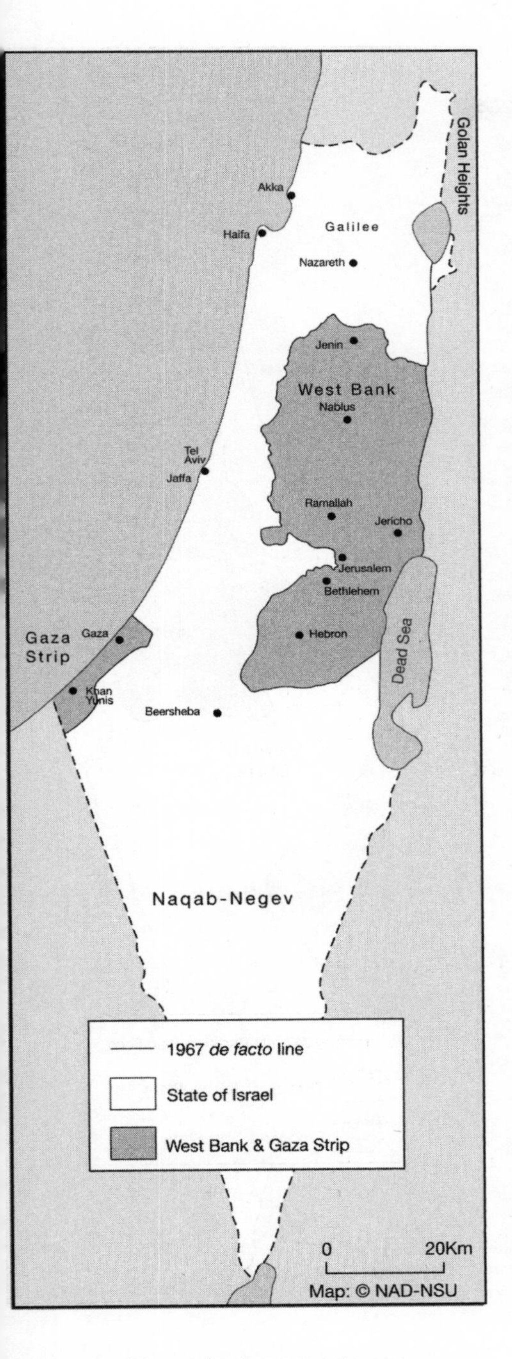

The 1949 Armistice Lines

Israeli Settlements on the West Bank

Introduction

The projection of their own evil impulses into demons is only one portion of a system which constituted the Weltanschauung *[World View] of primitive people, and which we shall come to know as "animism."*

—Sigmund Freud, "Taboo and Emotional Ambivalence," *Origins of Religion*

I am the son of two Holocaust survivors from southern Poland who, like most Polish Jews before 1939, shunned the Zionist call, supporting instead the socialist Jewish Labour Bund; like most other Jews, both considered Polish and Yiddish their languages and cultures. Both my parents were forcibly taken from the Nazi-controlled ghetto in Ostrowiec Świętokrzyski to the nearby Auschwitz extermination camp in June 1944, after the rest of their families were destroyed in Treblinka during 1943. Reduced to horrifying skeletons, they were forcibly marched to other camps in January 1945, as the Red Army approached Auschwitz. My mother was liberated from Bergen-Belsen by the British in April 1945; her weight at the time was recorded as thirty-four kilograms and she suffered from advanced typhoid. My father was liberated by the US Army from Gusen II, a subcamp of Mauthausen, on May 8, 1945. His recorded weight was thirty-two kilograms. They were married in a Torino Displaced Persons camp in October 1945. I was born, stateless, a year later in Rome.

Having failed to secure passage elsewhere, my parents decided to emigrate to Israel in May 1948, not a choice they would have otherwise considered. On the boat my father refused to undertake weapons training. After what he had experienced, he was not prepared to shed blood, his own or anyone else's. He was promptly arrested on arrival in Haifa as a draft resistor; he may have been the first, or one of the very first, conscientious objectors.

My mother and I were incarcerated in Athlit, a prison camp built by the Mandate authorities, then used to house immigrants. My father resisted for some weeks, but after realizing that he might spend years in prison, agreed to serve as an unarmed medic and was sent to one of the worst battles of 1948, in the Latrun area, at which almost 2,000 Israelis, mostly Holocaust survivors, perished; so too did a large number of Jordanian troops. Many were buried in mass graves; having just arrived, their identities were unknown.

How my father survived this hell I will never know. He never spoke to me about it or admitted that he had refused to serve in the army; later, when I became an officer in the Israel Defense Forces (IDF), he was ashamed to tell me about it. I only know this part of his story because his communist brother, who admired him for his stand, told me about it; he wanted me to appreciate my father's courage. This revelation affected me deeply.

I grew up in Jebaliya, a small modern town adjoining Jaffa, that was forcibly cleared of its Arab inhabitants by the Etzel (Irgun, the rightwing Zionist militia) even before the Mandate expired. Only a few Arabs managed to remain, becoming the unwilling and unequal captives of the Jewish State. The neighborhood was exclusively populated by Holocaust survivors in their twenties and thirties, and none of the many children had grandparents. We lived, like everyone else, in a flat that had been the home of a Palestinian family. Yosefa Loshitzky accurately describes this process:

> Many Holocaust survivors were, as a matter of government policy, settled in evacuated Palestinian homes in Arab towns like Jaffa, Haifa, Lod and Ramla, thus forcibly grafting the memory of the Holocaust onto Palestinian national memory, and symbolically linking the Holocaust of the Jewish people (mostly Polish Jews) to the Palestinian *Nakba*.[1]

This aptly describes our own situation. Jebaliya, and Jaffa itself, were hardly parts of Israel proper then—they existed in a twilight zone where Holocaust survivors were living cheek by jowl with Nakba survivors, their children studying in the same school, Al-Ahmadiyya, a green modern Bauhaus building within a copse of sycamore trees, renamed Dov Hoz after a Zionist apparatchik. We studied in Hebrew but also learned Arabic, and when later I was transferred to a religious school, I found that the Arab boys had to stay in for the Hebrew daily prayers—an odd punishment for the crime of being Other.

My parents, like so many other Holocaust survivors who came to Palestine/Israel after WWII, were hardly willing colonialists. But living as part of the colonial project, they were normalized into its ranks, and later also accepted its rationale and methods. When faced with such massive injustice, one either rises in opposition or, willingly or otherwise, joins in. By the time I was drafted at eighteen, in 1964, my parents had changed their relationship to military power; it had become the symbol of survival for them, as for most other survivors. I, on the other hand, was disinclined to join the IDF, having developed a naïve, instinctive gut pacifism but lacking the courage to follow in the footsteps of Giora Neuman, two years my senior, one of the famous draft resisters of Israel. He spent some years in prison for his principled stand, but I was not strong enough to emulate him or my own father (about whose courage I only learned later). I was selected for officer training, which I tried unsuccessfully to get out of or postpone.

I was placed in one of the few regular fighting units, the Golani infantry brigade, as a young second lieutenant, a role I held during the 1967 war. As part of the brigade command staff, I did not partake personally in the horrific battle in the Golan Heights, taking place a few hundred yards from us; I followed the battle through the communications system. When the battle was over, I heard the dazed voice of one of the battalion commanders asking the commanding officer standing next to me, a shaky voice emanating from the metal speakers: "I have 200 prisoners of war. What shall I do with them?"

He received no answer from the commanding officer, who snarled at us, "The idiot, doesn't he know what to do with them? Do I have to tell him? No one answer this idiot, do you hear?!"

After some further requests, the transmissions stopped. The penny dropped.

I was deeply shocked; throughout the officer training program we were told that the IDF was the most moral army; that we never harm civilians; that we never shoot prisoners of war. So, what was this officer, one I intensely disliked, trying to tell us? Deep bitterness grew inside me.

In the debriefing session after the war, it became clear to me that the battle fought by Golani had no real military objective: The men who had died like rats in a barrel had not represented a threat: their positions were isolated, their retreat was blocked, and the main force was getting around through other routes. I asked the commanding officer about the purpose of the attack. I was told that Golani had to earn its glory, like the paratroopers did in 1956, and that glory is only earned through battle and bloodshed.

For the first time in my young life I started to comprehend the deep gulf between reality and propaganda. I also grasped that as a young, white male of European origin, there may be some duties I am morally bound by and need to be committed to, as a past refugee indebted to the refugees in whose home I grew up. What could I do for them? I needed to find out. I also needed to get out of the Jewish State.

On arrival in Britain I was ready for a change. I studied for a graduate degree at the Royal College of Art, a progressive institution in the early 1970s, and soon enough met members of Matzpen, the radical organization of Middle Eastern anti-Zionists, mainly Israelis but also some from Arab countries, led by Moshe Machover, who by then had left Israel for London. In Israel it was at its zenith, with almost 2,000 members, while in London there were only ten of us at the weekly meeting, sometimes less. What followed was an intensive, political group study lasting months if not years. We read and discussed Zionist history and radical literature. Ironically, then as now, the main readers of key Zionist texts are anti-Zionists. At last, I started to understand the nature of Zionism and Israel. It was a painful experience of inner transformation. It allowed me to resolve my identity and beliefs and freed me from the all-powerful, stifling collectivities of Zionism.

In Israel, military service starts before birth. Take, for example, an advertisement in the rightwing newspaper *Makor Rishon*, depicting a mobilized Israeli fetus. The advertisement for Lis Maternity

Hospital, part of Ichilov Hospital in Tel Aviv, shows a fetus wearing a military beret with a caption reading: "Recipient of the Presidential Award of Excellence, 2038." Portraying an unborn child as a soldier is disturbing in its own right, regardless of country. In Israel, an occupying power whose military has for over five decades been primarily concerned with maintaining control over a civilian population, this advertisement is even more charged.

This example is by no means unique. Israelis perceive themselves as soldiers, whether during the long compulsory service that begins in the late teenage years for both sexes, or later, throughout adult life, in the reserves. Indeed, military training does not start at eighteen, but at the age of fourteen, when high school kids join the Gadna (acronym for Youth Battalions) and start their preparations for army life. By incorporating four years of frequent army training into high school, the army has already oriented the future soldiers before they even see service.

However, more important than physical and weapons training is the psychological and mental preparation of youngsters during their most impressionable years. This includes preparation for the violence they will need to effect and employ during their adult life. This psychological/ideological training does not start at eighteen or even at fourteen. Arguably, it is initiated even before school, in the societal envelope of family and friends. While the family lays the foundation, the school consolidates this uniquely Israeli mind-set, where history and collective memory are differentiated.

While children study (Jewish) history aged around eight, their socially mediated inculcation into collective memory is much earlier. Nurit Peled-Elhanan has studied the role Hebrew textbooks play in establishing collective memory at an early stage, when questioning authority is not only unlikely, but also impossible.[2] Her conclusions, based on the work of another Israeli scholar, Yael Zerubavel, are clear: "Studies show that Israeli schoolbooks aim to inculcate the collective memory created by Zionism which 'constitutes an entirely novel Jewish collective memory.'"[3] History, geography, civics, literature, and even science and grammar schoolbooks promulgate the Zionist grand narrative, directly or indirectly.

This narrative, according to Zerubavel, is made of two opposing parts: the account of decline from the "golden age" of antiquity through "exile" to its culmination in the Holocaust and, by contrast, a narrative of progress beginning with the Zionist return to the land of Israel leading to national redemption. This is told through nursery rhymes, religious and secular rites of commemoration, holidays—Independence Day, Holocaust Memorial Day, Hanukkah, Purim, Passover, Ninth of Av, Sukot, Shavu'ot—and " 'national heroes' " narratives. And these stories are absorbed at a very early stage, so that by the time children enter school at age six, they are already inducted into the mainstays of the Zionist narrative.

School further substantiates and broadens the reach of this narrative, adding "nature" and "wilderness" trips, poetry, literature, and Bible study. Thus, state institutions construct a personal, experiential underpinning for the foundation of collective memory. It can be argued that Israelis spend almost all their time in an ideological environment, and in this case the ideology is Zionism.

Both Zionism and its critics speak in absolute terms. Proponents of Zionism present it as the liberation movement of the "Jewish people" (whomever they might be). Critics understand it as a militarized settler-colonial movement of European (and later Arab) Jews in the Third World country, Palestine. Israel and Zionism have offered self-determination to some Jews, and Israel is, at the same time, an obvious colonial enterprise, a settler-colonial society that is understood and described as such by Theodor Herzl and early Zionist leaders.

Yet, definitions of Zionism as either a liberation movement or a settler-colonial project are both imprecise. Israel cannot be a real liberation site for Jews, because their conditions of control require the denial of rights of the colonized indigenous Other—the Palestinian. As Marx noted, "No nation can be free if it oppresses other nations."[4]

Israel is a settler-colonial project, but like many other instances of colonialism, it also diverges from the standard definition. The Zionist colony is not the vanguard of a specific colonial power in the way that the *pied noir* in Algeria was of France, or British colonists were in Australia. There was no single colonial power behind Zionism. Instead, it has been enabled by three imperial powers since 1918—Britain, France, and the United States. Israel is therefore a client state of the West, set up with some autonomy.

This complex nature of the Israeli-Zionist enterprise was pointed out by Maxime Rodinson just before the 1967 war.[5] He used an early definition of colonialism that has stood the test of time and analysis: "One can speak of colonization when there is, and by the very fact that there is, occupation with domination; when there is, and by the fact there is, emigration with legislation."[6] Rodinson also reminds us that early Zionists not only saw their project as a colonial-settler state but also described it as such, using the words "colonial" or "colony" to describe its institutions and habitations, seeing it as part of the larger project of Western culture:

> The creation of the State of Israel on Palestinian soil is the culmination of a process that fits perfectly into the great European-American movement of expansion in the nineteenth and twentieth centuries whose aim was to settle new inhabitants among other peoples or to dominate them economically and politically.[7]

Early Zionists originated in many countries and could not immediately rely on the support of a "mother country." Their support system was initially based on donations from Jewish supporters who wished to assist other Jews who were suffering antisemitism, mainly in Eastern Europe. They chose the colonial modus operandi because it was the normative European mode and thought little of it. (The severe critique of colonialism developing in leftwing and progressive intellectual circles was not shared by Zionists.) Even those who were aware of it either ignored it altogether or rationalized it away, using what will come to be termed *state of exception* argumentation.

Understanding this background is crucial for explaining Israeli feats of political contradiction—being simultaneously one of the world's smallest states and the fifth or sixth military power, a colonizing power that is allegedly a victim of terrorism.[8] An obvious element of any settler-colonial project is military violence, without which such undertakings are impossible.

As Wolfe points out, the specificity of settler-colonialism is the intention of removing the indigenous population, treating the land as *terra nullius* (empty territory): "Settler colonies were (are) premised on the elimination of native societies."[9] Palestine was no exception. Many Zionists find it difficult to publicly admit and accept the realities of such positioning.

The Myths of Zionism

Zionism is based on a system of myths and ideological truisms, in order to establish legitimacy as a "liberation movement." These myths make concrete the relationship between the army and the nation, placing the IDF at the heart of the national identity and function of the state.

First, Zionism concurs with historical antisemitism, the deadly enemy of Jews.[10] Both Zionism and antisemitism posit that Jews cannot (and indeed, should not even attempt to) live among non-Jews; both posit that only in a country without goys will Jewish life be safe. Antisemitism is crucially important to Zionism; it generates committed Zionists and operates as a recruiting sergeant, driving Jews toward Palestine. This "benefit" of Judeophobia was grasped by Zionist leaders beginning with Herzl. Without antisemitism, Zionism is a spent historical force; most Jews still abhor the idea of moving to Palestine unless they face a direct threat. Zionists operate fake "antisemitism" campaigns in many countries, whose goal is to target not actual antisemites (many of whom are now Israel's closest allies) but rather left anti-Zionist activists, most of whom are Jewish. Through these actions, resistance to Zionism had been relabeled *new antisemitism* and weaponized through the International Holocaust Remembrance Alliance "definition of antisemitism" as anti-Israel speech, convenient for silencing critics of Israeli atrocities and controlling the debate on Palestine.[11]

Zionists have argued that Jews should leave their European homes, in Germany, Britain, France, or Austria, where they are well established and have enjoyed hard-won rights and substantial advances following 1848, because of antisemitism by a minority of Gentiles; instead they were told to colonize a country where they had no legal standing and where their coreligionists made up a tiny portion of the population, living among a non-Jewish majority population and planning a gradual takeover in order to build a Jewish State. By the time of the publication of Herzl's plan, such projects were relics of the colonial past, seemingly no longer worthy of serious attention.

Such a risky objective is only possible if Zionists could also rid the country of its non-Jewish, indigenous population through a process of racist expulsion; otherwise it would have been suicidal.

Zionism, in fact, put Jews in harm's way. During the early period of his Zionist activity, Herzl in his naiveté (or the pretense of it)—typical of a Westerner of the period, supporting colonialism unquestioningly—believed this task to be relatively uncomplicated. However, achieving this goal was not possible without a major cataclysm. Having left the countries of Europe for the dream of living "without the hated and feared goy," Israelis now have to contend with a Palestinian population of over six million in Palestine and a similar number abroad.

To accept that Palestinians have the right to live in their own country runs counter to the historical tenets of Zionism, yet they are fully intent on staying put, despite the successes of Zionism and the high price wrested from them. This makes Israel the only country in which Jews live in mortal danger because they are Jewish (and colonialists, militarists, occupiers, and racists); no such danger faces Jews elsewhere. Thus, a double contradiction exists: Jews can and do live safely in other countries, while they cannot in Israel. This lack of security results from their political and military praxis, not their racial origin: as occupying, militarist colonials, they can never be safe, certainly not as safe as Jews living in London, Paris, or New York.

Another related contradiction is that despite its marked successes in over a century of political, settlement, and military action, Zionism has failed to empty Palestine of its indigenous population. Now it finds itself in a bind: it claims to be a modern and somewhat democratic state, but it operates a racialized society guided by the claimed, innate "Jewishness" of its projected democracy. It does not allow human or political rights to the very people it intends to destroy or expel so that "Jewishness" can flourish, untainted by the toxic presence of non-Jewish others. Such practices are deeply racist, mirroring the worst examples of institutionalised antisemitism in Europe, especially in the twentieth century, as well as other societies resorting to racialized genocide.

One therefore wonders about the realization of the Zionist colonization plan. How would non-Jews be removed from Palestine to enable a Jews-only democracy? Surely not by the methods foreseen by Herzl, who considered the local population fickle enough to leave their land and homes for a recompense of a few francs, or find employment elsewhere, as described in his diary. His followers knew that violence was required to "'cleanse'" the country of its indigenous people.

The concept and practice of "'Jewish democracy'" in Israel is, therefore, expressly antidemocratic, exceptionalist, and racist by its very nature. It is a democracy in which one must be Jewish to enjoy full civic rights. Zionism has marked the move from many small European Jewish ghettoes to a modern, large, and powerful ghetto setting itself apart. This particular ghetto has failed in its utopian project of building a Jewish existence without the goy. Clearly, such a democracy for Jews only is what sociopolitical scientists termed *herrenvolk democracy*,[12] or democracy of the master race, where others are not tolerated.

And here is the greater contradiction. Israel argued for decades that it wished to build a peaceful "Jewish democracy" and a "Jewish State" in Palestine; yet since June 1967, it has pointedly refused to withdraw from the Occupied Palestinian Territories (OPT), where more than four million Palestinians now live. Israel cannot be a "'Jewish State'" with millions of Palestinians under its control, unless it becomes a full-fledged apartheid state, justifying the lack of human rights for millions through racialized ideology, annulling their communal existence until such time as they can be expelled.

As a result, by continuing the occupation and its related iniquities, Israel has become an apartheid state. By making a two-state solution impossible, Israel is forcing the issue, as claimed by former Israeli prime minister Ehud Olmert. He foresaw dangers that such a situation might bring. His exhortation remains valid today:

> If the day comes when the two-state solution collapses, and we face a South African–style struggle for equal voting rights (also for the Palestinians in the territories), then, as soon as that happens, the State of Israel is finished ... The Jewish organizations, ... which were our power base in America, will be the first to come out against us because they will say they cannot support a state that does not support democracy and equal voting rights for all its residents.[13]

This would certainly reverse prospects of immigration of American Jews in significant numbers; hence Olmert wished to strongly support the two-state solution, even as such a solution was made impossible by Israeli politicians, including himself.

This distinction was recently abandoned by Donald Trump during a speech he gave on the occasion of Prime Minister Benjamin Netanyahu's visit to Washington in February 2017, when he publicly rejected the long-standing albeit insincere US support of the two-state solution in favor of a somewhat different formulation—the "happiness of both sides." "To be honest, if Bibi [Netanyahu] and the Palestinians, if Israel and the Palestinians are happy—I'm happy with the one they like the best."[14] During his visit to Israel in May that same year, Trump quickly retracted this statement in favor of one he believed Netanyahu might be happier with. He now supports anything Israel does, instead.

Zionism and its leaders use a mixture of legalistic, rights-based arguments alongside less universally impartial claims. Thus, the Jewish right for self-determination can be made with arguments based on religious beliefs, such as a biblical "promise by God" of the country to the Israelites in some unspecified future; Palestinians are not seen as having a right to self-determination, or to godly promises. That such a "promise" obviously overrides the rights of the indigenous population is not seen as a problem, no more of a problem than existed when the biblical Joshua reportedly exterminated the ancient peoples of Palestine. With these mythical claims atheists such as Herzl or David Ben-Gurion easily adopted biblical and mythological justifications not just for settling in Palestine, but for expelling its population.

For decades, Israel's leaders have claimed that "all Israelis want is to live in peace with their neighbors." Yet Israel has initiated war and conflict ever since its inception, refusing to bring the conflict to a negotiated end (for example, it rejected the Arab Peace Initiative of 2002). Agreement requires honoring UN resolutions and Geneva Conventions and removing Israeli colonies (Jewish-only settlements) from the OPT and Syria. Israel prefers the conflict with territories to a negotiated peace without them.

Once the system of Zionist maxims and achievements is examined with these contradictions in mind, many others emerge, redefining the whole project. One deep and searing contradiction is the very idea of Jewish democracy itself. After 1948, as some two thirds of Palestinians were expelled, Israel defied international law and UN resolutions and denied the refugees' return to their homes. Gradually, Israel managed to persuade many that with almost 80 percent of its population being Jewish, it was indeed a "Jewish democracy."

The war in 1967 changed all that. Millions of Palestinians again lost their freedom, becoming refugees in their own homes, in a country no longer theirs. Israel lost a Jewish majority in the country, by refusing to leave its military spoils and return to the pre-1967 Armistice lines, as demanded by the UN in 1967 and again in 2002 when lasting peace was offered to Israel, only to be rejected out of hand.

After 1967, the choice that faced Israel was simple though tortuous: to "protect its Jewishness" and vacate the territories as demanded by UN resolutions, or to hold on to them at the cost of illegally controlling millions of non-citizen Palestinians. Will it act to keep the "Jewish democratic state" much bigger, even though that would entail also making it distinctly non-Jewish and undemocratic?

Israel chose not to choose; it held on to the territory and tried to dislodge the people from the land, so as to achieve both aims of Zionism. The wish for a greater empire won over the need to stay "purely Jewish." The thinking, then as now, seems to be: "We got rid of most of them once before, so we just need to wait, and do it again."

And so, Zionism continues to delay, believing time is on its side, as it seemed to have been until now. The defunct two-state solution will not be accepted by Israel because that would terminate its mini-empire and confirm Palestinians' rights over part of their country. Despite paying lip service to the formula of two democratic states side by side, Israel has done everything it can to avoid that outcome. Such a prospect remains impossible but continues to be parroted by Western politicians as a demented mantra of an optimal goal.

IDF, the Crucible of the New Nation

The relationship between the state and the IDF, an army like no other, bears examination. Israelis, who incessantly speak of their wish for peace, are the greatest warmongers in the Middle East. To this end, they have developed an arms industry that is one of the largest in the world, built over 300 nuclear warheads and missiles to deliver them as far as central Asia or Europe, and have made the IDF the center of Israeli existence. Israel behaves as if peace is not only unobtainable, but positively undesirable.

This is not simply a case of *si vis pacem, para bellum*, as Israel often claims, but a real conceptual and deeply emotional politicocultural predilection. Here it may be useful to pose the possibility that having so excelled in military conflict throughout their short history, Israelis became what they know best; their consciousness has become formed by their praxis. If fighting is what one does, if weapons are what one excels at making and selling, if military adventures are what military professionals and politicians invest much thought and preparation in, then the national character must be deeply affected by choices from which it derives deep satisfaction and a sense of security. How could a nation so obsessed with "security" and "defense" be anything but deeply dependent on conflict and defined by it?

This heritage is as long as Zionist history itself. The irony inherent in Zionism, supposedly developed as an answer to European antisemitism, is that at the most basic level it accepts antisemitic tenets and descriptions of Jews. This shared understanding was transferred to Palestine after Herzl's death and has become the standard response among Zionists. The deep rejection of diasporic Jewish history, which Zionism saw as pliant and submissive to Gentile society (Herzl labeled it spinelessness), has coalesced into a dominant disavowal of the Jewish past by the Zionist movement. What is more dispiriting is Herzl's deeply seated antisemitism, expressed in the terms one could only expect from devout Judeophobes. In a little-known text, Herzl refers to Baron Edmond de Rothschild as "the Mauschel" (the German equivalent of "kike") because he resisted being lured to supporting Zionism:

> Mauschel is spineless, repressed, shabby—when the Jew feels pain or pride, Mauschel's face shows only miserable fright or a mocking grin—he carries on his dirty deals behind the masks of progress and reaction; with rabbis, writers, lawyers and doctors, who are only crafty profit-seekers.[15]

The deep rejection of the Diasporic Jewish experience has coalesced into dominant disavowal of the Jewish past by the Zionist movement. This was problematic. Ben-Gurion, the main author of this "negation of the Diaspora," understood that the identity of the new nation cannot be based on such a prominent negation; it required an affirmative center to hold it, and this he decided, could only be done by placing the army at the center. The new Jew, the sabra, was to be the soldier, the apotheosis of the Zionist project. Amnon Raz-Krakotzkin has identified the problematic results:

> Thus, the only link to collective identity remains the military service and the spilling of blood, and militarism remains the basic, definitional characteristic of collective identity [in Israel]. The characteristic "tribalism" which defines various aspects of Israeli culture is the clearest expression of the failure of an autonomic definition of Judaism, through the concepts of modern nationalism.[16]

The new Jew-Soldier would wipe the slate clean, through tilling the land and through military prowess.

Absent from this description is the Other of Israel—the indigenous Palestinian. Such absence clearly indicates the repressed status of the Palestinian, the turning of Palestinian identity into a taboo. Yizhak Laor is more conscious of this process of excising the Palestinian out of Israeli and dominant Western discourse. Through this practice, the Israeli sabra is distinguished from two inferior and repulsive stereotypes: the Ghetto Jew and the Palestinian Arab. These are the two racialized ghosts of the uncanny past who serve to provide a contrasting background for the Westernized New-Jew sabra, blond and blue-eyed.[17]

And here, the distance between "civic" and "military" collapses. Very early in the history of Israel, Prime Minister David Ben-Gurion outlined his model of "a nation under arms." He believed that while

Zionism had created both a state and an army, there was no nation proper. Instead, the task of nation building was to be allotted to the IDF, the radical instrument of transforming the "Old Jew" into the "Israeli"—using nationalist alchemy to turn the base metal of the Ghetto Jew into the refined gold of the sabra soldier.

The nation was constructed through military experience. Arguably, the task accomplished by the IDF is a unique success. Although Israel, like some other states, gives primacy to its army, Jeff Halper notes that

> in few countries, however, does the military play such a dominant role in government, the economy, the cultural life of its people, or in its international relations. Israel expends about $15 billion a year on its military, between 6.5–8.5 percent of its GDP. Finland, by contrast, with a comparable GDP, spends $3.8 billion annually on defense (1.3 percent of its GDP), Morocco, engaged in a long war in the Western Sahara, spends $3.4 billion (3.4 percent), while Uganda, locked into prolonged external and internal warfare, spends only $3.4 million (2.3 percent). Even the US, by far the largest spender on its military, devoted only 4.3 percent of its GDP for defense (though that added up in 2011 to a staggering $708 billion) ... Israel, with the world's fifteenth highest defense budget, spends far more on its army than [the much larger] Iran, North Korea, Pakistan, or Egypt.[18]

For reasons that have nothing to do with an objective threat to its society, Israel spends far more per capita on its military than any other society. This is even more surprising when one considers that Israel's Palestinian citizens, making up some 22 percent of its population, are hardly beneficiaries of this (in fact, money spent on the military is spent *against* them). If they are factored out, Israeli spending on the military is the equivalent of 8.2–10 percent of the GDP of its Jewish citizenry, a situation clearly without parallel elsewhere. By combining military experience with the Zionist narration, harking back 2,000 years to an idealized and fictitious "golden past," a New Jew had indeed been created, Frankenstein-fashion, out of the dead parts of Jewish history.

And out of this bellicose crucible, the civil society emerges. First, the IDF is created as an instrument of capturing territories beyond those apportioned for the Jewish State by the UN in the Partition Plan, then it acts as Ben-Gurion's transformational agency and creates the Jewish-Hebrew nation. Once the fundamentals of this fighting community are in place, its actions shape the polity.

The creation of the IDF was the most important task facing the nascent nation, and it still unites Israeli society today. The IDF and the social, financial, and cultural apparatus connected with it form the single most important institution in Israel—a politicocultural-economic military–industrial complex.

Separating the military from the civic in Israel is impossible. Yagil Levy notes that civilian control over the army is not a control over militarism, arguing that "a distinction should be made between two modes of civilian control over military affairs: control of *the military* and control of *militarism*."[19] Thus, the question of who controls the army, or whether the army controls the state is one simplification too far. By incorporating society at all levels and sectors in the military, it is not necessary for the military to control society directly: the whole social structure is militarized. Militarism has become Israel's default mode, its machinery of choice for resolving conflict and achieving national aims and objectives.

In societies where military leadership is sourced from a narrow elite, with its command and control structure confined to a ruling class or strata, as in Latin America, it became necessary at different points to impose overt military control. In Israel, this has never been the case. The nation was the creature of the army, which in turn became the crowning glory of the nation—democratically (as far as most Jews are concerned) relying on conscription—*Ha'am Kulo Tzava* (a nation under arms). If the whole nation is the army, then the question of who controls it is meaningless—the nation is in control. This type of control over the military leads to a militaristic society:

> Most interesting and relevant to the case of Israel is the situation in which a high level of control of the military encounters a low level of control of militarism. In such situations, effective civilian control may not only fail to restrain militarism, but, under specific conditions, actually encourages it.[20]

This paradox of growing militarism as a result of civilian control is a unique feature of the Israeli polity, not easily found elsewhere but in the United States, where the army controls much of the economy and political agenda, especially through the use of the military–industrial complex and its many appendages. It may well be that the great investment in creating the nation by using the IDF, as well as the over-determinative racialized arguments about a Jewish State/People, have driven Israel to this particular dead end. A social circle is closed and remains locked.

To operate such a militaristic society successfully requires the use of what Dalia Gavrieli-Nuri terms "war-normalizing discourse" (WND), which takes a variety of forms and is designed for habituating the population to continuous war and its negative effects.[21] In Israel, a variety of such WNDs are used successfully. Their efficacy can be assessed by examining the Israeli-Jewish public's reaction to various attacks initiated by the IDF. Such operations normally gain approval of at least 90 percent of the Jewish citizenry, a level of support difficult to imagine elsewhere. The 2014 attack on Gaza was one of the most brutal, as well as useless military actions taken by the IDF, yet some 95 percent of Israelis justified and supported the operation.[22]

Israel is using well-established modes to achieve such unanimity. The simplest but one of the more effective techniques is to use nonsense or "fun" names for especially brutal operations. Gavrieli-Nuri examined 239 titles of military operations since 1948 and found that the use of disguised names was ubiquitous. For example, the 1982 war that brought about the death of some 22,000 Lebanese and Syrians was called Peace in Galilee. In 2009, the IDF went even further, choosing a nursery rhyme as the name of a devastating attack on Gaza, which began on December 27, 2008. Yosefa Loshitzky explained the significance of the moniker:

> The name is borrowed from a Hebrew nursery rhyme which was (and may still be) very popular among Israeli children in the 1950s. In this song, a father promises to his child a special Hanukkah gift: "a cast lead *sevivon*." *Sevivon,* in Hebrew (a dreidel in Yiddish) is a four-sided spinning top, played with during the Jewish holiday of Hanukkah. Somebody, in the Israeli army, who apparently feels

nostalgic about his childhood, decided that if Israeli kids would enjoy a *sevivon* cast from lead there is no reason why Palestinian children would not appreciate it too. After all Operation Cast Lead is not the first (and unfortunately, will not be the last) of Israel's cruel war games.[23]

Gavrieli-Nuri argues that

> in WND terminology, the name Operation Cast Lead *justified* national initiatives by creating a historical link between the brave victorious Maccabees and the IDF soldiers. Its references to cast lead arouses faith in the IDF's power and, perhaps, in the nation's endurance.[24]

To date, roughly a third of the names chosen by the IDF have their source in the Bible, with another third referencing nature. In a similar manner the deadliest weapons, either Israeli-produced or imported, were named using biblical and lyrical names and references.[25] While this is not unique to Israel, its ubiquity and dependence on biblical references are probably without precedent.

The Hebrew Bible is the most familiar text for Israeli-Jewish children, studied throughout mandatory education, so biblical-sourced names resonate with the Israeli-Jewish public, suggesting a priori, mock-historical foundations. By referring to a commonly held cultural stock, the public feels related to the old/new name and acquainted with it, and it eases the public's acceptance of further bloodshed and brutalities.

Israel refers to wars as "operations," a practice that normalizes them and downgrades their horror:

> Put simply, the term "operation" dulls public awareness regarding an action's place in foreign policy and reduces potential criticism against the exploitation of military power. Worthy of note is the meaning of the term in the Israeli civil arena, where *mivtza* (operation) means the sale of goods for a short period of time, a "sale."[26]

Some of the wars, like the 1956 Suez War, or the 1982 and 2006 invasions of Lebanon, are in Israel dubbed *mivtza*. This linguistic insistence on demoting war to a mere "operation" is a deeply ideological, complex act for recruiting the social order. Only later, and in hindsight, did these "operations" gain the nomenclature of "wars," through media, academic and public discourse, a belated, muted admission of their horrific nature.

In addition, war was made into an exciting, libidinal experience of returning to "lost youth," especially in the novels of Moshe Shamir and later, A. B. Yehoshua and Amos Oz, where war and fighting have sexual overtones.[27] In a country where most cultural production was directly or indirectly under state/army control, flooding the cultural scene with romantic images of war and self-sacrifice and turning soldiers into masculinized objects of desire was a simple task. Authors assisted in shaping the fighting nation, the Sparta of the eastern Mediterranean.

Cultural production—music, theater, film, the press, radio (and later television), books, magazines, and newspapers, including children's weeklies—were part of the *hasbara* (propaganda) campaign, not by a secret cabal but by the open cultural, political, and economic collusion of Israel's elite. This was a coming together of all leading sectors of the polity to enhance the still-forming national consensus, a restatement of the militarized Zionist credo of "living on your sword." Love songs lionized the soldier. Children were militarized as young fighters by the iconic (and wildly popular) *Hasamba* books by Yigal Mosinzon in the early 1950s in which a secret group of children fought and won Israel's wars behind enemy lines.

It all served to redefine another normalcy, that of the "nation under arms," a Spartan society in constant, unremitting, merciless conflict, where the "best willingly sacrifice their life," with national identity made of such sacrifices and dependent on them. War, the only certainty, had to become as normal as sunrise. This was noticed by Hannah Arendt well before many others, as noted by Loshitzky in an article about Arendt's *Eichmann in Jerusalem*:[28] "The female soldiers who look alien compared to Arendt, dressed in European clothes (an uncommon sight in early 1960s Israel), may perhaps hint at Arendt's critique of the growing militarization of Israel, or modern-day Sparta."[29]

Even periods between wars were a time of fighting, if in different formats and formations. Israeli life was restaged as a heroic, unrelenting struggle, with the pain of loss elevated to what makes life worth living as well as worth sacrificing. Not only was this a success at home—it sold well to Western audiences.

But such narrative closure of the national "story" had a strong element of self-selection about it. It was, and still is, based on the normative exhortation by every Israeli leader: *Ha'olam Kulo Negdenu* (the whole world is against us). Such attitudes thus disqualified "the world" from judging or condemning Israel, as it could never properly understand Israel. This exceptionalist outlook is promulgated through mythical expressions in the Bible and later texts of the *chosen people*.

This in turn has been the inspiration for the development of the Israeli military–industrial complex, a commodification and commercialization of violent and hostile attitudes, made material and for material gain, hence twice justified: as an aspect of the mental defenses against the Other (a visible survival mechanism) and as a mechanism for economic survival and development, through producing actual instruments of violence for controlling the perceived or projected threat.

The Israeli military–industrial complex is a hi-tech enterprise using the most advanced technologies, enabling it to stay at the forefront of the global armament market, with a simple claim: its products have been tested "in action"—in Gaza, the West Bank, Lebanon, Syria, and elsewhere. This "tried and tested" label on its security products has turned the industry into a massive income generator.[30] Such success seems also to isolate the ruling elite from political fallout, despite the many corruption scandals that rocked the country during the last two decades.

The Israeli military–industrial complex should really be thought of as a military–industrial–academic complex, with Israel's seven universities and leading research centers collaborating with the IDF and armament production and training companies, creating a seamless security continuum. Les Levidow argues that the EU research funding agenda was heavily influenced by Israel and greatly benefits its universities and security firms:

> This agenda has been driven by and benefited Israeli partners of research projects for at least the past decade. From the standpoint

> of EU research chiefs, Israeli partners offer crucial expertise for enhancing the global competitiveness of European institutions. For Western elites more generally, the model is Israel's military-industrial complex, which has produced a world-leading security industry. As one of many examples, the RAND Corporation and the Hebrew University of Jerusalem participated in an EU project (EUSECON) aiming to establish a research network to provide "research-based policy advice on economic aspects of security."[31]

Incredibly, Israel's universities and research centers are the recipients of more EU research funding than the great majority of EU countries, and Israel is leading the world in research income per capita and research spending, mainly on military, policing, and security research—the main growth areas of the security industry, areas that are prioritized in the spending agenda of central and local governments everywhere.

This naturally sheds light on Israel's great political, intellectual, and financial investment in social control, conflict, war, and destruction. As opposed to investments in agriculture during Israel's early years, the current products are high-yield and highly sought, with no sell-by date and with high added value. Indeed, Israel is unable to keep up with the demand for its military wares, with some clients having to wait years for its products. Thus, Israel has turned adversity into hard cash, as well as into political and diplomatic power; countries that require such products cannot afford to oppose their supplier.

Hence, not only did Israel's economy boom, but its international standing has risen, while it continued to wreak havoc in the Middle East with impunity. Much may be explained by US support and the established powers of the pro-Israel lobby there,[32] but this does not offer a full explanation. Israel enjoys broad support beyond that offered by the United States, which constantly shields Israel from criticism and hostility. This is made up of a combination of a Holocaust guilt complex in European countries and the appeal of its security industries, not to mention its hundreds of nuclear weapons, built with the silent but crucial support of European nations.

The combined effect isolates Israel from criticism and censure, allowing IDF and Mossad free rein across the globe. Many assassinations attributed to Israel, such as those of Iranian military figures and scientists and Palestinian functionaries around the world, have gone unhindered and unpunished. Israeli officers and politicians are in the habit of taking credit for such criminal operations in the knowledge that no country is likely to confront Israel.

Speaking of the Holocaust

While trying to map Israeli attitudes to power, especially military power, one cannot refrain from examining that relationship historically, in the context of European antisemitism and the Holocaust. Jewish early history, during the Hellenic and Roman imperial highpoints, included long periods during which Jews served as mercenary soldiers in imperial armies. This type of intimacy with power and violence changed over the Middle Ages, as Jews came to live in well-defined communities in Europe, which later turned into ghettoes. Gradually, such communities, whose members were distinctly visible through clothing and facial features such as sidelocks and imposed signifiers such as distinctive hats or colors of clothes, have turned into the European Other, the butt of Christian religious animosity, and later, racist, antisemitic violence and vilification. Jews were severely limited in terms of occupation, settlement, movement, and ownership, their lives never safe from predatory social strata.

This fundamental insecurity, which typified Jewish life in Europe for more than a millennium, had what can be termed pathological effects on the Jewish psyche. The typical habit of searching for safety in finance and property ownership or through the protection of potentates was a flimsy and uncertain measure, proving irrelevant in the 1930s and later, during the Holocaust. Arguably, this has led to a deep-seated, understandable fear and to pathologizing security both by Zionists in Palestine and by Jewish communities abroad, especially following the Holocaust. One should not underestimate the part such attitudes played in the formation of Israel's security paranoia. Situations that would create worry and stress in other communities turn into communal paranoias for Israel's Jewish population, who lack the ability or willingness to see through pathology and propaganda and evaluate dangers accurately. Indeed, any doubt or a human need for amelioration is perceived as weakness, a throwback to the despised Ghetto Jew, weakening the national will to fight for total security.

Historically, the contours of Jewish distinction have been examined by a number of Jewish (and non-Jewish) intellectuals, none greater than Sigmund Freud, who spent his last years working on this topic in his groundbreaking *Moses and Monotheism*.[33] Freud treated the exodus as a historical fact, though without accepting the myths of the Book of Exodus.

Freud's reading of Jewish identity is complex. Its foundation is the thesis he advanced in the earlier (1912–13) *Totem and Taboo* articles and the presumed prehistorical episodes of the murder of the primal fathers by the oppressed sons banding together, leading to deep guilt and communal repression. In his later work, Freud returns to this thesis by introducing a reenactment of the murder of the primal father by the crowd of Israelites—the murder of Moses, their father/god figurehead, leading to further repression and the loss of the Moses/Akhenaten religion. Freud's Moses was an Egyptian high-ranking disciple of Amenhotep IV, or Akhenaten, a monotheistic zealot like the Pharaoh he outlived by escaping Egypt.

Crucial to our argument is Freud's understanding of Jewish identity as resulting from the great trauma of the killing of Moses by his flock—united by this trauma and harbored guilt.[34] This viewing of the group through a psychoanalytic perspective sees guilt in the group operating like neurosis in the individual, who represses knowledge of the crime, yet carries the guilt, which later becomes Jewish identity, through the return of the repressed. Freud notes that this had lasted hundreds of years; he also points out that Christianity, as a derivative of Judaism, is born of further enactment of the primal episode, this time through the killing of the son rather than the father. In Christianity, he argues, guilt is assuaged by blaming Judaism's refusal to recognize Jesus and be saved by his self-sacrifice. It is Freud's interpretation of the continued renewal of European antisemitic sentiment, and the deep guilt enmeshed in Jewish identity, which is at issue.

If this presumed murder of an overzealous, demanding leader has created a trauma that has lasted hundreds of years, what of the trauma of centuries of antisemitism that culminated in the Holocaust? Surely that might be a much deeper trauma?

The claim that this trauma was magically healed by militarily victorious Zionism is surely nonsensical. Any group experiencing such trauma would be deeply harmed, the repressed forming a pathology, returning in time to haunt it. Zionism has indeed repressed the Holocaust (at least until the Eichmann trial) as a shameful episode of Ghetto Jews' vulnerability, proof of the forever-present antisemitism of the non-Jew. Such ahistorical infantilization of this unique trauma was part of the New Jew epistemology—the warrior who dares and wins, is never afraid, asserts himself as an active agent of history. The trauma is exceedingly deep—the fear of physical extinction was real in the 1940s—and the Zionist movement leadership has exploited this fear for political gain.

This fear has been stoked by most Israeli politicians since Ben-Gurion as a mechanism of social control. Benjamin Netanyahu, for example, has spent more than two decades flogging the myth of the Iranian nuclear bomb, forever "around the corner," with bizarre visual aids at the UN and US Congress; Hezbollah and Hamas are routinely described as "existential threats," and in the last decade the Boycott, Divestment, and Sanctions movement has joined the list of existential threats to the Israeli body politic. (Interestingly, such threats to Israel's continued existence are periodically punctuated by loud threats by Israel against most of its neighbors, threats one should never disregard because most of them materialize.)

We can (and should) think of the New Jew ideal as an Israeli overreaction to Holocaust trauma and the fear it spread. The Israeli expectation for total security is totally unrealistic and results from an oversized group trauma; an expectation not affected by its impracticality or by the "reality principle." It is a truism that Jews are in danger only in the Jewish State, and while all Israeli Jews must somehow have realized this, they behave as if they haven't. Israelis are on a quest to decimate the last vestiges of opposition to their statehood and to achieve total security, through more missiles, mightier tanks, spy satellites, nuclear submarines, and all-seeing drones. The quest for total security is the pathological engine of Israeli society. To understand it is not to justify it.

A case in point is the first Gulf War in 1991, when Saddam Hussein lobbed 300 SCUD missiles at Israel, causing minimal damage to property and the death of one person from a stroke. Immediately at the onset of the war, fear of total destruction drove most of the population of Tel Aviv, Haifa, and many smaller towns to decamp to areas that were considered safer (like Jerusalem and Eilat), with some going further afield, to Europe and beyond. Normal life—education, production, commercial activities, services—came to a halt and was not renewed until the war's end. The few who stayed in the cities moved into communal shelters where they remained for the duration of the war.

None of us would like to live under traumatic bombardment. The people in London during the Blitz or the people of Beirut and Gaza, bombed repeatedly and massively, surely faced a more severe danger than the population of Tel Aviv in 1991. But this simple fact cannot penetrate the consciousness of most Israelis; in their minds, they have suffered the worst possible experience, which has remained etched in their memory and continues to affect their behavior. This heightened and unrealistic insecurity is striking: in the only Jewish State, with its mighty army, Jews who frequently rain death on Palestinians and others feel insecure, much more insecure than any Jews abroad. Zionism, which was allegedly intended to remove Jewish insecurity by building a Jewish State, has led to greatly intensifying insecurities in Israel, creating an existential paranoia.

The flip sides of this pathology are the militaristic fetish and mass neurosis of contradictory and complementary insecurity. Loshitzky notes the paradox:

> A displacement occurred in the course of Israel's effort to galvanise the Israeli/Jewish nation and construct its collective identity. The Arab, and particularly the Palestinian, was to become the container of Jewish fantasies of power and revenge. Jewish powerlessness and vulnerability, epitomized by the Holocaust, was transformed into a fantasy of absolute power exercised against the Palestinian as a substitute for the European goy (gentile).[35]

In the Israeli mindset, the only remedy to total fear is total security; Israelis realize this is impossible, but knowledge seems unable to affect the sentiment. Total security must be achieved by totally controlling the environs of the Jewish State. Like other *idées fixes,* this one is not attenuated by experience. Hence, the area that needs to be controlled must continuously expand, and so must the measures used. That this is further snagged by other requirements of Zionist security, such as the need to remove non-Jews from the society as well as from neighboring areas, makes this social and political project even more damaging and pointless. While there is no way to offer total security under the conditions of settler-colonialism, this is exactly what the Zionist state is demanding from the international community and promising its Jewish citizens; without it, what is the point of living in Palestine? But Palestine under occupation is the last place this security will be realized. In this way, Zionism perpetuates the deep insecurities etched on the Jewish psyche.

This basic contradiction at the heart of the Zionist project is forever troubling, but also constantly avoided—it is this disavowal of facing reality, which brings about the existential state of permanent crisis, of continuous and unchanging emergency as a way of life. This fundamental contradictory drive is the engine of the Zionist project, the secret of its fascination with military violence. The military is the way of reshaping the world and making it totally safe.

The deepest irony is the realization that war is not a solution but the *problem*. But such simple insight is well beyond the inclinations of a society specifically fashioned for war as the preferred method of dealing with conflicting interests, gearing all its social machinery to its conduct. As Chekhov noted of the theater, a gun hanging on the wall during the first act has to go off in the third; otherwise, why is it there? The hanging gun controls the mind, not the other way.

Overview of the Book

Examination of the various military conflicts that Israel has initiated or found itself involved in since 1948 is useful, because at those moments the dominant political credo of Zionism was tested on the battleground. And it prevailed. Israel has continuously "improved" its position by military occupation and genocidal expulsions, grabbing more territory. That these seven decades of armed conflict did not provoke serious opposition from the UN or its member states provides a key to understanding Israel.

The first chapter in Part 1 introduces the IDF—its background, history, "reality principles," operational methods, and developments. The six chapters that follow deal with the main "wars" in which Israel has been involved, wars mostly of its own making, in the seven decades since its inception. While each of these wars has its particular set of contextual conditions, aims, objectives, and concrete results, they all emerge from the sociopolitical status quo within Israeli society: that war is the best, almost the only means of advancing its settler-colonial agenda. In many ways, each war begets the next one, reinforcing the pattern.

Nations do not generally offer their land willingly to foreign occupiers, and Palestine is no exception. To continue and extend the occupation, Israel must use ratcheted-up force and violence, each war etching a higher level of suffering on its targets, which becomes normalized, a new "bottom line." This progression of violence is examined through Israel's six main wars to date, from 1948 to 2006. The wars and brutalities since 2006 are discussed in the context of other chapters.

Certain facts, so commonplace as to be almost obvious, nonetheless need to be examined to clarify underlying tendencies, as well as the resulting social and political processes that continue to shape Israel's social fiber. The simplest of these is the fact that this small nation has been involved in more conflicts and wars during its short existence than any other nation on earth; this fulfils two functions that Israeli hasbara has utilized throughout the period.

First comes the argument that Israel, never having had "peace," cannot count on politics, diplomacy, and negotiated solutions and must at all times be ready not just to defend itself but to preempt any likely attack by perceived enemies; this doctrine is the mainstay of the IDF, deeming such offensive acts as its absolute right, indeed, its duty, irrespective of normative international legal procedures. In effect, Israel claims that by virtue of its small size, the particular history of the Jews, and especially the Holocaust, it enjoys immunity from accusations of war crimes. Hence, Israel supposedly cannot commit war crimes, or rather, offenses defined as war crimes when committed by other nations are not war crimes when committed by Israel. This state-of-exception argument has been used successfully by Israeli hasbara and its many apologists.

The second argument is directed inward. Jewish-Israeli society is actively persuaded that there is no way of existing in Palestine without war and conflict, thus the necessity to recruit it toward the exceptional effort of sustaining a huge and brutal war machine. This persuasion is also useful for coaxing the society into the less-than-natural state of relentless, continuous war as a way of life—a normative process of not merely surviving, but of thriving through endless military conflagration. Without a perceived existential threat (a phrase much abused by Israeli leaders), society is less prepared to support such an overblown military machine, such a high level of the GDP spent on "defense," or the unusually high level of personal commitment of its citizens frequently asked to kill or risk being killed throughout their adult life.

It is impossible (as well as unnecessary) to cover here each instance of military action by Israel. Instead, the increasing militarization of Israel is examined through the main conflicts and social processes in the last seven decades. This explores the differences and similarities of Israel's armed campaigns against neighbors, as well as against its own Palestinian citizens, and obviously, the almost five million Palestinians now captive under its direct and indirect control, lacking civil, political, or any other form of human rights. By carefully observing the varying conditions pertaining in 1948, 1956, 1967, and 1973, a pattern of military-colonial involvement emerges, representing a chapter in Israel's history seemingly closed in 1979, with the various peace accords signed with Egypt and later with Jordan.

In the decades that followed, Israel's military might had been mainly used against nonstate actors—Fatah, Hamas, Islamic Jihad, and Hezbollah—in 1982, 2006, 2009, and 2014, in order to further Israeli and US interests in the region. Time and again, the two iconic targets of Gaza and South Lebanon are attacked, each time more devastatingly, to stoke fear and awe of Israel's military might. Internally, such attacks serve to reinforce the narrative of existential threat and cohere the nation, even though neither Gaza nor South Lebanon ever presented a mortal threat.

One of the interesting contradictions echoed in this history is that by definition, military action cannot always be successful. Indeed, since 1967 it was hardly ever successful even by the IDF's own terms or when examined against the political objectives that prompted aggressive action. Israel's clear edge over its adversaries has come to be questioned time and again, when small, relatively primitive bodies of popular resistance have managed to inflict heavy losses on the IDF and Israel, proving that military conflict is hardly the safest device in the political toolbox.

Equally important is that while such military adventures do not on the whole deliver their military or political objectives, they manage to unite Israeli Jews time and again. Israeli leaders can rely on their public when they go to war. In fact, the state had become more militarized and aggressive, more rightwing and extreme through conflict with nonstate organizations. The most obvious example is Netanyahu and his various campaigns, especially his "war on Iran"—a project carried over from the twentieth century that now looks close to realization, despite being totally unjustified, illogical, and highly dangerous for Israel and the whole volatile region.

In a sense, Israel's political aims have become clear since 1967. There was no good reason for Israel to partake in the 1956 Suez War, serving the declining British and French colonial powers. This war had certainly not advanced Israel's case, and it led to decades of conflict with Egypt and Syria. But, as Zeev Schiff noted: "The Sinai Campaign served as an expensive exercise in preparation for the Six-Day War."[36] As good a reason as any; thus, every war begets another.

After the 1967 victory Israel had achieved its ultimate goal—controlling the whole of Palestine—and the conflict with the Arab world became a distant memory. As long as Arab states did not challenge Israel's control of Palestine, it was hardly worth the effort to start a military conflict with any of them. The war in 1973 initiated by the Egyptian and Syrian armed forces, catching the IDF totally unawares, was the last time that Arab states were prepared to confront Israel militarily, claiming support for Palestine. A few years later, President Anwar Sadat flew to Jerusalem to speak before the Knesset, offering unconditional peace between the two nations, and was closely followed by King Hussein of Jordan. These moves removed any threat of armed conflict from the largest Arab state and Israel's immediate neighbor, allowing Israel free rein in its continued control of Palestine.

This becomes clear later on, with the advent of President Hosni Mubarak, after Sadat's assassination. Israel had no closer ally than

Mubarak, and his downfall in 2011 eventually brought to power General Abd al-Fattah al-Sisi, cut in the image of his predecessor and someone whom Israel can rely upon to support its aims of choking Gaza through a crippling blockade lasting more than a decade. This long-term development allowed Israel to concentrate on the settler-colonial aims in Palestine. While this conflict has clear international and human-rights aspects, it no longer involves Arab states, whose leaders ignore Palestine altogether. This has led to a radical change in the military balance of power as well as in the nature of the conflict, and hence, the nature and functions of the IDF.

The conflict, once referred to as the Arab–Israeli conflict (never an accurate title), is now being widely perceived as an Israel–Palestine conflict. As most states either support Israeli control of Palestine or are unwilling to challenge Israel beyond a periodic, symbolic vote at the UN, there is no threat of international or UN sanctions against Israel; the numerous UN resolutions on Palestine, demanding an end to occupation, and a return of the refugees of 1948, are ritualistic, ineffectual, and politically pointless. Israel has been free to focus its armed might on the Palestinian people and in all probability Iran and its allies in Lebanon are in its crosshairs. In these conflicts, Israel either represents the US position or influences it, the latter certainly applying under President Trump.

This crucial US support has allowed Israel to turn the large and powerful IDF into a policing and punitive occupational army, a role it has fulfilled for seventy years. That Israel's every move is determined by its relationship to the United States, as it once was hitched to France, may seem obvious, but the main machinery of serving the interests of such powerful nations (as seen by their rulers) is the IDF; Israel is not in any civic intercourse with the polities surrounding it, not even those countries with which it has peace treaties. Israel conducts relationships with the rest of the world, especially with its neighbors, through the IDF and the political and financial elites controlling it. This is as true now as it was in 1956, when Israel did Britain and France's bidding against Nasser; empires may have changed, but methods, military rationales, and modi operandi have not.

Israel should not be perceived through stereotypical and mechanistic models of imperial control; it is not a pliable or servile partner, blindly following orders. Indeed, the power relationship between the two may be seen in reverse by many Americans, who cannot see the use of Israeli's constant aggression for the US. Developments around the Iran nuclear deal during the Obama and Trump administrations are the best illustration of the substantial autonomy Israel enjoys for military action in the region. It is hardly a passive partner, and many times it is actively engaged in pushing its own agenda to the point of forcing it upon the United States. This complex relationship between the imperial center and a peripheral client state is dealt with in a number of the chapters in this book, offering a fuller understanding of these complex power relations and the IDF's role in such negotiated positioning. As the capabilities and reach of the IDF (and the related security apparatus) are Israel's main contribution to the US regional project, this military force has an effect not just on the tiny polity of Israel, but also on the political debate and priorities of the most powerful nation, through a vast network of open and covert channels of influence. Israel is the only country that is supported to the hilt by the United States and is also allowed to influence (sometimes even sabotage and divert) US policy, through its allies within the political elite in Congress and the Senate, not to mention the White House. One cannot think of any other army with comparable influence, not even the US military itself.

The role of the IDF as an occupation and settlement policing force, a task that engages the whole social structure in Israel, is not carried out very well, despite its many transformations. Arguably, the IDF is not equipped, inclined, or trained for the roles that have become its routine tasks. This is dealt with in the various chapters on each of the military campaigns (Chapters 2–7), which draw on Israeli official sources, as well as publications by prominent defense analysts and media correspondents. That Israel fails to consider political solutions should not surprise us; its inability illustrates the shortcomings of its political analysis, conducted for and through the IDF and its agenda; this has made a political solution nearly impossible. Ironically, this produces more problems than it solves and the longer it is allowed to prevail, the less likely is a lasting, just solution. Thus, the confusion between the military and civil society in Israel, which is a theme throughout this book, is the main difficulty of Israeli society. As the IDF navigates Israeli society into ever tighter corners, the stakes become higher and the situation a zero-sum game. While most of

Israeli society—brutalized, militarized, and dehumanized—may not be alert to its slide into the murky swathes of racialized and xenophobic policies and practices, it is evident to outside observers. History tells us that such internal processes within a militarized society are unlikely to lead to conflict resolution. I hope this book sheds light on the underlying sociopolitical processes in Israel and sends a sober warning about likely consequences.

The four chapters that make up the second part of the book deal with the state and its colonial appendages and actions—the vast armed colonial-settlement project, which has caused the intifadas of 1987 and 2000. Chapter 8 examines the early decisions taken by the cabinet to annex East Jerusalem and to start a settlement drive in the Golan Heights, West Bank, and the Gaza Strip as a way of securing the territories. It was clear to the government planners that such an illegal move and the incorporation of millions of Palestinians would make a "Jewish State" impossible. Despite this understanding, Israel's leaders preferred to build the settlements and used "security" arguments to justify this move, arguments that made no sense at all. The nature of the IDF had to change once settlement was pursued; instead of resolving security issues, settlements created security issues, and they predictably ignited Palestinian resistance to the illegal land confiscations and restrictions involved in the process of dispossession. In many ways, the settlements (the pet project of former defense minister and prime minister Ariel Sharon) not only changed the IDF and its practices (by requiring it to fight a civilian population rather than armies) but also transformed the nature of the Jewish polity of Israel beyond recognition, entrenching the settler-colonial experience and making it ubiquitous, as well as shifting the Israeli public sphere sharply to the right, toward apartheid.

Chapter 9, which focuses on the first intifada, examines the changes in the management of the settler-colonial project in the OPT and the developments leading to the Oslo Accords. The intifada brought about the end of a period in which the integrative vision of Moshe Dayan—many times a minister, especially minister of defense, 1967–1974—held sway. In this twenty-year period, Palestinians were used as a reserve army of labor, not only to build the new settlements but also to fill most of the unskilled and low-paid positions within the Israeli economy, as well as forming a captive market for Israeli goods.

This intimate contact with Israeli society has deeply affected and strengthened Palestinian society. The fear of Israel has been somewhat ameliorated, helping form a modern civil society with a clear position both on human and political rights. By 1987, the failure of the Dayan model was clear; Palestinians were rejecting an unequal and unjust partaking in Israeli society, demanding self-determination and equality, which was not something that Israel and the IDF were capable of or prepared to consider. The intifada delivered a deep shock to Israeli society, waking it from a daydream about the "success of the liberal occupation," as it was considered by both the left and the right. With the pretence of liberal occupation shattered, new methods of control and oppression were required and quickly developed. The great cost of such methods, both financially and socially, pushed Israel into the Oslo process, which provided a solution that would be safer for IDF personnel and attract less international animosity toward Israel and its occupation. Outsourcing security in the OPT to the PLO/Palestinian Authority was an ingenious solution: the cost of the system would be borne by the European Union and the United States, freeing Israel to further reinforce the settlement grid. While the IDF would grow smaller, as decreed by Barak's "small, smart army" dictum, leading to a more technological force, it certainly has not produced a more successful army; the major clashes with both Hezbollah and Hamas have proven that the fifth (or sixth) strongest army is unable to win against well-motivated guerrilla forces, though it certainly is capable of devastating noncombatant populations.

The chapter dealing with Israel's military–industrial complex—its global reach and financial and political functions (Chapter 10)—is crucial for understanding the way the IDF and its agenda shape and integrate the various social registers into a coherent militarized society. One of the benefits of this wide-ranging project has been to turn danger and difficulty into a business, crucial not just for the Israeli economy, but also for its diplomatic and political global outreach. The examples of collaboration with repressive and undemocratic regimes help to examine Israel's peculiar magnetism for such regimes, and the role it played in constructing a massive network of US bases of influence through Israeli arms and training contracts. The global role that Israeli cyber and information technology research now plays cannot be understood without mapping the gradual expansion into many more countries and areas of military and security arenas since the late 1950s. Indeed, Israel's successful economy, resistant to the shocks that overtook larger ones in 2008, could not have been built without the crucial role the military–industrial complex continues to play in its historical rising.

Chapter 11 describes recent developments in the IDF, as it turned from fighting armies to fighting civilians in the demoralizing environments of the refugee camps in Gaza, the West Bank, and Lebanon. This change is probably the most protracted and decisive in the long history of the IDF and has been also the most explosive politically. It had led to the blatant use of military attacks as election turners, and as stepping-stones for a number of political careers, notably those of Ariel Sharon and Ehud Barak. This new type of war has done much to normalize the agenda and methods of the IDF and inject them into the Israeli political agenda, a consequence of which is that there is more decisive public support for Israel's brutal attacks in Gaza and Lebanon, despite their clear inability to achieve a military solution to a political problem. The role of such conflagrations for the Israeli public has shifted from the functional toward the purely punitive and purgative. This development characterizes the current state of the conflict, and Israel's inability and unwillingness to search for political solutions is acceptable to most Israelis, who regard nonmilitary solutions as defeatist.

The third part of the book deals with Israeli society, its defining communities, and the tribal warfare constantly waged for control and influence. As explored in Chapter 12, the formative intra-Jewish conflict between the Ashkenazi leadership and the Mizrahi disenfranchised underclass has marked Israel deeply. While many of the worst examples of such racist conditioning have been transformed, the underlying sentiments and practices are unchanged, and the tensions and bitterness well-established and long-term. Such tensions were only deepened in the aftermath of the last large "European" Ashkenazi immigration from the former Soviet Union and its constituent parts after its dissolution in 1991. In a clear sense the favoring of the newcomers over the earlier Mizrahi populations has again inflamed the deep-seated fault lines in Israel. The later immigration of black Falashas from the Horn of Africa has inflamed the existing stresses, adding a color-based racialization to the existing othering methods. Official propaganda promotes the fiction that black Jews have been welcomed into Israel, but such window dressing cannot rid the society of its well-established racism.

But the deepest inequalities are those experienced by the Palestinian citizens of Israel. As citizens, they have never been considered equal to Jewish citizens in Israel, and for the first two decades of the state's existence, they lived under a military government. Yet their status as Israelis has never been under such a brutal attack as it is now. The chapter maps the legal and political efforts that have led to the disenfranchisement of this indigenous population, which constitutes 20 percent of the Israeli citizenry.

Chapter 13 examines and disproves Israeli claims of being "Jewish and democratic," as the society sheds any last democratic vestiges, turning into a fully fledged apartheid state not just in practice but de jure. In Israel as in South Africa, the antidemocratic nature of the settler-colonial regime was reinforced and battened by a tranche of racist legislation that was designed to institutionalize legal apartheid, to make the informal version sanctioned by law. The examination of Zionist and Israeli history proves that such a development was inevitable given the inherent inequalities within the Zionist project.

The final chapter is my attempt to assess the dangers and potentialities of the near future in light of the long-term trends examined in the book. I examine the claim by some liberal Zionists, such as Amos Oz, A. B. Yehoshua, and David Grossman, that another form of Zionism, a "humane" or "liberal" Zionism, is indeed possible or may be possible in the future. Such arguments are based on a denial or avoidance of the central tenet of Zionism—the perceived need of a settler-colonial movement to take over a territory and to dispose of its indigenous population by military force. By using tropes of "peace and coexistence" and "two states side by side," such apologists for Zionism are replacing rational, factual discussion of what is a most unequal power relationship between a militarized settler-colonial illegal administration and the indigenous population with a narrative describing a dispute between two equal sides about a piece of land. This pretence of an equality of claims has been the mainstay of liberal Zionism, providing cover for every war crime in the book.

Liberal Zionism has never existed and cannot exist by definition. There is no liberal colonialism, no more than there is a humane occupation. Both are creatures of the liberal Zionist imagination, where they live with another such phantasm, that of the "enlightened occupation" that supposedly followed the 1967 war. These forms of self-deceit are bizarre attempts to justify continued atrocities, allowing the perpetrators and their supporters to smear as traitor anyone who voices the slightest criticism. The concluding chapter examines such attempts at logic-defying intellectual legerdemain and puts them where they belong, with other acts of elite support for the illegal and immoral acts of criminal regimes.

The potential for change and the grave difficulties that bedevil any positive developments are examined in the context of the historical limiting factors—the settler-colonial nature of Zionism, the cultural foundation of extreme Zionism, the troubling traditions of Jewish zealotry, and the isolationism and separateness of Judaism that has arisen from historical antisemitism. Also examined are the tendency toward excommunication in small and threatened societies and the long history of militarism as the preferred method of negotiating reality, as well as the use of the "chosen people" myth as a tool for denying the claims of indigenous Palestinians.

The bottom line of this analysis is the need for international action on the scale and of the quality employed during the last decades of South African apartheid, in order to restore justice and the rights of the indigenous people of Palestine and to bring about a just solution to this settler-colonial conflict.

PART I

ISRAEL'S WARS

1

The Origins of the IDF

Jews wrested the Palestinian homeland from its people by force, and will—have to—ward off confronting that original sin. This, I believe, underlies the Israeli Jewish willingness to use extreme force against Palestinian civilians and the obscenely racist remarks about Palestinian parents willing to send their kids into the street.

Suad Dajani, "Yaffawiyya"

There is a well-known jibe about Prussia by Friedrich von Schrötter, who claimed, "Prussia was not a country with an army, but an army with a country."[1] This observation is, if anything, even more apt regarding the relationship between Israel and the IDF. Israeli militarism is not a fad of an extreme junta, but a firmly held belief among most Israeli Jews.

There is a fascinating relationship between military action and social attitudes toward militarism in Israel captured in the following expression: the more the medicine is used, the more it becomes the preferred method for curing the malady. Like some medicines, its efficacy might be doubtful, but that does not deter its usage; rather, it becomes an argument for a dosage increase. The prevalence of military action only makes the inclination for its use more automatic, with failure unlikely to reduce its appeal: when military action does not work, it is seen as an indication that insufficient force was used.

The more a society specializes in using force, the more the world looks like a dangerous place where one needs to use a big stick. As the saying goes: to a hammer, everything looks like a nail. These patterns of social addiction should not surprise us.

The Israeli state and society have been formed by militarism since the early stages of Zionism, without which controlling Palestine would have been unthinkable. The frequent and habitual use of the military at most junctures of Zionist history since the 1930s created deep furrows in the geopolitical landscape of the Middle East, ones consistently frequented, while offering no stable or sustainable existence. If in most states the use of military force is the option of last resort, after all other avenues have been exhausted, in Israel the use of military option comes first, negotiations used only after military options have failed.

The sheer primacy of Israeli militarism was, for decades, a taboo subject, repeatedly denied by Zionist historians and sociologists.[2] Baruch Kimmerling notes,

> As opposed to the prevalent position in most of the Israeli social science research, which strenuously denies the existence of Israeli militarism, and even the possibility of its creation, Israel has indeed developed a militarism, which, while changing in its form and intensity, has become one of the *central organising principles* of this society. Israeli militarism has mainly developed as one of the main reactions to the conflict created by the Zionist settlement project in Palestine, vis-à-vis the Arab environment which has perceived it as a definite colonisation project.[3]

Militarism has created material interests for the continuation of conflict in Palestine; it has become a crucial element of Israeli society.

The transformation of Israeli society from the fabled People of the Book to People of the Gun should not be underestimated. This was a complex and unique process of social engineering. It involved deep changes in the understanding, outlook, and culture of Zionist Jews and, after 1967, of Jewish communities around the world. This entailed creating a people from the divided followers of sacred texts and devotees of cosmopolitan idioms—the ultra-Orthodox Jews of the ghettoes and the liberal, radicalized metropolitan Jewish youth of the *fin de siècle*. Such a profound transformation was almost unimaginable; understanding its narratives, conditions, and complex development is crucial to understanding Zionism as theory, praxis, and historical phenomenon. This transformation was completed in a relatively short time.

While all modern states have some type of military force, most states could not be described as militaristic. Militarism is a deeply rooted social attitude—the idea that the prime vehicle for resolving national, international, and social conflicts is armed force results from complex sociohistorical processes.[4] Although militarism is socially constructed, there is nothing essentialist about it. In the Israeli context, Uri Ben-Eliezer argues that "the use of military force acquires legitimation, is perceived as a positive value and a high principle that is right and desirable, and is routinized and institutionalized within society."[5] We need to probe the specific conditions that created it.

To map Israeli militarism, one must examine Zionist colonization, a somewhat untypical enterprise that was launched when such projects were winding down. The colony had at least one empire protecting it and claimed a territory for itself not by overt conquest and military might, but by stealth and the gradual dispossession of the indigenous population and erosion of its rights. The move toward military force was used to expand its patrimony when other methods failed. Zionist narratives use tropes of "national liberation struggles" that have become the accepted history, occluding the colonial context.

This constructed liberation narrative is not at all obvious; after all, Theodor Herzl describes the project clearly as a settler-colonial enterprise.[6] Zionism set up a colonial bank, and its early habitations were called *moshavot* (colonies)—indeed, they are still called that today. Before the word *moshava* was coined, the Hebraized *colonia* was used. The colonizers themselves were fully cognizant of their methods, with Zionist historians later obfuscating it. That Zionism also offered East European Jews respite from antisemitism does not change the colonial contours of the project.

The Zionist project shares characteristics with earlier colonizing projects. In North America and South Africa, colonies were founded by small, religious groups of settlers escaping xenophobia and oppression in Europe. New England Puritans or Cape Town Dutch could hardly depend on the might of empires against indigenous populations. (In fact, many decades passed before the colonial bridgehead became a desirable asset of empires, one worth assisting.) The early Zionists of Khibat Zion (Love of Zion) were victims of religious oppression while they remained in the Russian Pale of Settlement. Like other European colonists, they settled in Palestine as farmers, buying or snatching land from local landowners, and employed the peasants (fellahin) as agricultural labor on the very land they had previously tilled.

Such exploitation led to deep-seated resentment against the dispossessors. In Palestine's old land-ownership system, villagers owned arable land in common (*musha'a*), sharing the labor of tilling it and the fruits thereof. Change was introduced by land parcellation and registration under the Ottomans in the eighteenth and nineteenth centuries, yet this hardly changed the fellahin's relationship to the land. While the land may now nominally "belong" to an absentee landlord, the fellahin treated the land as theirs, despite the obligation to pay a tithe. All this changed with the first Zionist colonies in the 1880s.

The incoming owners bought land not for investment purposes, but rather with the intention of living off it, despite their lack of farming experience. Like English gentlemen farmers, they had no intention of working the fields, a task thought unfitting for European owners; instead, they employed fellahin as day laborers. The fellahin, who until then had little use for cash, living in a rural, autarchic economy, were now exposed to a capitalist system, however primitive. During the first two decades of settlement, before 1905, tensions between the peasantry and the *colon* greatly increased, occasionally becoming violent, when peasants considered the colon especially unjust.

After the Russian government inspired pogroms in 1903–4, thousands of Jewish refugees arrived in Palestine. The newcomers of the second wave of immigration (referred to in Zionist history as the Second Aliyah) were mainly metropolitan workers from Russia and Poland, radicals purged after the failed revolution of 1905. The Tsar's reaction toward the society's most vulnerable—the poorest Jews living in the deeply hostile, antisemitic Pale of Settlement—was vicious. For many of these young workers, leaving Russia was the only way to survive. On arrival in Palestine, such ideologically motivated Zionists with a modernist leftist outlook (but no funds) were unable to purchase land and employ the fellahin like earlier colonists. They purchased land communally and tilled it themselves.

This innovative mode of Zionist colonization—the kernel of the collectivist Kibbutz movement—offered Zionism new opportunities. Until then established Western and Middle European Jewish communities had either ignored Zionism or actively shunned it as dangerous. For some, it threatened Jewish life in Europe (as well as Palestine) through adoption of antisemitic precepts (for example, claiming Jews did not belong in Europe). Denied support by European Jewish communities, Zionist leaders now looked to young and radical committed settlers, who were willing to face adversity en route to their nationalist social utopia. As a result, the movement seemingly morphed from a bourgeois-nationalist organization to one with a socialist agenda.

Though the first *kvutzot* (collective farming colonies) had to struggle with insurmountable difficulties such as severe shortages of food and tools, lack of agricultural experience, and a marshy floodplain infested with malarial mosquitoes, the newcomers' potential was soon realized. As the numbers of collectives grew, a pattern for the future Jewish State emerged. The new settlements were located in the periphery of Palestine, rather than the central coastal plain, where most earlier colonies were built. Both the Zionist leaders and the young settlers believed that settlements demarcated the boundaries of a future Jewish State. The new settlements also found the fellahin, who lost their land to the incoming "Muskovy" (*Muscovites* was used by the Arab population as an iconic shorthand for all newcomers), much more hostile than the earlier settlers. The dispossessed were no longer offered work, and the new settlements were closed to them, as the settlers wished to work the land themselves.[7]

As tensions increased and conflicts spread, the Zionist leadership understood that this animosity was likely to intensify. The fellahin, who were expected to appreciate the newcomers' modernity, machines, and medicine, proved to be belligerent. Other sources of conflict were the kibbutzniks' sartorial codes and lack of modesty, their refusal to learn Arabic, and their use of scarce water resources.[8] They did not bother getting to know the land and its inhabitants, whom they perceived as primitive rural brutes, treating the indigenous population in a way similar to colonists in North America centuries earlier. This could only lead to a violent conflict, and Zionists started preparing.

Beginnings of Military Organization

The indigenous Arab population, seeing itself as part of the Arab Mashriq (East), started organizing during the early stages of Zionist colonization. On June 24, 1891, Arab notables grouped together to petition the Sultan against Jewish immigration, without success. In their deposition, they claimed with surprisingly accurate foresight that if more Jews arrived in Palestine "the Muslims themselves will be the sufferers, as the European Jews being skilled in all kinds of trades, the Muslims could not compete against them."[9]

In September 1907, as tensions between Zionist colonies and neighboring villages mounted, a number of young Zionists set up a secret association of *Shomrim* (Watchmen) called Bar Giora (named after a Jewish zealot resisting Roman occupation).[10] This clandestine organization—illegal under Ottoman law—used the following revealing motto: "In blood and fire Judea fell, and in blood and fire Judea shall rise."[11] The increasing tension led in 1909 to the setting up of the larger *Hashomer* (Guardian) armed militia for defending Zionist settlements. Many of the founders of Hashomer later became key figures in Zionism, including Yitzhak Ben Zvi (Israel's second president), his wife Rachel Yanait, Israel Shochat, and David Ben-Gurion. Hashomer prefigured the Hagana set up in 1920.

The perceived Jewish demand for self-defense was influenced by the worldwide sense of horror after the April 1903 Kishinev pogrom, in which almost fifty Jews died and many more were wounded (and some were raped). This pogrom was instigated by the Russian secret police under the guidance of Interior Minister Vyacheslav von Plehve, a deeply antisemitic politician who supported Zionism because he believed it would rid Russia of its Jews. Following the pogrom, left Zionists and Bundists in Ukraine tried to set up self-defense organizations for protecting Jewish communities. This received much coverage in Oddesan journals and newspapers but was rejected by Jewish communities as dangerous radicalism. This was also Herzl's opinion expressed in his meetings and letters exchanged with von Plehve, which local Zionists followed.[12] This attitude prevailed until the uprising of 1905.

After 1905, the newly arrived young Palestine Zionists were mostly veterans of the anti-Tsarist uprising, having escaped Russia to avoid arrest. They turned their opposition toward the decaying Ottoman Empire, and as the distant Great War was brewing, it became clear that massive changes were under way. A number of young men from Zichron Yaacov, one of the older settlements supported by Baron Edmond James de Rothschild (and named after his father), settled in nearby Shune forming a protomilitary organization called *HaGid'onim* (the Gideonites). This assembly laid down the foundations for a Jewish military force. The group translated its opposition to the Ottomans into active treason by setting up a covert spy network known later as NILI to assist the British.[13] This was the beginning of the special relationship between Zionism and Britain and would help to bring about the Balfour Declaration.

Many Russian socialists in Palestine had military experience, having served in the Russian Imperial Army during the Russo–Japanese War. A leading figure was Joseph Trumpeldor, who joined the army in 1902, returning to serve after losing an arm in 1905, only to be captured by the Japanese at Port Arthur. In captivity Trumpeldor discovered Zionism, forsaking his Russian nationalism. On his release in 1906, he was one of the first Jews to be decorated in Russia, the first Jewish officer of the Tsarist army. Nonetheless, he left for Palestine in 1911, ending up in Degania, a new kibbutz beside Lake Tiberias. At the outbreak of World War I, Trumpeldor ended up in Cairo, together with fellow countryman, Vladimir Jabotinsky, a prominent Odessan intellectual.

Jabotinsky and Trumpeldor not only believed that Palestine was theirs by right, but also that the area now forming the state of Jordan was destined to be included in the future Jewish State. They knew that such a large territory could not be legally purchased and so they planned a Zionist army under British command for conquering Palestine from Turkey. Jabotinsky was certain that this would involve force, as he later wrote: "That the Arabs of the Land of Israel should willingly come to an agreement with us is beyond all hopes and dreams at present, and in the foreseeable future."[14]

A modern army had to be set up, equipped and trained without delay, and Trumpeldor was directed to this task. The outbreak of World War I was a godsend for Jabotinsky and Trumpeldor: here was the chance to bargain with political powers, offering a force of committed Jewish soldiers. Jabotinsky, Chaim Weizmann, and other Zionist leaders managed to persuade the British command in Cairo that a Jewish military force might be useful, and the Zion Mule Corps was formed in 1915, as the British were gearing up to fight in Gallipoli.

The battalion saw action in the Dardenelles and was evacuated with the rest of the British Expeditionary Force, ending up in Britain where it was disbanded. Trumpeldor served as an officer in this battalion and got wounded again; he returned to Russia, where he trained Zionists preparing to leave for Palestine. He returned to Palestine after the Ottoman defeat and he and Jabotinsky then spent months persuading the British that Jewish battalions would assist victory. Eventually three Jewish battalions were established in 1917, dubbed by Jabotinsky rather grandly as the Jewish Legion. Jabotinsky himself served as a lieutenant in the 38th Royal Fusiliers in Palestine in 1918, during the end of the war against Ottoman Turkey. The Battalions (ancillary units that saw no action and disbanded) were a great disappointment to both friends but led to the establishment of the Hagana in 1920.

Not all Zionists shared these militaristic sentiments—some actually resisted militarism. One leading political figure criticizing the need for a military force was Eliezer Yoffe, one of the leaders of the Yishuv's (the Jewish community) second-largest party, Hapoel Hatza'ir (Young Worker). Yoffe attacked the removal of workers from productive agriculture, for military training:

> We have as yet neither a nation nor a land, yet already we aspire to an army. Is this a primary need? Is it necessary to take people who are now in the Land out of it in order to guard it? It is like uprooting the tress from a garden to make a fence around the garden. But if the trees are uprooted—where is the garden and why a fence around it?[15]

But his was a voice in the wilderness.

The Hagana (defense) was conceived as a national force of regional units with a centralized command, training, supply and intelligence. This force was illegal under the British Mandate regulations, which forbade the setting up of militias in Palestine. Nonetheless, the first high commissioner for Palestine, Herbert Samuel, was an ardent

Zionist who closed his eyes to its existence. The Palestine Mandate authorities understood the need for a military force if the Zionist movement was to control the whole of Palestine. Thus, the Hagana was treated with a wink and a nod by the administration, satisfying the Zionist Federation, the newly set up Jewish Agency, the Histadrut (Hebrew Federation of Trade Unions), and other Yishuv bodies including the political parties and cultural institutions, set up after the British takeover. The military force, though clandestine, was part of future state machinery. The Zionist leadership, functioning like a state-in-waiting, was distinct from the Palestinian leadership.

Despite the high hopes invested in the Hagana, it was an odd organization. The centralized command structure was countermanded locally by regional commanders from different factions, more committed to political interests than "national" ones.[16] Hagana units were placed in kibbutzim, which, while representing a tiny portion of the Yishuv, became crucial to the militia's ideology and doctrine. This meant unnecessary duplication, causing frequent conflict. Although established by the rightwing Jabotinsky, the leftwing parties soon started vying to control it.

After splitting from mainstream Zionism in 1923 due to differences with Chaim Weizmann (then chairman of World Zionist Organization), Jabotinsky worked on building the Alliance of Revisionists-Zionists and its youth movement, Betar. This would lead to the IZL (Irgun Zvai Leumi, National Military Organization) known in English as Irgun, in 1931. The British expelled Jabotinsky from Palestine in 1930. He never returned. For almost two decades Irgun fought both the British and Palestine's Arabs. Jabotinsky, close to Mussolini in his outlook, made use of his links with the Italian regime during the 1930s, getting the Italians to set up a training camp in Bari for IZL's naval wing in 1934.[17]

Jabotinsky forced the Zionist movement into its deepest crisis by dividing it between left and right in 1923, and the echoes of this bitter split resound even now. (However, there was little difference between Jabotinsky and his arch-rival Ben-Gurion on the central tenets of Zionist colonization.) Both factions were united on the main problem of Zionism—the Palestinians. Both agreed that force was necessary because the Palestinians would not give up their country willingly, but the two men disagreed on timing and manner. Ben-Gurion saw

Jabotinsky as a game-spoiler who revealed future moves before it was strictly necessary or wise. According to Ben-Gurion, Zionists would be unable to overcome both the Arab population and the British Mandate authorities by force during the 1920s and 1930s, so they should bide their time. For them, Jabotinsky was a necessary evil, the ideal man to set up and inspire the Hagana, after his arduous battle to set up the Jewish Battalions.

Nonetheless, the conflict illustrates some key points. While Ben-Gurion and the left of his party considered themselves socialists, and Hapoel Hatzair members also thought of themselves as internationalists and anti-imperialists, they never supported a common struggle with Arab workers, whom they did not consider partners. Theirs was a narrow settler-colonial, racialized, nationalist platform. Despite homilies of "brotherhood of man," no Arabs were accepted as members in the Kibbutz movement or Histadrut.

Most Zionist historians—from Nahum Sokolow, Adolf Böhm, and Ben Zion Dinur to Anita Shapira, Walter Laqueur, and Hillel Halkin —avoid dealing with these realities, which Jabotinsky never shirked; he understood the opposition of Palestine's indigenous population to Zionism and expected it to increase, hence a strong army was needed for the Zionist takeover. This was accepted by the left but fudged in order not to negate the humanistic tradition they projected.

For example, during the Arab Rebellion of 1936–39, Poale Zion and Hashomer Hatzair (the two leftwing Palestine Zionist sections) agreed to assist the British forces in defeating the Palestinians, even though they considered themselves to be anti-imperialists.[18] In *A History of Zionism*, Walter Laqueur acknowledges the difficulties faced in the 1930s: "The British administration in Palestine was by no means totally sympathetic towards Zionism, and the Arabs were actively hostile."[19] Did he expect the Palestinians to be more helpful to their dispossessors and disappear?

This tendency to obscure the broader picture in favor of a limited, racialized perspective of Zionism is evidence of a deeper problem, lasting over a century. According to Ben-Eliezer, such a limited political and conceptual framework is the result of Zionism's ongoing process of militarization.[20] When the only option considered is military force, the Other can only be conceived of as an enemy to be eliminated. According to Ben-Eliezer, the historian of Israeli militarism, the contradiction between the leaders' moderate language and their use of military force was crucial; they employed "moderate rhetoric" even as "they acted aggressively."[21]

Building the Nation

The intention by the Zionist movement to establish a large and powerful military force was anything but obvious. Even following the Balfour Declaration in 1917, Jews (mostly non-Zionists) only made up less than 8 percent of Palestine's population. How could a small minority take over the whole country and dispossess its indigenous population?

That it could not happen in the manner described by Herzl in his diary was obvious.[22] While some effendis were willing to part with land for large sums, the peasants were unwilling to do so, despite Herzl's plan to "spirit them away." After 1921 it seemed clear that the growing Palestinian national movement would oppose the Zionist project with all its might. Even with the substantial support of Mandate authorities, who were committed to the Balfour Declaration, there was no way that Zionism could create a national home in Palestine without use of military force. This was openly admitted by Jabotinsky but avoided by left Zionists, led by Chaim Weizmann, David Ben-Gurion, Moshe Shertok, and Berl Kazenelson. They recognized the need to build a military force but also the necessity to keep quiet about it.

After the establishment of the Hagana in 1920, it was clear that the process would take decades. The urgent task was to use the British to expand the Zionist base in Palestine. The Zionist left realized that any mention of military options was dangerous, so they preferred lulling the Palestinians with talk of cooperation and coexistence.

But this was not the only reason for the growth of Zionist militarism. Jabotinsky, and later also Ben-Gurion, understood the need to turn the refugees congregating in Palestine into a nation. Using armed force to transform Jewish society in Palestine became the main tenet of Ben-Gurion's political program in the 1940s and 1950s, the formative period of Israel. His conceptualization of Israel as a "nation-in-arms" also contained the belief that the nation itself was yet to be forged. Even during the 1951 election campaign, three years after the Declaration of Independence of 1948, he noted: "I see in these elections the shaping of a nation for the state; because there is a state but not a nation."[23]

The Zionist project concentrated on securing the land—building state machinery and an army to occupy the country and expel its indigenous inhabitants—but neglected the creation of the nation itself. Despite claims to the contrary, Judaism was not a nation, certainly not for the last two millennia—the nation was a very recent phenomenon. Zionists persuaded themselves that Jews were a nation, despite evidence to the contrary. Ben-Gurion and Jabotinsky understood the difference between propaganda and realities on the ground and knew there was an urgent need to construct a nation, now that the state, army, and national institutions were in place, creating a semblance of a long-standing entity while occluding a lacuna.

The army appeared to Ben-Gurion, as to Jabotinsky before him, the ideal social mechanism for constructing the elusive Israeli-Jewish nation. Both believed a distinction needed to be drawn between the Jewish Diaspora experience (the "old Jew") and the Zionist pioneer (the "new Jew," a term coined by Herzl).[24] This required a massive social engineering process, *kur hituch* (crucible), which would melt together the different ethnicities, cultures, languages, and beliefs, forming a hybrid Jewish community in Palestine—a new nation.

In order to claim Palestine, therefore, Zionists knew that they had to invent a long history rather than a recent "new nation," a problem encountered by all "imagined communities."[25] The problem was this: how to reject and deny the Ghetto Jew—that despicable stereotype of the antisemites—yet claim Jewish historical continuity with Palestine.

The solution was nothing short of ingenious. Zionism rejected the old Diasporic Jew while embracing the mythical biblical Jew, selecting a gallery of military leaders of past revolts against the Hellenic and Roman empires. They invented an unambiguous trait of Jewish militarism, building an ideological foundation for the new army destined to become the state's central institution. Thus, Jewish history underwent a radical transformation—2,000 years of tainted Diasporic history were excised, in an unstoppable advance toward national bliss in Palestine. The new Jew became the imagined heir of genocidal Joshua, Bar Kochba, and Judas Maccabeus—a rogues gallery of militarized heroes who formed the foundation for the conquest of Palestine—suggesting methods and requisite mental attitudes.

Employing the IDF as a grand social deus ex machina for this transformation was almost obvious. After all, the new state had a large, effective army, its most important organ. At the height of the long 1948 war, the tiny nation-in-formation numbered around 625,000 Jews. At the same time, the army mushroomed from 80,000 in mid-1948, to more than 110,000 by the end of the year, or some 17 percent of the total Jewish population.[26] Taking into account women, children, and old people, it is clear that all able-bodied Jewish men were under arms. Hence, the newly created IDF (whose title belied its function) was the ideal instrument of social change.

Despite the difficulties involved, Ben-Gurion cherished this miraculous process, translating the Prussian *Das Volk in Waffen* (the Nation in Arms) into the Hebrew as *Ha'am Kulo Tzava*, a term accurately descriptive and deeply ideological. Thus, the new community, the nation, is an army, and the army is the nation. By strongly opposing the left Kibbutz movement (which demanded a smaller, militia-based volunteer force),[27] Ben-Gurion created the army of his dreams—a centralized, authoritarian, ideological, and professional body—an instrument of occupation, expulsion, and subjugation. This was a popular army (if one overlooks its racialized base), creating and fusing the nation, constituting an effective war machine whose task was to deliver the old country to the new nation.

For such a huge muscular instrument to be effective, it needed to be constantly active and based on offensive rather than defensive strategies. Moshe Dayan, speaking in April 1967, put it succinctly: "The Israel Defense Force is a decidedly aggressive assault army in the way that it thinks, the way that it plans, the way it implements. Aggression is in its bones and its spirit."[28] It is safe to assume that Dayan was not aware of the inherent irony of his statement. All Israeli leaders since Ben-Gurion have used the army for the same ends and in similar ways. The many wars and instances of military aggression since 1947 became junctures of social cohesion, of the tribe gathering around the bonfire of combat, a coming together of warriors.

This also explains the great popularity of battle in Israel. Each war is always supported by huge majorities within the Jewish population and is a crucial instance of national fusion and identity avowal, seen as "great and positive moments"[29] of cohesion and human compassion. This "human kindness" is highly ethnocentric, excluding the victims, as they do not belong to 'Am Habchira (chosen people); rather, they are dehumanized by racialized propaganda and hence can be killed with impunity, as *homo sacer* (accursed man).

Elsewhere, the IDF use for social cohesion and identity formation had some real achievements, including offering very limited equality to women and Mizrahi Jews. The role of the IDF has been contradictory, on the one hand praising equality, and on the other (due to prevalent racial and sexual harassment within the IDF) enshrining inequalities. Women were initially not allowed into the IDF's combat units, though they did serve as fighters in the prestate Palmach militia. According to Nira Yuval-Davis, women were "formally barred from the front zone and/or combatant roles" because of patriarchal traditions concerning the "proper" roles for females and males.[30] What many Israeli male officers think about women was elucidated in an infamous lecture given to high school students by Colonel Gershon HaCohen in January 1995: "From the earliest times men were warriors and women were whores, and most women in the IDF fulfill administrative duties, that a computer and answerphone would replace with similar results."[31] Such views are far from exceptional. Similarly, due to Israel's culture of racism, Mizrahi Jews were initially allocated support and junior roles within the IDF; this has changed, though full equality is a long way off. The same approach was used for the inclusion of lesbian, gay, bisexual, and transgender individuals and vegans, for example.[32]

The role of the IDF has been contradictory: it praises equality even as it enshrines inequalities, as seen by prevalent racial and sexual harassment within the IDF. Thus, the IDF can be seen acting in contradictory fashion at specific junctures; examined over longer periods, clear changes can be observed in the social positioning of both women and Mizrahim. The fact that IDF claims to treat most Jews equally has strengthened its social standing, even if realities are far from stated claims.

Other groups of IDF soldiers are also interesting in this regard. Men of the Druze, Bedouin, and Circassian communities are expected to serve in the IDF, though the percentage of serving men varies between the groups. Bedouins' readiness to serve has sharply declined over the last decade, with Naqab Bedouin communities targeted by the state, like Palestinians in the OPT. The repeated destruction of "unrecognized" Bedouin villages and the state's refusal to provide services of any kind has affected the relationship between Bedouins and the IDF.

The Druze and Circassians serve in the most vicious IDF units such as the Border Guard and treat Palestinians in the West Bank brutally, reflecting Israel's divide and rule policy. Recently, after the passage of the 2018 nation-state bill, the Druze community has been deeply offended by this further disenfranchisement, after decades of falling readiness to serve in the IDF.[33] As a result, some Druze officers have given vent to their anger; a twenty-three-year-old officer, Shady Zaidan, has been suspended from duty after announcing in a Facebook post, as reported in *Haaretz*,

> his intention to leave the army due to the passage of the controversial legislation. "I'm a citizen like everyone and gave my all to the state ... And in the end, I wind up a second-class citizen." Zaidan described himself as a deputy company commander in a combat unit, in which he served for the past five years. "I'm not prepared to be a part of this. I'm also joining the struggle, I've decided to stop serving this country," said Zaidan's post.[34]

Zaidan was the second officer to make such an announcement. In all likelihood, Israel will not be able to sustain recruitment from the three communities for much longer; it is safe to assume that the passage of the racist nation-state bill in 2019, turning Israel from a *de facto* to a *de jure* apartheid state, will continue to reverberate, damaging the social structure in Israel.

Transforming people from disparate cultures, with little that binds them, into an efficient, modern fighting force was (and remains) a slow and complex process. If at the start of the 1930s the Hagana was a large, mainly underground militia equipped with light arms, ostensibly trained for defense, the following decade created a unified, Spartan,[35] and disciplined force, with modern armaments and trained for assault. A number of factors enabled such swift transformation.

Key Phases of Militarization

The 1930s saw Palestine change beyond recognition, due to the concatenation of social and political processes. The rise of Nazism brought about a sudden official Judeophobia and harsh measures against Germany's Jews, forcing many to consider emigrating. Faced with the opportunity of capitalizing on this human disaster, the Zionist organizations signed the Ha'avara Agreement with the Nazi Party, allowing German Jews to take part of their property when immigrating to Palestine. This ensured that Palestine became the only practical destination for German Jewish migrants leaving with capital.[36] This would have been impossible without the collusion of the British Mandate authorities, who changed entry requirements by adding a new category, "capitalists," which allowed entry for anyone possessing £500 (later £1,000). An enormous number of people entered Palestine on top of the immigration quotas, which were calculated each year according to the "economic absorption of the local economy."[37]

Ironically, Mandate authorities calculated this elusive condition based on population figures, so that the more refugees came into Palestine, the larger its economic capacity became, doubling the quotas from an annual average of a few thousand. Between 1930 and 1939, the Jewish population climbed from a mere 11 percent of the population of Palestine to 33 percent. This required capitalizing Palestine's economy, building a modern port in Tel Aviv and extending the others, as well as developing a manufacturing sector. All this massively increased the Jewish middle class in cities and towns, where a building boom supplied housing for hundreds of thousands from Germany and eastern and central Europe.[38] The British Mandate authorities justified this influx by listing supposed benefits: the creation of new jobs, improvement in services, and economic modernization.

Though the benefits were real, they did not apply to Palestinians, but only to the Zionist community that had no intention of sharing either economic benefits or the country itself. An efficient campaign against the employment of Palestinians quickly emerged, based on the racializing *Kibbush Ha'avoda* (Conquest of Labor) and *Avoda Ivrit* (Hebrew Labor) campaigns, strongly supported by Ben-Gurion since the 1920s.[39] Noncomplying employers were punished; most relented.

That this policy was presented by left Zionism as a campaign against cheap labor did not fool anyone, least of all the Palestinian workers. It was a nationalist and racist campaign against the existence of Palestinians in their own country, and it was assisted by Zionist industrial muscle, the Histadrut and Zionist youth movements, the substantial cultural and artistic sectors, academics and teachers, health organizations, and the kibbutzim and moshavim. The Arabs of Palestine had no chance against this European influx of modernist nationalism, assisted by compliant British administration bent on building a Jewish State in Palestine, euphemistically known by the quaint phrase "Jewish National Home."

Many absentee landowners of Palestinian agricultural land also assisted the Zionist project by selling lands on which thousands of Palestinians subsisted. By so doing, they removed their countrymen's livelihoods, turning them into an impoverished subproletariat[40]; they lacked the means of organization and struggle built over decades in the developed world and available to Jewish workers.

Within a decade, Palestine was lost to its inhabitants. The British administration, supposedly there to guarantee their rights, combined forces with the Zionist colon, leaving Palestinians exposed and helpless against the European influx. This was an innovative pattern of colonial occupation, built upon the empire's gunboats and League of Nations Mandate.

This resulted in a general strike of Palestinians, culminating in the 1936–39 Arab Rebellion. The British were unprepared for the intensity of Palestinian resistance and had to ship much of the British army there, despite trouble brewing elsewhere. More than 25,000 British soldiers were needed to brutally suppress the Arab Rebellion. (This contrasted radically with the earlier years of the Mandate, when less than 1,000 British soldiers sufficed.) The British empire was in flames, and subduing Palestinian resistance proved most difficult, despite the small size of the country, the small numbers of resistance fighters, and the additional and enthusiastic support of Zionism.

In the breach, the Jewish *notrim* (guardsmen) militias were recruited by the British, as they knew the countryside, as well as the intimate details of each Arab habitation, through large, secret operations started at the end of the 1920s—mapping and recording the Arab population of Palestine, conducted by the Sherut Yedioth

(Intelligence Service) of Hagana. A huge database called Tikey Ha'-Kfarim (Village Files) was created and used to identify, capture, and kill activists in each Arab neighborhood, decapitating the rebellion.[41] The Hagana intelligence service was superior to that of the British, and it was crucial for the eventual suppression of the uprising in 1939. This information once again became crucial during the 1948–49 war, allowing the IDF to round up and kill most of the active nationalists in the Palestine countryside.[42]

An added benefit of the suppression of the Arab uprising was that the British garrison in Palestine was supplied almost exclusively by the growing industries of the Zionist project. The financial siege of Palestinians was not less significant than the military fight against them, and in a sense, continues today.

The 1930s were rife with new modes of military organization, mainly in countries such as Japan, Spain, the Soviet Union, and Germany, which "innovated" the means of military cohesion by combining them with popular youth movements. Other types of groups were also developed, such as the SA (*Sturmabteilung*, Storm Detachment), Brownshirt thugs used by the Nazi Party since 1919 against the Left and labor unions, and the SS (*Schutzstaffel*, Protection Corps) which, despite its name, was an aggressive paramilitary organization developed by the Nazi Party of Germany in the 1920s. Under Himmler's leadership during the period 1929–45, the SS became the regime's shock troopers. Another model was the Spanish Falangist forces of Franco, which defeated the International Brigades fighting in support of the Republic.

Such models, the hallmarks of militarizing societies, had an effect on similar formations in Palestine. In 1936, Hagana established Fosh (acronym for Field Battalions), squads fashioned on European models, intended for mobile, swift offensives against Arab villages during the uprising. By early 1938, the Fosh included some 1,500 highly trained commandos, organized in regional groups with modern arms stolen from or donated by the British forces.[43] These bands acted under cover of darkness, presaging IDF later raids. Training included Arabic classes and patrols into villages, sometimes in Arab garb.

This was closely followed in 1938 by establishing Plugot Ha'Layla Ha'Meyukhadot (Special Night Squads, SNS) by Captain Orde Wingate, an eccentric British pro-Zionist.[44] These covert units were made up of mostly British but also some Zionist fighters, with support from the British army. The corps included the young Moshe Dayan and Yigal Allon. While the units' official task was protection of the Iraqi Petroleum Company pipeline in the north of Palestine (a target of Palestinian resistance), their operations soon expanded. For example, one operation included shooting every tenth man in a village to force others to volunteer information.[45] The brutality shocked even hardened Fosh veterans who joined SNS, and a number of them complained (unsuccessfully). Wingate had no compunction about killing scores of fighters in cold blood, as reported by the Israeli novelist Yoram Kaniuk,[46] and this would have a powerful influence on the future direction of the IDF. Zionist SNS officers became the military leaders of the 1947–49 war, and SNS methods formed battle doctrines of the Palmach, discussed below, and later of IDF.

Following these early developments, the Hagana secretly established the Special Operations branch in June 1939.[47] By that time the Palestinian uprising against the British had been suppressed, and the unit was intended for the next phase of the conflict. Inspired by the SNS, the unit used "unauthorized" brutal actions,[48] killing villagers to elicit collaboration and indiscriminately destroying houses in "uncooperative" villages. Such brutal revenge became ubiquitous, prompting criticism within the Hagana, leading to a protocol "for maintaining discipline in the security force, to prevent pointless killing."[49] Complaints were also made to the British HQ but never resulted in any action against the SNS and its commander.

The defeat of the Arab Rebellion with over 9,000 Palestinian casualties was inevitable, bearing in mind the combined forces of the British Army, the Royal Air Force, the Palestine Police and the well-equipped Hagana Notrim (Auxiliaries). Most nationalist political and military cadres were eliminated, and others were incarcerated. The conflict had forged a close relationship between the British military and Hagana, which in turn led to the Mandate authorities turning a blind eye to the massive arms imports by Zionist militias. While Palestinian fighters were shot on sight if caught armed, the British were training and arming the Hagana auxiliaries and SNS.

The Opportunities of WWII

The last vestiges of resistance in Palestine were extinguished by the summer of 1939. The timing was crucial for Britain, which needed to extricate itself from Palestine to prepare for the coming war. As it happened, Palestine became a garrison country, holding the line against Rommel's Afrika Korps. This force needed provisioning: the Arab population, which lacked modern industry, could not do it. In contrast, the Jewish middle class, reinforced by German immigrants, rose to the challenge and built a war-supply economy, feeding and equipping the British forces.

Hence the war gave a great financial, political, and military boost to the Zionist project. The same industry would later supply and equip the IDF.[50] In this manner the state-in-waiting was built; Zionists received crucial training by serving in the British forces and (by 1944) in the Jewish Brigade.

Though they could see what was happening, the Palestinians could not stop or delay such developments. Despite the British call for volunteers, few Palestinians considered fighting alongside soldiers who so recently brutally butchered their brethren, let alone serving an empire that made their dispossession its mission. Serving in the British army sealed the pact between Britain and Zionism, despite the 1939 White Paper.

This White Paper, emerging from a Royal Commission chaired by Colonial Secretary Malcolm McDonald, resulted from the 1936 uprising. The fierceness of resistance caused great unease in Britain and elsewhere, forcing a rethink of Mandate aims; this dealt a blow to the Partition Plan (never acknowledged by British politicians after the Balfour Declaration, but implicit in it). With the growing Jewish population (from 9 percent in 1917 to around 33 percent in 1937), change was due and the White Paper limited Jewish immigration into Palestine to 75,000 over five years, until 1944. After that, Jewish immigrants would be allowed only by agreement of the Arab majority. This was a corollary of abandoning the Partition Plan;

the new plan was for a binational state, preserving a reduced Arab majority.

After three brutal years, Britain was eager to conclude the Mandate while protecting their interests in the region, greatly threatened by any policy perceived by Arabs to be hostile. Britain also needed to protect its "soft belly" in the Middle East—Egypt, Palestine, and to a lesser degree Iraq and Transjordan—and so it used a conciliatory move to garner Arab support in the coming conflict. The Palestinians never consented to the partition of Palestine; the White Paper, an attempt to satisfy both sides, managed to anger both.

The Palestinians saw the addition of 75,000 Jewish immigrants as British betrayal. The Zionists, on the other hand, perceived it as meager in light of the European refugee crisis. Zionists saw Palestine as their patrimony and "losing" any of it was anathema. The contradiction built into Zionism—coveting the whole country, but with no Arabs—made the White Paper unacceptable. The British Mandate was doomed.

Indeed, Britain's entry into World War II was a complex affair in Palestine. If Jews had no other option but to support Britain against Nazism, most Palestinians eschewed volunteering to fight, and very few did—6,000 Palestinian Arabs compared to around 40,000 Zionists. While Palestinians understood the dangers of Nazi Germany, which caused mass Jewish immigration into Palestine, few were lured by the Nazi Judeophobia and considered the Nazi threat to Britain in North Africa an opportunity for ending the hated Mandate.[51]

This was the position of the Grand Mufti of Jerusalem, Haj Amin Al-Husseini, who was deported by the British because of vocal opposition to the suppression of the Arab Rebellion and his tendency to incite violence against both the Mandate and Zionism.[52] Husseini, based in Beirut after his expulsion, left for Berlin, spending the war under Nazi patronage. Zionist apologists frequently use this isolated episode as proof of "antisemitic tendencies" in Palestine, even though most Palestinians, though hostile to the Mandate, did not collaborate with or support Axis powers:

> An overwhelming majority of liberal pro-independence activists, a majority of progressive nationalists and all Marxists—in a word, all those Arabs who shared an allegiance to the values of political

> liberalism that issued from the Western Enlightenment, particularly the idea of human rights ... rejected Nazism and alliances with Hitler's Germany in the name of those values.[53]

In contrast, while the Palestinians were nonhostile captives of the British during the war, the Jews of Palestine supported British war efforts, making the most of this political choice, anticipating their future brutal struggle against both the British and Palestinians. To that end, Zionism engaged in crucial military training, reequipping and reorganizing, coupled with industrial and financial preparations for war. Many more Jewish immigrants were required for the fighting force to win Palestine, and their hopes were pinned on the millions of Jewish refugees that might flock to Palestine, before the full horror of Nazi extermination was revealed.

The clearest expression of this was the 1942 Biltmore Program, through which the Zionist movement abandoned support for partition, though this was never openly stated. The plan demanded immigration of one million Jews to Palestine, to force a Jewish majority. Views on the plan were varied but all agreed that statehood could not have been achieved without armed conflict with Palestine's Arabs, expelling most if not all from the territory. Views on the plan were varied, but its direction was clear: "Ben-Gurion championed immediate statehood, the establishment of a Jewish state in all of Palestine, and armed resistance, if necessary, to achieve Zionist goals."[54]

Britain could not publicly reject the Biltmore Program. The United States assisted Britain in its fight for survival, keeping Britain afloat since March 1941. Even before it entered the war, US support for Zionist sentiments was an important part of political calculations in both US political parties, so that the Zionist delegates in May 1942 could assume solid support for their radical program. There was a quid prop quo between British support for the Biltmore Program and US support for Britain.

The understanding of the Biltmore Program as a proclamation of the intention to establish a Jewish Zionist state by force was widespread. The aggression and threat involved in this approach were quickly recognized by some liberal Zionists, who believed it imperiled the future of common life in Palestine. In opposition, staunch Zionists such as Henrietta Szold and Judah L. Magnes formed a new party,

Ichud (Unity) whose political program was the formation of a Jewish-Arab Federation, an overture toward a binational state in Palestine. Ichud never managed to garner support in the strongly nationalist Jewish sector in Palestine, and it became the exception proving the rule; Zionism chose to act by force in 1942 and ever since.

The Biltmore Program was followed by more practical steps. As opposed to the Arabs of Palestine, pegging their hopes for a future Arab state on surrounding Arab countries, Zionists were realistic. They set about building a military infrastructure at a time of relative calm, while much of the rest of the world was being brutally destroyed.

The future formation of Israel as a war economy originated in this period. Some surprising personalities stood out in its transformation into a war economy. Europe's largest munitions and armament manufacturer before and during World War II was a Hungarian firm owned by the Jewish billionaire Baron Manfred Weiss, the main munitions supplier to the Third Reich. Weiss handed over the firm (and his prodigious art collection) to the Nazis in 1944 in return for the family's escape to Portugal on a Nazi military plane. Some Jewish experts from the Weiss plant came to Palestine, reinforcing the industrial base of what would become Israel's military–industrial complex.

Beyond this the war had obvious influences on Zionism. The concept of blitzkrieg, developed by Nazi Germany, guided Zionist military planners. Bearing in mind the small numbers of Jews in Palestine, and the threat by large Arab armies surrounding Palestine, fighting a defensive war was dangerous. Despite the Israeli myth of "the few against the many," Zionism had numerical advantages from the very start. This case is made by Benny Morris even after his conversion to rightwing politics:

> One of the most tenacious myths relating to 1948 is that of "David and Goliath"—that the Arabs were overwhelmingly stronger militarily than the Yishuv. The simple truth—as conveyed by Flapan, Shlaim, Pappe and myself—is that the stronger side won.[55]

Nonetheless, Zionist habitations were too small and scattered across the country. They were vulnerable to attacks and being cut off from Jewish population centers. What was required, Zionist leaders reasoned, was a dynamic, offensive battle plan, using well-equipped modern units exploiting fault lines of enemy deployment—commandos penetrating and destroying enemy rear positions. Tactics, developed by T. E. Lawrence during World War I, in which small dynamic forces were employed against large formations, were carefully studied.

The result of such deliberations was the establishment of the Palmach (acronym of *Plugot Machatz*, Strike Battalions) by the kibbutzim and Hagana in May 1941 (a year before the Biltmore Program publication), proving that taking Palestine by force was the result of long-term planning.[56] While ideological considerations were important in the creation of this commando force of 5,000 leftwing fighters, these were quickly forsaken in the complex realities of Palestine. It was a reaction to depletion of the Hagana manpower as a result of thousands of its troops joining the British forces and a political manifestation of the Kibbutz movement.[57] The complex history of Palmach is dealt with elsewhere[58] and is clearly beyond the scope of the current volume; however, it is important to examine here the main characteristics of this force as it coalesced during World War II.

Palmach was conceived as a national force with regional autonomous units; it enjoyed autonomy under its commander, Yitzhak Sade, who persuaded the Yishuv leadership to create the force as a defense against the much-feared invasion by Rommel's Afrika Korps. Initially it was a militia of Ha-Kibbutz HaArtzi, an arm of the leftwing Mapam (United Workers Party). This common background of the volunteers created an esprit de-corps and a "uniform outlook"[59] among the force, that of a military elite. In most kibbutzim, it was *de rigueur* to join Palmach.

The early years were devoted to military professionalism, in contrast to other Zionist paramilitary organizations that acted as citizens' militias. This move flowed from the SNS, and the serving of some forty volunteers with undercover British units in Syria early in the war. Training was initially conducted by the British Army, which had a stake in developing such units against a possible Nazi invasion.[60] With training over, permanent bases were built, rather than returning recruits to their kibbutzim. This was the beginning of a permanent, professional Zionist army. The recruits underwent physical and arms training, learned Arabic for undercover work, and hiked in order to learn the lay of the land and become hardened soldiers. The hikes were not limited to Palestine—they included areas beyond the Jordan River, where Palmach was planning to expand the patrimony of Zionism, so the hikes were not merely physical training bouts but had ideological and political functions.

Set up by a leftwing party, the Palmach was notionally committed to socialism, and most members shared the kibbutz ethos. However, tensions developed between them and Irgun and Lehi volunteers, belonging to rightwing organizations. Nonetheless, such class consciousness was quickly displaced by nationalist sentiments after the end of the war, as the British became targets for military action. By then, Irgun and Lehi had agreed to coordinate their actions with the Hagana. Despite ideological disagreements, the forces now shared the same outlook—a military solution.

For example, in July 1946, when Ben-Gurion ordered the building of twenty-two settlements to scupper a plan by British minister Herbert Morrison for a federal Palestine, all three organizations collaborated.[61] The plan was to bring about a revised version of the Peel recommendations, in which Jews and Arabs would live in quasi-cantons. Like many later Israeli actions that punished "unruly" international leaders who proposed peace plans, Ben-Gurion chose aggressive settlement as a fitting answer to the "threats" of peace. Conquering the land and settling it were regarded as two points on a continuum; indeed, the Hebrew term for settling the land was *Kibush Ha'adamah*, conquering of the land.

The coming together of the three forces meant synchronizing military activities, arms smuggling, and supplies. The Palmach was part of this new reality, and its command structure was intact. Thus, the second step was taken toward the formation of a professional army. The ethos and experience of Palmach would form the new IDF post–1948, even though the actual force was disbanded in 1950 by Ben-Gurion, who was afraid of a coup. (This suspicion was unwarranted, as argued by Ben-Eliezer in his iconic work on Israeli militarism.)[62]

Between 1945 and the end of 1947, the combined forces conducted steadily increasing military actions against Mandate forces and Arab communities. The collaboration of the Zionist forces meant that the British could no longer deal with Zionist terror, and were forced to enlarge the number of soldiers in Palestine beyond 100,000.[63] The famous acts of brutal terror are well-known; the pivotal terror act was the bombing by the Irgun, with Hagana's support, of the British forces headquarters at the King David Hotel in July 1946. Ninety-one Jews, Arabs, and British nationals were killed, and many more were wounded and maimed. All were government employees.

It soon became clear that the Mandate was no longer operational and could not guarantee the safety of anyone in Palestine. That the King David Hotel bombing occurred only a few days after the publication of the Report of the Anglo-American Committee of ministers, which was an attempt to resolve the conflict of interests in Palestine, was a reminder that Zionism did not countenance diplomatic solutions. The decision had been taken to ratchet up the military conflict with the British to force their departure. Britain was no longer useful for the Zionist project and was a hindrance to seizing the country. Zionism was confidently preparing itself for the real conflict—with the Palestinians.

The British, for their part, had little energy left for the Mandate. The enormous cost of the Palestine garrison and the growing effects of Zionist terrorism as Britain was rebuilding its infrastructure and services were becoming insupportable. It was understood that Britain could no longer afford the Mandate as a lynchpin of its Middle Eastern policy and was preparing for its termination, while trying to leave Britain in virtual control, with reduced risks and costs.

Some politicians had realized much earlier the need to leave Palestine. Winston Churchill, writing in disgust just before losing the election in 1945, described Palestine as "this very difficult place," saying he "was not aware of the slightest advantage that has ever accrued to Great Britain from this painful and thankless task. Someone else should have their turn now" of administering the Mandate.[64] For the Zionist Churchill, this was a painful turn. However, others within his government thought differently: General Ismay was convinced that "Palestine is the key to the security of the Middle East area." Harold Beeley of the Foreign Office was more emphatic: "Abdication in Palestine would be regarded in the Middle East as symptomatic of our abdication as a great power."[65] Leaving Palestine became an urgent topic by the end of the war, even before the Zionist terror campaign against the British.

The IDF's and Israel's Impunity from Censure

The terror campaign against the British forces, as well as ransacking bases for weapons and ammunition, had by now intensified. It involved all Jewish militias, especially Palmach. Some operations included occupation of land and its settlement by military force, to guarantee its inclusion in the future Jewish State; such settlements (*Heahzuyot*) were established in the Negev (Naqab) on September 1, 1947, as a strike at the British authorities and Palestinians. Mandate authorities, nominally in control, did not react to such illegal and aggressive moves. This sent a clear message to the Zionists that there would be no opposition to their moves to occupy the rest of Palestine. Nothing could stop the coming genocidal expulsions.

Even more important than the military and political advantages thus gained by Zionism was another advantage being established—an immunity from public censure. Every Israeli government since that day has invoked that advantage. Following a long history of European antisemitism, which, especially after the horrors of the Holocaust, were still fresh in peoples' minds, Zionism was in a unique position: it could act illegally and immorally without danger of international censure. Immune from punitive measures, the IDF could inflict terrible damage on the Palestinians.

Thus, a pattern was set—Israelis enjoyed international extrajudicial impunity, based on what will later be defined as *exceptionalism*, a concept originally developed by Herder and Fichte.[66] That meant that Israeli acts came to be judged not by normative legal frameworks, such as UN resolutions and international treaties and agreements, but instead were considered *exceptional,* free of normative yardsticks and regulations.

The question of the racialized expulsion of Palestinians is a case in point. By the end of the Mandate, May 14, 1948, more than 375,000 Palestinians had already been made refugees, expelled by a combination of force, atrocities, and a terror campaign. (By the end of the war, the number of refugees would double.) Thousands were already dead, and many more wounded, before a single soldier from the Arab Salvation Army entered Palestine on May 15, 1948, a fact that refutes Zionist claims that the Nakba resulted from the Arab states' attack on Israel.[67] It is quite extraordinary to realize how pervasive such lies still are.

The 1948 war developed and shaped not only the IDF and Israeli attitudes to fighting the Palestinians, but also its deep-seated approaches to the discourses of history and communal memory. A combination of denial—not new to Zionist thought—and a certain economy with facts has been established, to become the dominant mode within the Israeli polity.

Walter Laqueur's rendering of the period between November 29, 1947, and the end of the Mandate in May 1948, during which over 350,000 Palestinians were expelled by the IDF or its militia forerunners, is a case in point. This occupies just three pages in his 640-page *A History of Zionism* (1972, 2003). The book makes no mention of the forced expulsion of the Palestinians; instead, he writes,

> By the end of April, 15,000 Arabs have left Palestine. What impelled them to do so was debated ever since. The Arabs claim that the Jews, by massacres and threats of massacre, forced them out and that this was part of a systematic policy. The Jews asserted that the Palestinian Arabs followed the call of their leaders, believing they would soon return in the wake of the victorious Arab armies.[68]

Thus, the 750,000 expelled refugees were practically erased, conveniently replaced by "15,000" who "left Palestine." The claims that Arab leaders have called upon the Palestinians to leave their land were never substantiated, either in Laqueur's tome or elsewhere; despite this, it had entered the canon of Zionist history, and as such, also into argumentation of its Western supporters. In such ways mythistories (to use a term coined by Shlomo Sand) are created.[69]

The salient facts of this tragedy are not only that it was allowed to happen even as the British were preparing to leave, but that it went through without any serious reaction by the international community —neither at the time, nor since; it was displaced by the "rebirth" of Jewish statehood:

> The Palestinians were excluded from the unfolding of this history. Their catastrophe was either disregarded or reduced to a question

> of ill-fated refugees, similar to the many millions around the world—those who wandered in Europe following the end of World War II or those forced to flee the violence that accompanied the partition of India.[70]

The Israeli mythistories were adopted by the West and played a crucial part in the normalization and tolerance of the Nakba. Israel established its ability to act illegally with impunity; it has managed to remove itself from international jurisdiction, and insert itself into the small group of nations protected by the exceptionalist argument.

What was further grounded is the protocol of "shoot first, lie later." Many future illegal actions by Israeli forces would draw on this precedent. The politicians—Ben-Gurion and Moshe Sharett during the early years, and all Israeli leaders since—lied confidently, in Hebrew and in English, knowing they were quite safe in doing so.

By the end of the war, in January 1950, patterns of military, political, and social conduct were being settled; today, they can no longer be questioned. It is through such perspectives of military doctrine that Israeli identity was conceived and formed.

It would take many decades, and many wars, for Israelis to evaluate IDF activities in 1948 and later through a self-reflective, critical historical discourse. By the early 1990s, as the world was changing beyond recognition, Israelis briefly experienced a moment of contemplation. An untypically liberal attitude to the archived materials of the IDF seems to have overridden state doctrine and allowed unprecedented access to a younger generation of Israeli historians. Drawing on newly available archival research and documentation, a group of historians came to challenge established myths of Zionism, especially the events in 1947–49. Referred to as the "new historians," they include mainly the Jewish Israeli historians Benny Morris, Tom Segev, Ilan Pappe, Avi Shlaim, and Uri Ram; to this list I add Nur Masalha, the main historian of the Nakba and an esteemed historian of Palestine and Zionism, who is normally excluded from this list because he is not Jewish. His exclusion is clear evidence of racialized academic exception.

Long-term Patterns Emerging in the IDF

The new historians have covered the many aspects of battles, policies, and military and civil illegalities involved in Israel's "War of Independence," which lasted from late 1947 until early 1949. This surprising access to the archives did not last long. The secrets in the archives remain locked today more firmly than ever, under new rules by the Netanyahu government. Patterns noted by the new historians in the early stage of the IDF's history have become iconic and formative in the following decades. Many of the documents quoted by the new historians are no longer available in the archives; they were removed in order to cast doubt on the books discussing 1948, as openly admitted by Yehiel Horev, head of the defense ministry's security department from 1986 until 2007 who also oversaw control of the archives:

> I don't remember the document you're referring to, but if he quoted from it and the document itself is not there [i.e., where Morris says it is], then his facts aren't strong. If he says, "Yes, I have the document," I can't argue with that. But if he says that it's written there, that could be right and it could be wrong. If the document were already outside and were sealed in the archive, I would say that that's folly. But if someone quoted from it—there's a difference of day and night in terms of the validity of the evidence he cited.[71]

Israeli policymakers were inclined ever since 1948 to act through military means to resolve political differences and conflicts. Aggressive military action would become the first organizing principle of Israeli diplomacy/politics. Whether this meant that politicians used the army to advance political aims or that army officers using the politicians to advance military aims is a moot point; both groups had identical aims. It may suit politicians at different junctures to present controversial military actions as army (or even "concerned citizens") initiatives, but this was never the case—no conflict of interests ever existed.

War, unsurprisingly, was used for Zionist expansion. Zionism has learned in 1948–49 that it would be allowed to hold on to occupied territory, despite international law and UN resolutions. Thus, military conquest was used to extend the boundaries of the Zionist project as opportunities arose or were carefully manufactured. This did not end in 1967, with Israel controlling the whole of Palestine, but continued in Lebanon, Syria, and Egypt, as Israel extended its power and controlled its boundaries militarily. Forcing any Israeli withdrawals demanded intense international pressure; the only examples occurred during 1956 (the tripartite aggression against Egypt) and 1979 (under the Camp David agreement). The withdrawal of Israel from Southern Lebanon in 2000 and from Gaza in 2005 was occasioned by IDF difficulties in controlling such territories, rather than by negotiations or agreements. In that sense, there is a difference between control by the IDF, as is the case in the West Bank, and direct political control by the state, as in the Golan Heights, annexed in 1981; for political reasons, the West Bank annexation has been deferred, with military control imposed in 1967. Hence the settlements were at first excluded from Israeli law, and then Israeli jurisdiction was extended to include them. The last phase of the process is legal annexation, and Israel is currently preparing to annex most of the West Bank, the inhabitants considered nonpersons.[72]

The second organizing principle concerns the nature of the IDF and its doctrines. The IDF has adopted the earlier lessons of the SNS and Palmach in 1948 and has preferred the offensive doctrine ever since. This is partially explained by Israel's lack of "strategic depth"; thus, the IDF prefers attack to defensive measures. However, the reasons are also sociopolitical; IDF is one of the region's largest forces, but keeping men permanently drafted is not economically sustainable. Hence, the IDF is based on a core of conscripted soldiers. The main force is the reserves, called up at short notice; reservists cannot be used to implement defensive strategies, because it would take too long to mobilize them. Israel has always preferred to strike first, and its military philosophy and practice is attuned to this need. This was also the basic conception established before 1947, requiring battles to be staged on enemy territory, occupying ever larger tracts of Palestine. "Transferring the battle to enemy territory" (*le'haavir et ha'krav le'shetach ha'oyev*) has become the first tenet of IDF's doctrine, inherited from the offensives of 1948.

An important characteristic of the new state is the increasing blurring of civilian and military spheres: Israel and specifically, the IDF, do not separate the civilian and military spheres of activity, but instead blend and integrate them. Ben-Gurion has promoted the notion of the nation under arms, opposing a professional army. This became more pronounced over time, as Israel built a war economy as its financial mainstay. The armament and security industries of Israel are world leaders, attracting enormous income from sales and training across the globe. The military–industrial complex is Israel's largest industrial sector, employing (directly and indirectly) hundreds of thousands of highly skilled scientists, engineers, and researchers. Israel transformed the conflict from a serious difficulty into a lucrative operation; it lives off the conflict and depends on its continuation.

The main functions of the IDF have continuously morphed from an army fighting state actors toward a force of occupation and settlement. Military force was used for the settlement of Palestine even before 1948. After that war, soldiers were organized in special units of the Nahal (Fighting Pioneer Youth), an elite force of the IDF, building military settlements that are later pronounced "civilian."[73] A continuity was established: the IDF occupied, financed, administered, secured, settled, and then "civilized" the territories of Palestine, controlling legal, social, military, industrial, religious, civic, cultural, educational, and communications sectors, as well as water, electricity, building, and transportation.

To be able to complete this complex process of taking over the territory, the Israeli political elite had combined lies, denial, pretense, playing dumb, and delaying tactics in conjunction with illegal military actions. While this is done elsewhere and is not in itself unique, it is employed by Israel so methodically and with such brazen disregard that it is clearly exceptional. This approach employs a wink and a nod toward the Israeli population: those who need to know, do; those who do not need to know, don't.

The need to get the nation behind questionable IDF actions was simpler in 1948–49, when most Jewish men were mobilized. It became more complicated later when military actions were taken against returning Palestinian farmers and their families and against communities of Palestinian refugees across cease-fire lines, as was increasingly the case in the 1950s. New measures were required—a permanent war economy and militarized culture were needed to bind the nascent nation to its maturing military. To guarantee discipline it was crucial to ensure ignorance, so an Office for Military Censorship was created in 1948. To this day, every Israeli publication must be approved by the censor. Keeping the population in the dark has worked miracles.

A war economy was instituted after 1945 when Zionism laid the foundations for an industrial base through meeting the needs of the British garrison, and it was further developed during the early years of Israeli statehood. The mind behind the military–industrial complex was Shimon Peres, director-general of the Defense Ministry during the 1950s. As early as 1947, the twenty-four-year-old Peres was appointed by Ben-Gurion as the official responsible for Hagana personnel and arms procurement; he soon turned his role into the most powerful in the ministry, a key post in every government since. In 1952 he became the deputy director-general of the Defense Ministry and a year later the director-general. Through the network of personal contacts he built with foreign visitors, he managed to befriend and influence many European leaders, some of them Jewish, mainly in France, laying a foundation for Israel's military–industrial complex and nuclear capacity. According to Beit-Hallahmi, Ben-Gurion decided to develop nuclear weapons in order to "avoid the fate of the crusaders,"[74] and Peres was the one to realize his wish. Through the ultimate control held by the government party, Mapai (acronym, Land of Israel's Workers Party) and over the Histadrut, Peres has managed to turn this powerful and affluent body into the main industrial player in Israel, owning and operating some two-thirds of Israel's economy in its first two decades. Thus, instead of being a workers' representative body, the Histadrut became part of Mapai's control apparatus, the major industrial and social player whose wage negotiations and workers' rights section was one of the smallest and most insignificant subdivisions.

With virtual control over the Histadrut, Peres and Ben-Gurion could direct its development and investment policies, creating a powerful instrument for building a successful armament industry. Ben-Gurion supported his ambitious protégé, affording him unlimited powers through the various control channels. The young nation was led by those who learnt the crucial lessons of recent European history—the importance of control of the three bases of popular power: party, state, and labor unions. By buttressing such control across social strata and effectively blurring distinctions between the three branches, fascists in Italy, Nazis in Germany, and Stalinists in the Soviet Union controlled their complex economies and societies. Such undemocratic conflation of social and political boundaries, with representative assemblies devoid of power, was crucial for the total control that no democratically elected government could achieve. Ben-Gurion certainly understood the need of exerting control over society to hammer it into a nation.

Despite propaganda claims about the Jewish people returning to an abandoned homeland, Ben-Gurion was aware that no Israeli nation existed. This was especially true as the Jewish population of Palestine doubled twice, between 1930 and 1939 and between 1948 and 1951;[75] the incoming 700,000 Jews from various Arab countries were neither committed Zionists nor welcomed or assisted on their arrival, and could hardly be said to constitute a nation.[76] The shameful way in which such immigrants were treated—as human trip wires along the new frontiers, protecting the soft belly of Zionism where Ashkenazi Jews were the majority—created deep resentment in the racialized Mizrahi (Arab) Jewish community, which suffered from the unequal and unjust nature of the Zionist project. They resented the racist rejection by the Ashkenazi middle class but were unable to do much about it; neither could they change the demeaning conditions in what were called development towns, a euphemism for clusters of tents or prefabs with no infrastructure.[77] (The lucky ones were housed in depopulated Arab villages.)

Ben-Gurion realized that his policy of attracting Mizrahi Jews to Israel was not conducive to social cohesion in the racist society controlled by European Jews—dangerous in such a small polity "surrounded by enemies," as he never tired of repeating. During the 1951 election campaign, he said, "I see in these elections the shaping of a nation for the state; because there is a state but not a nation."[78] This insight led him to create an army, which then formed its own state, for which the construction of a nation was an urgent task. The nation he was to create would be a nation at arms, in a state of neither peace nor war. To make this mode of existence (exceptional by any yardstick) into Israel's modus vivendi, a major social engineering project lasting decades would follow, requiring constant renewal. As late as 1954, Ben-Gurion was worried about the nation's nonexistence; in the Government Almanac, he noted, "For thousands of years we were a nation without a state. Now there is a danger that Israel will be a state without a nation."[79] Putting aside the absurd notion that Jews had been a nation for thousands of years, it is fascinating that the "missing nation" narratio was used for such a long period. It combines cynical notions of influencing the Zionist narrative, as well as his deep-seated belief in the urgency of engineering an Israeli-Jewish (or as some called it, a Hebrew) nation.

That the nation did not exist was a real difficulty for Ben-Gurion and many of his generation, especially those in the military field: "The problem [being] that we are not a nation ... [There should be] a desire to fight, and an ability to fight. In order to want to fight, there must be a nation, and we are not a nation."[80] A circular argument is being made: In order to fight, one needs a nation, and in order to have a nation, one needs an army and war. Thus, it is hardly surprising that the ideal instrument for fashioning the nation was deemed to be the very army that preceded and created it. Hence, the Israeli army was not an instrument for the exceptional moment of war, but the foundational social institution of the new state, its guarantor of identity and existence. This army "came to stand for all that was good and noble within the nation."[81]

Ben-Gurion's generation of Zionist politicians, although not Marxists, were well aware of the tenets of Marx's thinking about identity formation and its mechanisms. They understood that existence forms consciousness rather than the reverse. Hence, they set about creating the conditions for a society of soldiers, whose consciousness is shaped by military praxis. The most important role for the "civilian" aspect of society was given to the military–industrial complex, a condition still pertaining; Israel's armament and related industries are not just the largest sector of the economy, but also one of the largest anywhere, regularly achieving a leading position among global armament exporters. Israelis are living proof that one *is* what one *does*—the specialization in "security" and conflict gave rise to a society that perceives reality though the gunsight, metaphorically speaking. Such considerations have led to an army that is like no other in the post–World War II period, with the possible exception of North Korea. From its very inception, the IDF has been a wide-ranging, social network of enormous political and cultural complexity. It will be important to outline below some of the patterns of this unique institution and its social positioning in Israel.

The level of involvement of army officers in all parts of the legislature and executive in Israel, central and local, is without precedent in any society with democratic aspirations. This was true throughout Israel's history; if anything, it is even clearer today. If in some societies private schools may form the elite, in Israel the IDF serves this role; it forms the military, industrial, academic, scientific, media, cultural, and political elites. The prevalence of Israeli generals as executives in all sectors is more pronounced at the political upper echelons: the composition of various governments and cabinets, the Knesset, the various ministries. There is always a place of honor and influence for a retired general in political parties. Indeed, a military career seems to be the required preparation for Israeli politicians. Those who, like Benjamin Netanyahu, were not senior officers find it necessary to highlight their military achievements, as demonstrated in this excerpt from his Wikipedia page:

> Netanyahu joined the Israel Defense Forces during the Six-Day War in 1967, and became a team leader in the Sayeret Matkal special forces unit. He took part in many missions, including Operation Inferno (1968), Operation Gift (1968) and Operation Isotope (1972), during which he was shot in the shoulder. He fought on the front lines in the War of Attrition and the Yom Kippur War in 1973, taking part in special forces raids along the Suez Canal, and then leading a commando assault deep into Syrian territory. He was wounded in combat twice. He achieved the rank of captain before being discharged.

Obviously, inflating Netanyahu's limited military career is necessary exactly because he had not attained a status higher than captain; the normative bottom line for Israeli politicians is that of a lieutenant general, with some colonels also slipping in. Each party packs its ranks with "security experts." Writing in *Haaretz*, three months before the 2015 March election, Yossi Verter noted about the Labor Party:

> Next month, a retired major general will also join the party, the de rigueur "security expert" in Israel. It might be MK Shaul Mofaz,

"Mr. Security" in terms of experience and his record, or possibly former director of Military Intelligence Amos Yadlin.[82]

A major party without a pack of generals is doomed. Most areas of elite control in Israel are also peppered with army officers, who tend to retire from the military before they reach age fifty, allowing for a second career in politics, business, industry, finance, or academia. The relationships forming in the IDF tend to parallel ties formed at the public school system in Britain. Ubiquitous army officers on appointment panels guarantee that the "right candidates" are selected.

A fascinating case in point is the recent scandal over the Eilat-Ashkelon Pipeline Company (EAPC), which has turned into an exemplary event combining corruption, secrecy, and the illicit culpability of senior ex-officers. It all came to light after a devastating breach of the pipeline, causing total destruction of a prime site of natural beauty. This environmental disaster contributed to uncovering the secretive EAPC which is beyond the reach of Israeli law: the company is governed by a group of retired army officers, pays no taxes of any kind, does not release information about its finances or ownership, and for all intents and purposes is ex-territorial, not subject to Israeli or any other law. *Haaretz* editor Aluf Benn published an article trying to map, for the first time in Israeli history, the ownership and control network of this elusive body (by no means the only one), which cannot even be approached when it is responsible for a major environmental disaster, but even this long article does little more than scratch the surface of this saga of corruption, lawlessness, and secrecy.[83] Military censorship has ensured that the picture remains incomplete and confusing; the secret deal with the Shah of Iran in 1968, in which the main investors were the government of Israel and a group of shadowy Iranian businessmen, has stayed under the radar of public scrutiny for five decades, a model of the strange fruits emerging from blurring the lines between the military and the civic sectors. The ongoing corruption allegations against Netanyahu in connection with the submarines' deal with Thyssen-Krupp is another example, with no prosecution likely.[84]

The IDF has served Israel's needs by forming the nation and then by serving as the hothouse for the elites. During the early years, immediately after the 1947–49 war, the elites were typically kibbutz members trained in the SNS, Fosh, and the Palmach, with right-wing underground groups shunned by the "socialist" Ben-Gurion administration. The IDF was a powerful mechanism of entrenching elite control over the polity. On leaving the army, most generals become central players in other sectors; an example is Yigael Yadin, who on retiring started a career in academia, as the leading archaeologist in Israel, with a central role in "confirming" foundational myths through spurious digs and questionable finds, all in the service of affirming Israel's ideological readings of the Bible as real-estate certification: "The founding fathers of Israeli archaeology explicitly set out with the Bible in one hand and a pick in the other, seeking findings from the biblical eras, as part of the Zionist project."[85] Yadin's career did not end in academia, though; in the late 1960s he set up a centrist party and started a new political career. He was not the only one; other veterans of 1948 were also tunneling toward politics, including Moshe Dayan, Ariel Sharon, and Yitzhak Rabin. A pattern was established that still holds: with the economy very much based on the conflict, with the military–industrial complex growing in size, importance, and revenue generation, many army figures became industrial managers, a career well-suited to officers lacking academic qualifications.

This preponderance of Ashkenazi kibbutz and left-of-center officers in the higher echelons of Israeli society continued after the 1967 war, the last war that saw the old elite in full control of the IDF, industry, the economy, media, culture, and the arts, through the double stranglehold of the party (Mapai) and the Histadrut. This control created a deeply divided society, where sectors such as the Palestinian citizens of Israel, ultra-Orthodox Jews, Mizrahi Jews, and most women were effectively excluded from positions of power and influence. In 1977 the rightwing Herut Party (the political organ of Jabotinsky and the Menachem Begin-led IZL, which was incorporated into the IDF in 1948) gained power in what was termed the *Mahapach* (upheaval). With Begin in power, a period of "liberalizing" the economy began, from a command economy controlled by government and Histadrut (like the Indian economy under the contemporaneous Congress Party) to an aggressively privatized laissez-faire market economy. With political change came a change in elite makeup. The kibbutz elite started its decline, especially in the

IDF, while a new elite of financiers, bankers, and the controlling clans of the Israeli economy gained ascendance, especially those identified with the "fighting family" (a term used by Begin for the few families at the top of the Betar/Herut/Etzel ranks).

The new elites also emerged from the illegal settlements, as well as two groups previously excluded: the Mizrahi Jews and the ultra-Orthodox. As many of the settlers came from these two communities, they gained a foothold in Israeli politics. Again, it was the IDF—through its role in establishing settlements and guaranteeing the occupation—which had shaped the new elites of the right, thus stabilizing the "upheaval." The IDF continues to be the social space where the young members of the elites form their contacts in the many elite units (a euphemism for the death squads unleashed on the Palestinian civilian population).

To understand the wide penetration of the IDF into all strata of life in Israel, it is instructive to look not just at the elite, but also at the margins. It is hardly surprising then that the IDF and earlier militias have played a major role in the development of Hebrew slang, which radiated outward to wider society, as it was the only confluent location of all Jewish (and some other) Israeli males and most Jewish women. New expressions coined in one IDF unit quickly circulated, and if successful, were normalized by societal use. The evidence for this process is massive, and in 1982 the first dictionary of Israeli slang was published, recording the military birthplace of such expressions.[86]

Another crucial characteristic of Hebrew slang is that most expressions came from other languages. The largest group came from Arabic, followed by Yiddish, German, and Russian, and to a lesser extent English and Spanish. This pattern is evident even before 1948 and hardly surprising.

In the five decades before 1948, hardly any incoming Zionists spoke Hebrew with proficiency. Hebrew was, like Latin, a dead language, used exclusively for praying and religious rites by Jews of all communities, including in Palestine. It was called Leshon HaKodesh (Sacred Language) and was banned from daily use by religious edicts. This meant that it did not develop since the destruction of the Second Temple in 70 AD, and probably earlier, as Aramaic was the lingua franca of Palestine. Hebrew lacked words required by the enormous changes to human experience, and such lacunae had to be bridged. This meant that books and periodicals published in Hebrew between 1860 and the 1930s used archaic and unwieldy expressions, filling gaps with words adapted or imported from European languages in which authors were proficient. Thus, foreign idioms impinged on the language, forcing ancient Hebrew into contortions it was unfit for. The poetic structures of ancient Hebrew were useless in trying to express modern content, and biblical forms were inappropriate for modern discourse.

The fast development forced upon Hebrew was necessary in order to serve the settler-colonial community needs. The Committee of the Hebrew Language (Va'ad Halashon) that was set up by Eliezer Ben-Yehuda in 1889 was formalized in 1905 by a group of Zionist writers, grammarians, and linguists. By 1953, the Israeli state institutionalized it as the Academy for the Hebrew Language.[87] The committee was responsible for modernizing Hebrew, with thousands of words added annually; in its work, linguists used other languages —mainly Arabic, Aramaic, Greek, and Persian—spoken at the time of Hebrew's demise. The Zionist movement in Palestine eschewed the advice of Herzl in both his diary and his polemical novel *Altneuland*, published in German in 1902, and later in Yiddish and Hebrew versions; Herzl believed the Jewish community in Palestine would bring its languages and cultures with them, continuing to speak, write, and read them. He had no command of Hebrew, which he considered a dead language.

But Herzl did not contend with the new generation of Zionists, exemplified by Ben-Gurion, who refused to speak any language but Hebrew and silenced even Yiddish, the language of European Jewry. Ben-Gurion understood the role of language as a national and identity agent, forcing it on the Palestine community through social agencies—parties and Zionist institutions, educational systems and cultural bodies. It became unacceptable to speak any other language in public, despite the great limitations of contemporaneous Hebrew. Modernizing Hebrew was seen as an urgent national task, attracting enormous social efforts as well as resources. This haste ironically meant that the development of the unofficial Hebrew was fast—and this was mainly done through the militias in the mid-1940s.

Ironically, the official suppression of foreign languages has led to a fascinating linguistic return of the repressed; languages mostly used by the Zionist militias before 1948—Yiddish, Russian, Arabic, and English—have all re-emerged in slang. It was easier to adopt and adapt words from such languages than wait for the august committee to add urgently needed words. Thus, a car exhaust was named *egzoz* (exhaust), and an open truck was christened *tender*, from the English (still in use); the cops-and-robbers game was called *Enzap* (hands-up), and one with a cushy job in the IDF was called *Jobnik*. Most words used for military equipment came from English. Ben-Gurion's own moniker became Bee-Gee.[88]

Terms for swearing, sexual functions, or private parts were adopted from Russian, Arabic, or Yiddish, with the prefixes such as *Abu* (father, Arabic) and *Um* (mother, Arabic) in great demand; *Abu Ali*, for example, was used for a hefty male, following a local Palestinian expression involving a famous strongman. Yiddish, a language rife with humor and expletives, was used liberally—*gurnisht* (literally "nothing" in Yiddish) was used to denote someone of little or no consequence. The word *bardak* (Turkish for bordello) was used for describing mayhem or pandemonium. These and many thousands of other single and multiword expressions have survived seventy years of Israeli Zionism and are used daily by millions. Ella Shohat tells us in a memoir: "Arabic was also the language in which I learnt to curse so well, much to the chagrin of the adults around me."[89] This gives us an important clue as to the function of slang in Israel. The vigorous development of slang in the context of the toxic and aggressive campaign against Diasporic languages and for exclusive use of Hebrew, led by Ben-Gurion, was nastiest against Yiddish and Arabic, the two languages spoken by most immigrants. By taking over Hebrew slang, the repressed languages proved their richness and vivacity, contrasting with the aridity and sterility of most invented Hebrew. This minor cultural victory is evidence of a difficult cultural struggle against ideological, racialized social repression by Zionism.

Historically, the IDF has heavily invested in cultural production for molding national identity. Thus, it set up a radio station in 1948. (It had had illegal stations during the Mandate years.) At the time, the only two radio stations in operation were the national channel, Kol Israel (Voice of Israel) and Galei Zahal (IDF Waves), both serving as overt propaganda instruments on the models established during the 1930s in Britain, Germany, and the Soviet Union. Kol Israel was controlled by the prime minister, while Galei Zahal was under control of the minister of defense. For many years Ben-Gurion filled both functions, so he controlled all media channels for the nation he created. Galei Zahal, which now includes a range of stations, is the most popular radio channel in Israel and still operates in the same manner. It has also served for almost seven decades as a nursery for media tyros. Aspiring professionals compete to serve on the channel's staff during the obligatory military service, because it has become an elite gateway for media careers.

Another important media outlet controlled by the IDF/ministry of defense is the weekly *Ba'Machane* (At Camp), historically the most popular weekly in Israel, inducting journalists into press careers. Various forces also have their own publications, the best known of which is the Air Force journal (*Bitaon Khel Ha'Avir*), a mass selling publication. The army is also a massive publishing entity; its publishing arms Ma'arachot (Battles) and Carmel are major players in Israel's cultural life. All the organs mentioned (and many others) receive the largest portion of Israel's state budget, which explains their lavish production and protection from market forces.

In addition, the IDF has set up and financed a large host of dramatic and musical troupes, used not only to entertain the troops but "civilian" audiences, too. These became virtual training grounds for actors, directors, producers, singers, art directors, designers, and playwrights of Israel's theater, film, and television. The best known of these was the Nahal Troupe, which plays an important role in forming Israeli identity. As an example, take its very famous 1956 ditty "*Totachim Bimkom Garbayim*" (Guns instead of Socks) as Israel prepared for the 1956 war, collecting a special "defense tax" from its citizens. To spur the public into parting with meagre resources, the stanzas implored:

> Guns instead of socks,
> Instead of shoes, a tank
> And instead of underwear,
> Give us two armoured cars
> And in place of Sunday best
> Let us get—a fighter jet!

> It goes to show the front is also at home
> It goes to show the winter shall be cold
> And when I asked for warm clothes, they retorted
> The main thing is—get more guns![90]

This was a successful campaign. Over and above the tax, many children were persuaded to part with paltry savings to replace underwear with guns.[91]

It will be important to describe IDF's financial functions. The beginnings of the IDF commercial enterprise arose from its canteens, called *Shekem* (acronym for Special Canteen Services), but it has long since become the largest department-store brand in Israel, with some fifty branches in all cities and towns. Because of inbuilt subsidies, the Shekem network is easily the most successful of Israel's commercial ventures, with a large captive market, offering reductions to serving soldiers. Because most Israeli-Jewish men serve in the army every year as reservists, it is basically an unofficial discount service for Jews (and the few Druze and Bedouins who serve in the army), acting as part of the commercial arm of the apartheid system.

Many retired officers have set up commercial companies in the security, armament, and information technology (IT) areas, creating a continuum between the IDF and commercial, financial, and industrial Israel. The IT industry is owned mainly by such officers. The main financial feature of the IDF is the existence and growth of the Israeli military–industrial complex, a major player in the global armament and security market (covered in Chapter 10). The consequences are far-reaching. Take, for example, the NSO company, which developed the hijacking software that has affected millions of global users of WhatsApp in 2019.[92]

Another area of involvement is academia. IDF is deeply involved in all Israeli academic institutions, including eight universities and some twenty colleges. Most institutions employ ex-officers as administrators, and most presidents of Israeli academic institutions are retired senior officers. Academia is a racialized part of the apartheid machinery; it offers special deals and arrangements for IDF veterans, thus excluding most non-Jewish citizens. After each of the recent attacks on Gaza (in 2008–9, 2012, and 2014), institutions vied with one another, offering soldiers special arrangements and waiving tuition fees. As most Israeli men and some Israeli women serve not just the universal conscription but also in the reserves, special arrangements are made for them, like flexible exam deadlines. Because most of the higher education system is state-funded, it is easier to associate with the IDF—these are tentacles of the same beast.

Most Israeli universities are built on the ruins of Palestinian towns and villages conquered in 1948. For example, Tel Aviv University is built over the destroyed large village of Sheikh Muwanis.[93] Large parts of the Jerusalem Campus on Mount Scopus are built on the lands of Issawiyya and Azariyya, two large Palestinian villages adjacent to East Jerusalem.

All universities and colleges make up much of their income from running training programs for the IDF and security services, and much of the research income is from IDF-related budgets, which use and develop research that serves its functions. Israel's academic institutions are major recipients of European Union and US research funding and contribute toward presenting an academic front for security and defense projects financed in whole or part by either the European Union or the United States, further subsidizing the lucrative military–industrial complex, removing the burden from Israeli taxpayers. Much of the cutting-edge research—in IT, encoding, web security, drone technology, and surveillance systems—is developed either by university researchers or with their collaboration, making them academics partners of the military–industrial complex.

The links are not limited to higher education. The IDF has been directly involved in all levels of education ever since 1948. The typical role for female recruits was teaching in *yishuvei sfar* (border settlements), or "development towns," a euphemism for neglected towns in remote border zones, home to the underclass Mizrahi Jews. Women soldiers taught in full military uniform while bearing arms, the first intimate contact for young children with the IDF. The IDF was involved in teaching Hebrew and basic skills to new immigrants, including ideological initiation into Israel proper.

As mentioned previously, all fourteen-year-old Jewish high-school students go through paramilitary training in the Gadna and are readied for fighting units at the age of eighteen after a period of basic training. This preconscription preserves the IDF as the major instrument of social identity and value formation.

While these are the main characteristics that separate the IDF from other armed forces in the West, they are not the only ones. From the above it should be clear that the IDF (and its predecessors) not only functioned as the major force in building Zionism since 1920, but also as the social machinery of nation and identity formation, supporting it ideologically, economically, industrially, and culturally, not to mention militarily. This was true even before 1948 but it intensified in the decades that followed, as the new nation made conflict and war its essence.

2

The 1948 War

In all of Palestine, Ben-Gurion did not know of a single Arab of political significance who would consent to continued immigration or a Palestine in which Jews were anything but a minority. A political settlement was impossible.

Shabtai Teveth, *Ben-Gurion and the Palestine Arabs*

What happened in Palestine was by no means an unintended consequence, a fortuitous occurrence, or even, a "miracle," as Israel's first president Chaim Weizmann later proclaimed. Rather, it was the result of long and meticulous planning.

Ilan Pappe, "The 1948 Ethnic Cleansing of Palestine"

The 1948 war, called in Israel by other names (the three most commonly used being *Milhemet Ha'Shihrur*, War of Liberation; *Milhemet Ha'Atzmaout*, War of Independence; and *Milhemet Ha'-Komemiyut*, a "poetic" rendering of Independence) actually started in December 1947 and did not end officially until January 1949. It was the foundational war of the settler-colonial project. The result also shaped the nature of the Israeli polity, in ways that few could have predicted at the time. The conflict saw the conquest of four fifths of Palestinian territory and the expulsions of some 750,000 people (two thirds of the total Palestinian Arab population) from their homes. The war was the definitive move to establish a settler-colonial state with a Jewish majority, the ultimate aim of the Zionist project.

But if the war was one of liberation, who was the land supposedly liberated from? The popular understanding in Israel is, rather curiously, that it was liberated from Arab dominion. This is counterintuitive. After all, it was the Palestinians who expected liberation from British colonialism and Zionism, which aimed to build a Jewish "national home" at their expense. (As the Balfour Declaration had put it, Palestinians merely resided in Palestine.) Thus, the Hebrew titles for the war are deeply ideological, even mythological, denoting a change from a position of partial control to that of total control of the territory in question. This was liberation both from the British Mandate, which had outlived its usefulness, and of the indigenous population of Palestine, "ethnically cleansed" to allow the building of a settler-colonial Jewish *herrenvolk democracy*.

The Palestinian nomenclature is more accurate. In Arabic the war is called *al-Nakba*, literally "the catastrophe," an apt description of what happened to Palestine and its people: the loss of the country, the cities, villages, homes and fields, the deaths of thousands and expulsion of two thirds of all Palestinians. Its closest conceptual parallel, curiously enough, is the Hebrew name given to the destruction of European Jewry, *Ha'Shoah* (Holocaust). Both names of these events denote a catastrophic event of biblical proportions in the life of the people who have suffered it. While both events are obviously of a very different nature, what makes them closely related is that the group that suffered one event (Shoah)—Jewish refugees and settlers mainly from Europe, before and after the Holocaust—inflicted the other (Nakba). Israeli intellectuals have noted the affinities between both events, most famously the poet Avot Yeshurun, who even used the Hebrew Shoah to denote both events: "The Holocaust of the Jews of Europe and the Holocaust of the Arabs of *Erets Yisrael* are one Holocaust of the Jewish people. The two gaze directly into one another's face. It is of this that I speak."[1]

As I point out elsewhere, the connection between the two traumas, the Holocaust and the Nakba, is crucial to the understanding of Israeli identity.[2] That Israelis see themselves as victims in the first is inaccurate but understandable—it was the Jews of Europe rather than the Israelis in Palestine who were destroyed, after all. That they see themselves as victims in 1948, frequently evoking the war as an attack by the "seven armies of Arabia" (a war of the few against the many), is altogether rather obtuse.

~

The three years after the Holocaust, 1945–48, were a crucial period for the Zionist project. Many Jews were changing their minds on Zionism, and the number of volunteers who wished only to assist the military campaign but not to settle grew substantially. Many were Jewish soldiers demobbed from Allied units, with battle experience crucial for the ensuing war, and they complemented the nucleus of the IDF.

The period was typified by open conflict with the British, who were clinging to vestiges of the Mandate even though they realized that they could no longer control the entity they had nurtured for three decades. Other parts of the empire, more important than Palestine, were also in revolt. India was inflamed by decades of struggle against the British and by the internal conflict between Hindus and Muslims, which would eventually culminate in a bloody partition. The empire was disintegrating, after six debilitating years of intense struggle against Nazism and fascism. The Mandate was doomed and Zionism was simultaneously doing all it could to terminate it and to build a state-in-waiting.

The same could not be said for the Palestinians.[3] After years of battling the Mandate authorities during the 1930s, the population was dispirited, leaderless, and unprepared. Thousands of resistance fighters were dead and many others were in prison. Palestine lacked a modern fighting force. As Arabs, Palestinians looked to Arab neighbors for assistance, but it was not forthcoming. The regimes of neighboring countries, and some remote ones, were under the influence of Britain and France, reluctant to help the Palestinians in their plight. They did not support the setting up of a Palestine force, which was left to local initiatives. Local militias had no unified command, no training or modern facilities, and resistance was built on the vestiges of the failed rebellion.[4]

Only a well-armed, large, and modern force of the Arab nations could possibly save Palestine, but such a force did not emerge. The only modern and well-trained armed force in the region was the Arab Legion of Transjordan, commanded by the British officer Glubb Pasha, but it was miniscule in comparison to the large force built by the Zionist Yishuv. Zionist military leaders were clear:

> In summing up the situation in October 1947, Galili [Hagana leader] foresaw an initial pattern of hostilities resembling that of 1936–9. The Palestinian Arabs, he noted, had made no preparations for war on a large scale "such as the training and conscription of forces and staff planning."[5]

Zionism had prepared for war.

During this final phase of the Mandate, Zionist leaders concentrated on bringing in many young, ideologically trained Zionist settlers and volunteers, legally or otherwise. The main reservoirs of Jewish young people in Europe were the Displaced Persons camps, where some 250,000 Holocaust survivors lived in rudimentary conditions, hoping to resettle and restart their lives. Most lost whole families, surviving slave labor in the camps; not surprisingly, many survivors were deeply depressed.

Different factions within the Zionist movement identified this potential and competed over the survivors, each wishing to enlarge its own political camp and militia. The country (which they called Eretz Yisrael) was described to these survivors as a virtual paradise, the grim realities of the conflict minimized or unmentioned. They were not told that they might have to risk illegal entry and that they would be deported by the British if caught.[6] The displaced persons had little choice; most did not consider returning to the small towns they came from, where they were likely to face antisemitism and even death.[7] Most of them had heard of the pogrom in Kielce, in southern Poland, during July 1946, in which forty-two returning Jews were killed by an antisemitic mob.[8] Most displaced persons wanted to leave Europe altogether and emigrate to America or Australia, but few could obtain visas. The rest had to stay in the camps, probably for years, with hardly any rights and no possibility of working for a living. Or they could go to Palestine.

The paths to Palestine were arduous in the postwar chaos, but the same conditions enabled Zionist emissaries in trafficking thousands of Jews through war-ravaged zones and onto small and unfit craft, on the dangerous sea passage to Palestine. They had to dodge British naval patrols or risk being caught and exiled to Cyprus, Kenya, or Mauritius. Not all displaced people were welcomed. The Zionist recruiters looked for young, strong males with ideological training —in other words, cannon fodder.[9]

The displaced people were a bargaining chip for Zionism. In 1940, hundreds of refugees from war-torn Europe managed to get to Haifa in a rickety old boat called *Patria*, only to be refused entry by the British authorities, who planned to tow them away. To frustrate this plan, Hagana sent a diver to sabotage the vessel and the bomb capsized the ship almost immediately with 267 of its human cargo, mostly refugees; most of the crew and some British guards, trapped in the hold, also perished. Another boatful of refugees aboard the *Struma* was refused berthing in Istanbul in February 1942 and was towed to the Black Sea, where she was torpedoed by a Soviet submarine, with the loss of some 780 passengers and ten crew. Some smaller craft sunk without a trace and without media attention, in mid-sea. When assessing their options, displaced people had to take into account the risk of the journey as well as what they might find once they arrived in Paradise.

They kept coming, and many thousands managed to defy the Royal Navy and sneak into Palestine, there to join the armed forces of the Hagana, Irgun (IZL or Etzel), and Lehi (Stern Gang). As the build-up to war intensified, the number of soldiers mustered by the militias grew, as did the materiel. Despite the arms embargo, much arrived illicitly. In most cases, the arms supply grew by a factor of ten or even more,[10] and the number of mobilized troops rose by a factor of twenty, from 4,000 to 80,000;[11] some authorities place the figure at around 120,000. A substantial air force was built consisting of some sixty aircraft, many of which were Messerschmidt ME-109's (Israel bought forty-nine) and twenty-five Spitfires, mostly from the Soviet bloc, as well as a small but efficient navy.[12] "The Yishuv's military capabilities improved significantly during the immediate postwar years. One element was the establishment of a clandestine arms industry."[13] Such were the beginnings of the mighty military–industrial complex.

No parallel increases took place on the Palestinian side. Even after the fateful November 1947, when Resolution 181 passed by a large majority, assisted by the United States and the Soviet Union, Palestinian civil society was ill-prepared for war, with no army, no command structure, no intelligence or military hardware capable of facing Zionist forces. The reasons for this failure are complex.

The main obstacle to Palestinian national unity was the decimation of the leadership during the Arab Rebellion and the role played by the urban, land-owning bourgeoisie, sometimes referred to as the *'ayan* (notables); this part of the Palestinian population became affluent during the Zionist settlement period, through selling land to Zionist organizations even as they made the requisite nationalist noises. Such action by many absentee landlords and the Palestine elite meant that it was far from invested in the struggle against the growing military might of Zionism. Many preferred to leave with their money to an Arab capital, rather than reap the storm they have helped to sow. They were spurred on by Zionist military actions targeted at regions and cities that were identified for ethnic cleansing.

While the countryside was inclined to resist, as were city workers, the elite never properly united behind the resistance to the Zionist onslaught, nor was it prepared to invest by supplying and equipping a military force. The effort of resisting Zionist forces fell to a small, nationalist segment of the elite,[14] which staked its hopes on local, ramshackle militias and on the promised assistance by Arab regimes. It was clear that no defense was to be proffered by the British, whose policy during this period, according to Avi Shlaim, was "essentially an exercise in damage control."[15]

After the war, Zionist propaganda claimed that Arab leaders issued calls for Palestinians to leave their country; such myths were dispelled by historians, proving the opposite: "In late December 1947, the AHC [Arab High Committee] apparently issued a general, secret directive 'forbidding all Arab males capable of participating in the battle to leave the Country.'"[16] Simha Flapan argues that the AHC did all it could to stem the tide.[17]

Another reason for the success of ethnic cleansing before the end of the Mandate was the equivocation of the Arab regimes—mostly client kingdoms supported and controlled by Britain. While much vocal protest was heard in Arab capitals about the UN resolution dispossessing Palestinians of most of their country, precious little was done to prepare militarily. The Arab armies were equipped to defend corrupt rulers against their people, but were badly organized, inexperienced, poorly equipped, and lacking in motivation and morale. Wealthy Palestinians who knew the Arab capitals could clearly conjecture the likely results of such unpreparedness. For them, the safest step in the circumstances, they surmised, was to leave Palestine until the Arab armies come to reclaim it.

The third most decisive element enabling the ethnic cleansing were attitudes of the British authorities in Palestine, placed there to guarantee the Mandate provisions.[18] Despite many successful, murderous operations carried out by Zionist militias against the British armed forces, Mandate authorities still played a role in the process of clearing Palestine of its indigenous inhabitants, with many officers supporting Zionism. In the spirit of the Balfour Declaration, they assisted the setting up of the "Jewish national home" in Palestine. In most instances, government forces stood aside as Zionist forces attacked Arab positions. With their superior training and firepower and larger numbers, the Zionist forces could easily defeat the Arab forces.

British abstention was, therefore, a major factor in the defeat, before and after May 1948. Half of the total number of Palestinian refugees (almost 350,000 of the total nearing 750,000) had already been expelled or fled from their homes before a single unit from the Arab countries arrived in Palestine.[19] The Israeli claim that the war and expulsion followed the invasion of Palestine by Arab armies is simply a convenient propaganda canard.

But clearly the overriding factor that ensured Zionist success was that the settler-colonial forces had been trained, armed, and prepared through a modern regime of military build-up for three decades before the war. Both numerically and in terms of equipment (thanks mainly to Soviet arms supplied by Czechoslovakia but also of morale and esprit de corps), Zionist forces were clearly superior to the small, disorganized, and badly equipped and untrained fighters in Palestine. This was true even when one takes into account the armed forces of neighboring countries, which, with the exception of the British-trained and officered Arab Legion, were hardly comparable to the combined forces of Hagana, IZL, and Lehi.

The three military forces that were to form the IDF were boosted by European Jews with military experience gained in World War II. And by July 1948, the IDF had some 63,000 under arms.[20] By the end of the war, Israel could muster some 110,000 soldiers, a much larger force than the combined Arab forces. Indeed, it was easy for this armed force to take not just the 55 percent of Palestine proposed by the UN, but 78 percent of Palestine.

The Arab Liberation Army—a grand title for the ramshackle collection that crossed into Palestine—numbered less than 2,000 badly prepared fighters: "The ALA commanders immediately proved both mendacious and incompetent."[21] Their rudimentary training, nonstandard and obsolete weapons, and lack of battle experience undoubtedly contributed to the decision of many families to flee fighting zones. The Arab Legion, an instrument of British policy, had no intention of resisting Zionist forces: "Abdullah explained that he would not fight at all, unless pushed to it by other Arab countries, and then do so only for appearances sake."[22]

The exodus from Palestinian urban centers started almost immediately after the UN vote. A flow of Arabs from Haifa was noted days later, in early December, by the British intelligence services. The richest families of Halissa left first, as early as December 4, 1947.[23] The continued fighting in border areas of Haifa (and other mixed cities), including barrel bombs, sniping, and hostile demolition of large areas by the embryonic IDF, was extremely unsafe for the thousands of trapped families. Such intimidation by Zionist forces, designed to expel Arabs, was very effective, as the IDF controlled the higher ground in the Haifa and were able to fire down on the communities below.[24]

The main offensives of Zionist forces did not start until March 1948, when Hagana and the other militias received large arms supplies. By April and early May, ethnic cleansing campaigns were widespread. The British authorities turned a blind eye to these shipments, mainly from Czechoslovakia and the United States; the new equipment and munitions enabled Zionist forces to clear the large towns and cities of their Arab inhabitants. The process started in Tiberias in March and continued with Haifa, the second largest Palestinian city, during mid-April.

During this period, the IZL (Irgun) bombed the Haifa oil refinery, where Jews and Arabs worked together (and perished together). This incited a revenge attack by Arab local forces on Jewish refinery workers, in turn leading to Hagana reprisals in Balad al-Sheikh, which only multiplied the number of victims and escalated the violence. The

British forces adopted an indifferent posture; they increased patrols around Haifa, their port of departure for the end of the Mandate in May 1948 but made no attempt at controlling the violence.[25]

All moves for a truce in Haifa were brushed aside by the Zionist forces, because the continued fighting was leading to mass evacuation, the outcome they wanted.[26] By opposing a truce, they were able to drive out 70,000 Arabs from Haifa, controlling it without hindrance. By May 1, 1947—well before the end of the Mandate, and before a single Arab soldier ventured into Palestine—most of the Palestinians were already refugees, not just from Haifa, but from most of Palestine.

The pattern established in Tiberias was repeated elsewhere,[27] marking the start of the ethnic cleansing of Palestine. The larger cities and towns—Haifa, Jaffa, Beersheba, Tiberias, Lydda, Ramla, Majdal, Acre, Safad, Nazareth, Afuleh, and Kastina—fell before May 14, 1948; many smaller communities, including hundreds of villages, were also overrun and their inhabitants expelled. By the official start of the war on May 14, 1948, its result was a fait accompli.

Many years after writing his book about the Nakba (though without using this term), Morris returned to the topic of ethnic cleansing. No longer intending to obfuscate the issue of "intentionality" behind the actions of the IDF and the orders given by Ben-Gurion, he justified ethnic cleansing:

> Ben-Gurion was right ... Without the uprooting of the Palestinians, a Jewish state would not have arisen here... There are circumstances in history that justify ethnic cleansing. I know that this term is completely negative in the discourse of the 21st century, but when the choice is between ethnic cleansing and genocide ... I prefer ethnic cleansing.[28]

Such statements by Morris are enormously revealing; he knows, and we know, that the Zionists never faced genocide in Palestine. Thus, the genocide he refers to must be that of Palestinians. That he should even air such an option is indicative of the debate held by Ben-Gurion and his closest allies. Disturbingly, after decades of denial, Morris admits that the intention of Ben-Gurion was to expel all Palestinians from their land, while harsher options were also considered. Later in the interview he goes on to mourn the failure to totally clear Palestine, while threatening further ethnic cleansing.[29]

The main question that exercises historians about the 1948 war is the issue of intentionality on the part of Israeli leaders. Zionist historians typically argue that the Nakba was a result of the Arab states' attack on Israel and deny deliberate expulsion or ethnic cleansing.[30] This line has been taken up by Israel and its apologists for almost seven decades, and many accept this claim, put forward by one of the most sophisticated propaganda machines in modern history. After all, if the Nakba was not an intentional or a preplanned process, then it is difficult to attach blame to Israel. As the saying goes, *à la guerre comme à la guerre* (in times of war, act accordingly). Zionists had committed a perfect crime—one without a perpetrator.

However, the movement's intentions can be gleaned from the Biltmore Program, released by the Zionist Organization of America on May 11, 1942. After rejecting the White Paper of 1939 and blaming Britain for reneging on the Balfour Declaration, the document ends with a simple message: "The Conference urges that ... Palestine be established as a Jewish Commonwealth integrated in the structure of the new democratic world."[31] How might the whole of Palestine become a "Jewish Commonwealth" when less than 30 percent of the inhabitants were Jewish and the majority were opposed? No such plan could ever be enacted without first expelling the indigenous population.

Like Palestinian historians Walid Khalidi, Rashid Khalidi, and Nur Masalha, the new historians in Israel used Israeli sources to question this official version of history. Most of these historians agreed that the process used in 1948 by the IDF could only be termed racialized genocidal expulsion. Morris accepted that ethnic cleansing took place but argues that it was not planned, organized, and single-minded, but a result of local initiatives—an "organic" process resulting from the "dynamics of war." While writing *Birth of the Palestine Refugee Problem*, Morris was equivocal about the process, and his book cannot be seen as a justification of this illegal act, an essential war crime. A few decades later, in the infamous interview quoted above, he justifies the expulsion and regrets it did not go further, due to the growth of the Palestinian population since 1948.

Elsewhere, one can learn about Ben-Gurion's own position on ethnic cleansing indirectly, from a neglected paragraph in his diary, quoted by Teveth:

> I have never felt hatred for Arabs, and their pranks have never stirred the desire for revenge in me. But I would welcome the destruction of Jaffa, port and city. Let it come; it would be for the better. This city, which grew fat from Jewish immigration and settlement, deserves to be destroyed for having waved an axe at those who built her and made her prosper. If Jaffa went to hell, I would not count myself among the mourners.[32]

One has to take his statement that he "never felt hatred for the Arabs" with a large pinch of salt; his views of Palestinian Arabs were tinged with deep racism and great animosity. It could hardly be otherwise: "The Arab is a political creature who is unable to withstand the pressures of his environment, or the emotive and collective drives of his people."[33] Writing about Ben-Gurion's public positions toward the Arabs between 1916 and 1936, during which period he professed friendship and even brotherhood with the Palestinians, only to flip his position later, Teveth pulls no punches. It is important to quote him at length:

> Ben-Gurion's claim that he knew of Arab opposition to Zionism as early as 1915 raises serious questions about the sincerity of his professed positions on the "Arab question." In fact, as early as 1910, Ben-Gurion recognized that a conflict existed between Arab and Jewish aspirations, and later in 1914 he asked, concerning the Arabs, "Who hates us as they do?" In 1916, he openly spoke and wrote about the "hatred" of the Arabs for the Jews in Palestine. Only in the years between 1917 and 1936 did he avoid mention of the conflict and even deny its existence.
>
> But this was his public position. In his diary, and behind the closed doors of party fora, he showed himself alert to the problem of Arab rejection. A careful comparison of Ben-Gurion's public and private positions leads inexorably to the conclusion that this twenty-year denial of the conflict was a calculated tactic, born of pragmatism rather than profundity of conviction. The idea that

> Jews and Arabs could reconcile their differences through class solidarity, a notion he championed between 1919 and 1929, was a delaying tactic. Once the Yishuv had gained strength, Ben-Gurion abandoned it. The belief in a compromise solution, which Ben-Gurion professed for the seven years between 1929 and 1936, was also a tactic, designed to win continued British support for Zionism. The only genuine convictions that underlay Ben-Gurion's approach to the Arab question were two: that the support of the power that ruled Palestine was more important to Zionism than any agreement with the Arabs, and that the Arabs would reconcile themselves to the Jewish presence only after they conceded their inability to destroy it.[34]

Two important historians of the intentionality of expulsion and ethnic cleansing are Ilan Pappe and Nur Masalha, who have widely published on the period with each devoting more than one volume to the question.[35] Unlike Morris, who deals only with documents from the time of the events and the late 1940s, both Masalha and Pappe use a larger historical canvas, examining the development of the concepts of transfer (a euphemism for ethnic cleansing) and a "pure" Jewish State within the history of the Zionist movement, ever since Herzl published his iconic text in 1896.

For Masalha, the development of a racialized, settler-colonial Zionism is part of European nationalism of the nineteenth century and was based on ideas of ethnos, land, and blood (kinship and racial purity).[36] This was informed by Herzl's attachment to Prussian ideas on nationality, race, and land, later influencing Nazi concepts of *lebensraum* (living space) and *blut und boden* (blood and soil). For liberal Zionists, such as Brit Shalom, this connection to Prussian nationalism and its antecedents was a difficult and worrying phenomenon, especially as the right wing of Zionism, led by Jabotinsky, had adopted not only the concepts and terminology but also the visual trappings of this toxic brand of German nationalism during the early 1930s.

Such political thinking could only lead to forced expulsion in favor of the colonizer. The plans for controlling the whole of Palestine by force were presented by Ben-Gurion already at the 20th Zionist Congress, in Zurich, during August 1937: "After the formation of a large army in the wake of the establishment of the state, we will abolish partition and expand to the whole of Palestine."[37] For Zionism, partition was the first stage toward getting the whole country and expelling the indigenous population. This was further elucidated by Ben-Gurion in 1942, when speaking about the Biltmore Program: "This is why we formulated our demand not as a Jewish state in Palestine but Palestine as a Jewish state."[38] Again, speaking four days after the partition resolution was passed at the UN, on December 3, 1947, Ben-Gurion tells his audience that partition "does not provide a stable basis for a Jewish State,"[39] because it would include Arab citizens, which was unacceptable.

Additional sources that illuminate the issue of Zionist intentionality are: Plan C (*Tochnit Gimmel*) and Plan D (*Tochnit Dalet*). These major operational plans by Hagana were published for the first time in 1972[40] and commented upon in an important article by Walid Khalidi in 1988,[41] following his original articles about elements of Plan Dalet in 1959 and 1961.[42] Khalidi struggles against the then prevalent and dominant versions of the 1948 war written by the British brothers Jon and David Kimche, the latter being an active Mossad agent and later involved in the Irangate scandal.[43]

In his original article, Khalidi comments on the question of intentionality, assessing the plans by Zionist leaders to depopulate Palestine:

> We know that as early as Theodor Herzl they had decided that the answer was to be found in the theory of "the lesser evil": in other words, that any hardship inflicted on the Indigenous population of the land chosen by them was outweighed by the solution that the Zionist possession of the land offered to the Jewish problem. The yardstick of the lesser evil (consciously or subconsciously applied) became the moral alibi of the Zionist movement, dwarfing and finally submerging the anguish of its victims.[44]

The attitude in both plans is graduated, with Plan C concentrating on harming Palestinian defenses and infrastructure in order to cause lasting expulsions of the inhabitants of towns and villages through sabotage and covert operations by the Hagana and Palmach. While much damage was indeed inflicted upon the Palestinians until

March 1948, it became clear that Plan C did not achieve its aims: "Up to 1 March not one single Arab village had been vacated by its inhabitants and the number of people leaving the mixed towns was insignificant."[45] (Khalidi published a correction in 1988: At least ten villages out of the four hundred that fell in the period 1948–49 were captured by Zionist forces by March 1, 1948.) The small error causing the qualification to be added to the text in 1988 hardly changes the article's drift or argument: Plan C has failed to dislodge enough Palestinians from their land and homes. A further, more aggressive approach was clearly needed and was supplied by Plan D in May 1948.

A proper, wide-ranging analysis of Plans C and D requires a separate monograph. Here I quote small excerpts, relating to the question of the intentionality of the Nakba expulsions. Although illegal orders are usually not put in writing, especially after World War II, Plan D is an exception; it is one of the most detailed and lengthy military orders ever written. While military orders are merely the basis for action rather than its delimitation, Plan D goes well beyond the norm in terms of clarity of aims. It describes the mission of forces operating in population centers in the following terms:

> 4. Mounting operations against enemy population centers located inside or near our defensive system in order to prevent them from being used as bases by an active armed force. These operations can be divided into the following categories:
>
> —Destruction of villages (setting fire to, blowing up, and planting mines in the debris), especially those population centers which are difficult to control continuously.
> —Mounting search and control operations according to the following guidelines: encirclement of the village and conducting a search inside it. In the event of resistance, the armed force must be destroyed and the population must be expelled outside the borders of the state.
> —The villages which are emptied in the manner described above must be included in the fixed defensive system and must be fortified as necessary.[46]

The Nakba—Plan and Execution

Exactly fifty years after Herzl wrote about transfer, it would become a reality. Between these two junctures, others have promoted and argued for transfer as the enabling mechanism of the Zionist settler-colonial state: Leon Motzkin, Nahman Syrkin, Menahem Usishkin, Chaim Weizmann, David Ben-Gurion, Yitzhak Tabenkin, Avraham Granovsky, Israel Zangwill, Yitzhak Ben-Zvi, Pinhas Rutenberg, Aaron Aaronson, Vladimir Jabotinsky, and Berl Kaznelson.[47]

The planning and monitoring of the genocidal expulsions during 1947–49 was a complex task, shared across a number of parallel, competing agencies—a practice not invented by Zionism, but pioneered by totalitarian states during the 1930s.

Ben-Gurion was the main (and for most of the prestate period, the only) arbiter of action, his power base being Mapai (Workers Party of Palestine) and organizations that he either set up or controlled were peopled with party apparatchiks: World Zionist Federation, the Hagana, the National Committee (Vaad Leumi), Zionist Congress, the Trade Union Federation (Histadrut), Jewish Agency, the Palestine Bank, Jewish National Fund (Keren Kayemet Le'Yisrael), the Land Commission, the Representative Assembly (Va'ad Hazirim), the Zionist Executive (Ha'executiva Hazionit), the Security Committee (Va'ad Ha'bitachon), the Transfer Committee, the Consultancy, and many others. By controlling this administrative labyrinth, Ben-Gurion held the real power in the movement. His was the casting vote and controlling voice; all bodies were either chaired by him or directly or indirectly reported to him. He controlled the budget, military, land, transfer, labor, external relations, negotiations with Mandate authorities, and relations with the Diaspora.

After May 14, 1948, Ben-Gurion retained most of these roles and control mechanisms, but added two crucial new titles to the many he held before: prime minister and minister of defense. Such a combination of power is almost unique; there are few states, if any, where the elected chief executive is also directly responsible for defense. He also determined the financial and foreign policy, to the point that both the finance and foreign ministers were constantly embittered about their political impotence. This introduced the pattern of military interests navigating the state, rather than the opposite.

The 1947–49 war was methodically prepared; for more than a decade, the Intelligence Service of the Hagana (Shai) as well as some workers of the Jewish National Fund worked tirelessly on collecting and assembling data and information on every Palestinian village and town. This enormous database, complete with maps, population information, details of political activists, village leaders, numbers and types of weaponry, military training, defensive arrangements, would prove invaluable in the coming war.

Each village was carefully studied in order to determine the best way of occupying it.[48] The data, dubbed the Village Files (though towns were included), and the attack plans were safely filed. Some of the early data was used to assist the British during the Palestine Rebellion, 1936–39. This period helped to perfect the Village Files and map the Palestinian nationalist/activist base, in order to target fighters. Hagana used the events as training for the coming battles.

Ethnic cleansing was the topic of a special group, the Transfer Committee, which first convened in November 1937. Weizmann's Transfer Plan of 1930 provided a foundation for its urgent deliberations, as the situation in Palestine was by then very precipitous (an added benefit from the Zionist perspective). The British government had eventually abandoned Peel's proposals, due to their explosive potential; the Transfer Committee's work (carefully mapped by Masalha[49]) would not be wasted. The resolution of the Palestine conflict was merely relegated to a future date by the brutal suppression of the Arab Rebellion in 1939 and the imminent world war.

Zionism had won more time for consolidating military capabilities, training tens of thousands of soldiers in the British ranks and expanding the Yishuv with Holocaust survivors; by 1947 the stage was set for the decisive phase of the conflict, and Zionism was much better prepared. A decade later, Yosef Weitz (whom Ben-Gurion implicitly trusted) was named chair of the Transfer Committee. The secret cabal had carefully studied the Village Files, intelligence by Shai, and the planning activities of the regular army and Palmach. Some members of the Transfer Committee were also members of a parallel body, the Consultancy (Ha'vada Ha'myaetzet), which did much of the detailed planning and met weekly with Ben-Gurion.[50]

Minutes were not filed, due to their damaging potential; however, Pappe located a number that survived (probably through error) in the Hagana archive. Even from the little that has evaded the redactors of the IDF, the efficacy of the Village Files as a tool of genocidal expulsions is clear; it enabled preparation and guidance of the process, starting in December 1947. Plans were passed through Ben-Gurion, the commander-in-chief, to generals in various fronts, for planning operations based on Consultancy proposals. Not only were villages attacked, occupied, and expelled, the Transfer Committee and Consultancy had also planned the repopulation of whole regions, replacing the expelled with new immigrants who had flooded Haifa after May 1948 and required resettlement.

The villages, homes, orchards, fields, and contents were "recycled" as Israeli outposts. This made impossible any return of Palestinian refugees who managed to escape the military squads. Thus, the military and civilian aspects of the war were impossible to untangle; the pattern was to be repeated in later decades.

The actual process of expulsion differed from place to place, as the army was made up of somewhat incompatible elements. Deep conflicts between Mapai and Herut, played out by the personalities of Ben-Gurion and Menachem Begin (who replaced Jabotinsky in 1940), also framed the conflict between the two militias they headed. When opposition and animosity suited one or both sides, that became the rule, but mostly, purported conflict was used to conceal close cooperation.

The bombing of the King David Hotel, the conquest of Jaffa, and the murderous massacre in Deir Yassin were carried out by IZL with the assistance of the Hagana; this did not prevent Hagana from defining them as "irregularities" beyond its control. Indeed, the Hagana was crucial in the Deir Yassin massacre; its officers and soldiers removed and hid corpses before international observers arrived.[51] This type of cooperation existed ever since the end of World War II and grew stronger after November 1947. At that point, the Hagana assumed overall command, with IZL serving as special forces for "arm's length" operations.

Later revelations clarify Ben-Gurion's role in directing and commanding this process of ethnic cleansing. Traces of the process were carefully redacted from minutes, records, and written orders either by avoiding mention at the time or by careful editing after the fact. More than thirty-five years later, the archives were partially opened to researchers, with crucial documents still unreleased.

The various diaries of politicians included those of Ben-Gurion and Sharett, written in coded language with some details redacted. It is with later publications, like Rabin's diary *Pinkas Sherut* (Service Record), that this question resurfaces. Rabin's first draft included a mention of Ben-Gurion's famous hand wave, answering (Rabin's and Allon's) repeated inquiries "What shall we do with the Arabs?" raised during Operation Dani, concluding with the expulsion of some 60,000 from Lydda and Ramle. While not replying during the meeting, when the query was repeated afterward, Ben-Gurion gestured unmistakably: "Out." They got his message and executed it brutally and efficiently.[52] Thus, they had their expulsion order from the supreme leader of Zionism but managed to avoid a (written) record of it.

The cynical Zionism of Ben-Gurion and the IDF officers responsible for the ethnic cleansing is pointed out by Yitzhak Laor, interpreting Ben-Gurion's gesture as avoidance of written orders.[53] One may clearly infer the methodical nature, not only of the ethnic cleansing process, but of the systematic political denial from Ben-Gurion on down. During a debate in the Provisional State Council, in the middle of the fiercest battles, Zvi Lurie, a Mapam representative, asks Ben-Gurion pointedly:

> Do you know that Arab communities which did not take part in military activities against the state of Israel and which did not shelter enemy gangs, and whose inhabitants are leaving the areas under our control—are being destroyed? Is it known to you that these include communities far from the front, and that it will be difficult to find a military-strategic justification for such destruction, and which government authority is deciding on it?

Ben-Gurion replied: "I do not know and there is no government authority given to this. Such activity is clearly prohibited outside actual fighting times."[54]

In this manner, Ben-Gurion was responsible for the ethnic cleansing, complemented by the systemic silencing of debate through lying, denial, censorship, bureaucratic prevarication, and conscious abjuration.

This cozy relationship has endured ever since; it is best to avoid written evidence where a wink and a nod will do. Israeli society has become used to this operational mode, and it still pertains. Only gradually was evidence released by official sources, memoirs, diaries, journalistic inquiries, and later, historians' careful digging in the massive archives in which the documentation is buried. In fact, in 2018, Netanyahu has decided to close the archive to researchers for another two decades.[55]

We are unlikely to have the full record of massacres committed in this period by the IDF, but disturbing evidence keeps emerging. Some has recently come to light in an article by Yair Auron in *Haaretz* on February 5, 2016, describing a series of massacres. Typically, it was *not* translated for the English edition, so it could not be accessed by non-Hebrew-speaking readers. An analysis of the issue is available on another website, quoting at length from the Hebrew article.[56] The original article quotes a contemporary unpublished letter reporting on the al-Dawayima massacre in October 1948, written by a participating soldier and sent to a Mapam member, S. Kaplan. Shocked by the soldier's testimony, he sent the letter to Eliezer Peri, the editor. The letter makes for disturbing reading.

> A testimony provided to me by an officer which was in [Al] Dawayima the day after its conquering: The soldier is one of ours, intellectual, reliable, in all 100%. He had confided in me out of a need to unload the heaviness of his soul from the horror of the recognition that such level of barbarism can be reached by our educated and cultured people. He confided in me because not many are the hearts today who are able to listen.
>
> There was no battle and no resistance (and no Egyptians). The first conquerors killed from eighty to a hundred Arabs [including] women and children. The children were killed by smashing of their skulls with sticks. Is it possible to shout about Deir Yassin and be silent about something much worse?
>
> One soldier boasted that he raped an Arab woman and afterwards shot her. An Arab woman with a days-old infant was used for cleaning the back yard where the soldiers eat. She serviced them for

> a day or two, after which they shot her and the infant. The soldier tells that the commanders who are cultured and polite, considered good guys in society, have become vile murderers, and this occurs not in the storm of battle and heated response, but rather from a system of expulsion and destruction. The fewer Arabs remain—the better. This principle is the main political motive of [the] expulsions and acts of horror which no-one objects to, not in the field command nor amongst the highest military command. I myself was at the front for two weeks and heard boasting stories of soldiers and commanders, of how they excelled in the acts of hunting and "fucking" [*sic*]. To fuck an Arab, just like that, and in any circumstance, is considered an impressive mission and there is competition on winning this [trophy].[57]

The letter supplied graphic evidence of war crimes. On its publication in *Haaretz*, seven decades after the events, the main public reaction was of disinterest—Israelis have become so inured to worse crimes by the IDF in Lebanon, Gaza, and the West Bank that such early atrocities raise no eyebrows. Despite numerous denials, it is clear that the IDF has committed a series of massacres and other war crimes between December 1947 and February 1949 as well as later.[58] These have been presented as "deviations" from the supposedly high moral code of the IDF, but they were in fact the necessary steps of wide-ranging, intentional ethnic cleansing.

Expelling two thirds of all Palestinian Arabs was not a simple operation and could not be done without planning and committing wide-ranging atrocities. During the war but after 750,000 Palestinians had already become refugees, the UN recognized the atrocious blunder it had committed in November 1947, which ignited the war and sealed the fate of the Palestinians. By December 1948, Israel controlled most of Palestine and denied the refugees a return to their homes. Only then did the UN General Assembly try to redress this earlier mistake by passing Resolution 194 (III), which stated in part:

> The refugees wishing to return to their homes and live at peace with their neighbours should be permitted to do so at the earliest

practicable date, and that compensation should be paid for the property of those choosing not to return and for loss or damage to property which, under principles of international law or in equity, should be made good by the governments or authorities responsible.[59]

But the expulsion and ethnic cleansing were a fait accompli; Resolution 194 became one of many UN resolutions totally ignored by Israel, establishing a pattern of exceptionalism that still holds true. This was not the first UN resolution Israel ignored with impunity. Resolution 181, passed in November 1947, recommended partition, but Israel by then controlled most of Palestine and refused to vacate areas allocated to the Arab state. The UN had not reacted to the thrashing of Resolution 181 by Israeli conquests and then avoided naming Israel in Resolution 194 (III) above; likewise, when Israel refused to allow the refugees to return, it did nothing to enforce its resolutions.

Thus, early in the history of both the UN and the Israeli state, a pattern emerged: while the UN may occasionally pass resolutions demanding action or censuring Israel, it remains oblivious to Israel's habitual rejection of international law. In all such cases, Israel was buffered by the support of Western nations led by the United States, which afforded it immunity against punitive action. The UN was allowed no role in trying to resolve the conflict it had helped to escalate; its refusal to act on Israeli rejection was proof of its culpability and feebleness.

After the signing of the armistice agreements in 1949, there were additional "adjustments," one of which meant both an achievement and a problem for Israel. Israel had forced Jordan (earlier Transjordan) to cede the Wadi 'Ara villages, in what was termed the Small Triangle; it comprised some sixteen villages and a section of main highway connecting Jerusalem to Haifa, later replaced by a costal road. Israel forced King Abdullah to cede the territory. "The British *charge d'affaires* in Amman, Pirie Gordon, compared Abdullah's cession of territory under military threat to Czech President Hacha's capitulation to Hitler in March 1939."[60] This large area of prime agricultural land was a boon, but it came with a "mortgage" of 15,000 Arab inhabitants within the ceded area. A specific provision in the agreement prohibited Israel from ethnically cleansing these Arab villagers.

This suspicion was justified: Israel turned several thousands of the remaining citizens of the town Majdal (later renamed Ashkelon) and the village Zakaria into refugees after signing the armistice agreements. Without the support of international law and the UN, Jordan had few options when Israel threatened it with military attacks if it refused to cede territory; its army had suffered badly in 1948. Nor was Egypt capable of stopping the continued ethnic cleansing into the Gaza Strip, which it nominally controlled. Thus, Ben-Gurion managed to establish not only UN irrelevance on Palestine, but an international climate of denial and collusion whenever Israel threatened or attacked its Arab neighbors. For seven decades since then, this pattern of collusion was maintained and reinforced, affording Israel total impunity from international jurisdiction, through its diplomatic and political Western tutelage.

While it is undeniable that this political impunity/legal immunity has its roots in the Holocaust and the destruction of European Jewry, such an explanation is inadequate when used today to explain Israel's relationship with the United States. Unlike many nations in Europe that collaborated with the Nazis to a greater or lesser extent in the process of identifying, incarcerating, and deporting Jews to the death camps, the United States and the United Kingdom were hardly prone to guilt feelings over the Holocaust; yet both governments played a crucial role in defending Israel from international justice. It is time to come up with a better explanation for the continued support they and other Western states offer Israel today.

To continue to use the Holocaust to justify this nefarious support of Israeli war crimes is to deny much of the political agenda of Western nations after World War II. If Israel did not offer a real and crucial service to old and new imperial powers in the Middle East, their support would not be forthcoming.

The 1947–49 war produced a new terrifying reality for the Palestinians, the severity of which was not fully recognized at the time. The Palestinians lost their homeland, their homes, their livelihood, and their human rights, none of which have been restored in the decades that followed. An indigenous people became refugees in their own land overnight and are still living in camps set up in

1948 and afterward, as Israel continues to empty Palestine. Most Palestinians today live outside Palestine, while around six million still live in it, refugees in their own country, with few or no rights of any kind. More than four million of them live in the West Bank and Gaza, under brutal military occupation with no end in sight. This terrifying fate was set in motion in 1948, being reinforced and intensified since.

For Israelis, the war and its associated crimes have established a terrifying pattern, which Moshe Dayan outlined in one of his famous speeches in 1956.[61] Israel has chosen the path of the pariah, the exceptionalist Spartan war machine,[62] the ever-ready, belligerent militarized society, surrounded by a supposedly forever-hostile Arab and Muslim world. In Dayan's words:

> We are a generation that settles the land and without the steel helmet and the cannon's maw, we will not be able to plant a tree and build a home. Let us not be deterred from seeing the loathing that is inflaming and filling the lives of the hundreds of thousands of Arabs who live around us. Let us not avert our eyes lest our arms weaken. This is the fate of our generation. This is our life's choice—to be prepared and armed, strong and determined, lest the sword be stricken from our fist and our lives cut down.[63]

The battle of David vs. Goliath became the foundational myth fixating Israeli society for the next seven decades. From the Zionist perspective, Palestine in its entirety had to be taken and its population expelled to make way for a Jewish State without a hostile minority. Not only had the country to be conquered and controlled, but so had the national and historical narrative. The Zionist narrative of a small helpless state attacked by seven Arab armies, fighting back to conquer most of Palestine, with Palestinians obeying their leaders who told them to run is a clearly functional fiction that covers up malevolent intentions: the ethnic cleansing of Palestine.

The 1948 war was only partially successful, leaving parts of Palestine unconquered and some Palestinians unmoved. The reason for this failure was clearly described by Yigael Allon, commander of the Palmach:

> If it wasn't for the Arab invasion [the Arab armies entering Palestine after May 14, 1948, to defend the Palestinians] there would have been no stop to the expansion of the forces of the Hagana who could have, with the same drive, reached the natural borders of Western [Eretz] Israel [Palestine; Zionists also considered Jordan to be part of their god-promised latifundia, calling it Eastern Israel, or Eastern Eretz Israel], because in this stage most of the local enemy forces were paralyzed.[64]

It is beyond doubt that both the intention and the practice of the IDF and its politician-commanders had been to expel the Palestinians and control the whole country. That this illegal and immoral act had partially failed is not even a result of the patchy Palestinian resistance, but of the small Arab forces entering Palestine in order to frustrate the Israeli plan. Israel learned an important lesson: the international community would never act to restrain it, and it could keep whatever areas of Palestine it could grab by force. The 1948 war had not achieved all its objectives, but those that were not achieved were not abandoned; they were just deferred.

Toward the end of the war (mid-October 1948), the IDF was on a roll. Breaking the truce, the IDF successfully moved against the Egyptian forces to take the Negev, despite all the UN injunctions.[65] Until the end of the year, Britain and the United States quibbled about the proper way of stopping the IDF offensive but reached no decision. After neutralizing the Arab Legion by threatening Aqaba, the IDF attacked the Egyptian army and got as far as al-Arish.[66]

It became clear that unless forced to stop and withdraw, Israel might cause the rout of both Arab armies and take much more territory. This was a red line for the Western powers. Britain would have to come to Egypt's defense under the 1936 Anglo–Egyptian Treaty, and Ben-Gurion caved in and agreed to retreat from the Egyptian territory captured. All sides agreed to hold peace negotiations, leading to the cease-fire agreements in Rhodes in early 1949. It was clear that the supremacy of the IDF allowed it freedom to reshape the territories under Israeli control.[67]

The new historians in Israel covered the policies and military and civil transgressions involved in the war lasting from late 1947 until early 1949, so there is no need to detail their distressing findings.

Here I briefly list some of the emerging patterns that would persist for the next seven decades.

- War as the preferred option for action: Israeli policy-makers in this war (and ever since) used military means in order to make political gains. Military action would become the default instrument of Israeli diplomacy/politics. Only under intense pressure would Israel be prepared to negotiate.
- War as a tool for territorial expansion: As a corollary of the preceding observation, offensives and conquest were used to extend the boundaries of the Jewish State as opportunities arose. This did not end in 1967, with Israel controlling the whole of Palestine, but continued in Lebanon, Syria, and Egypt later on, as Israel extended its boundaries by military conquest. Only intense international pressure has in the past succeeded in forcing limited Israeli withdrawals.
- Offensive "security" and military organizing: The IDF has adopted the early lessons of the Special Night Squads, the Palmach, and Fosh in the war, preferring offensive to defensive strategies. This is partially explained by Israel's lack of strategic depth due to the narrowness of Palestine, hence preferring to attack rather than defend. However, the reasons for this preference are social and financial—keeping a huge reserve force mobilized is impossible. Thus, Israel has always preferred to strike first, and its military philosophy and practice are tailored to this reality.
- Blurring and combining of the civilian and military spheres: Israel (and more specifically, the IDF) does not differentiate civilian and military spheres of activity; it blends and combines them. David Ben-Gurion formed this credo in 1948, and this became more pronounced as Israel matured, building a war economy as its financial mainstay. Israel's military–industrial complex is world-leading, producing immense wealth through exports and training; it is also the largest industrial sector in Israel, including thousands of high-tech companies and state organs, employing hundreds of thousands of academically trained scientists, engineers, and researchers. Israel transformed the conflict from a difficulty into a lucrative solution—it lives off conflict and depends on its continuation.

- Using the army for settling and controlling occupied land: The military has played a central role in the settlement of Palestine ever since 1948. Schoolchildren are militarily trained through the Gadna at age fourteen. Before joining the IDF at eighteen, many are organized in special units of the Nahal (Fighting Pioneer Youth), an elite force whose soldiers spend some of their time in military settlements tilling the land, many staying to turn the strongholds into "civilian" settlements[68] after their obligatory service. Thus, the army occupies, clears, administers, secures, settles, and then "civilizes" the territories of Palestine. IDF controls legal, social, military, industrial, religious, civic, cultural, educational, and academic matters, as well as communications, water, electricity, building, and transport in the territories.
- A policy of ambiguity and lies: Ever since 1948 Israel's political elite have combined lies, denials, playing dumb, and delaying tactics when dealing with illegal, politically sensitive, or unpopular military actions. With the Israeli population, the attitude is: those who need to know, know; those who do not, don't. This allows leaders to hint at military excesses they wish to take credit for (in a country where military prowess is de rigueur a condition of political high office) while never admitting responsibility. Thus, Ben-Gurion could take credit for the ethnic cleansing he ordered with a hand wave and that Rabin then carried out.[69] Benny Morris, who first uncovered ethnic cleansing evidence (and later justified it in a *Haaretz* interview in 2004) notes: "Ben-Gurion always refrained from issuing clear or written expulsion orders; he preferred that his generals 'understand' what he wanted done. He wished to avoid going down in history as the 'great expeller.'"[70] It was understood that there are certain things one must do but never speak of.
- Neither peace, nor war: The constant state of emergency from which Israel emerged was adopted by conscious decision. This state of military flux, termed "positive militarism,"[71] was a doctrine that blurred all distinctions between war and peace, army and civilians, party and government. It created a sociopolitical sandbox, where the population was allowed minimal freedoms and kept under constant vigilance and perpetual emergency. This has been well presented in the famous *Davar* (Ben-Gurion's

media organ) article,[72] which for the first time describes and details this shadowy concept, which has formed the mainstay of Israeli political and military thinking. This allowed the creation of a permanent state of emergency—an existential condition of Israel and its governing principle.

- The endless war concept: An extension of the above is that war is a necessary instrument of the Israeli state, its kernel of identity. War becomes the existential condition of the nation despite occasional lulls.[73] That is also why it is difficult to distinguish between soldiers and civilians in Israel. Shimon Peres, Ben-Gurion's adept follower, explained: "Nowadays, soldiers and civilians are exchangeable. Today's soldier would be tomorrow's civilian and vice versa, just as today's civilian settlement would be tomorrow stronghold."[74] This formed the base of the settlement policy since the 1920s and continues to operate on the same foundational assumptions.

Such foundational tenets have served both as Israel's political bedrock and the IDF working norms, becoming the sine qua non of Israeli identity. A pattern was established; under some liberal pressure, military and political leaders may be forced to issue half-hearted exhortations against "pointless" brutalities. Such noises were never sincere or efficacious—they were not meant to be. The pattern has become part of Israeli military doctrine: Ben-Gurion would from time to time make poker-faced announcements about "angry citizens taking the law into their own hands" after an especially brutal attack, denying that any IDF units were involved. The truth would often be revealed by IDF commanders bragging in the Hebrew press. What was developed as a deflective temporary measure has turned into a well-tuned machinery perfected over decades. Lying became a safe way of deflecting blame and international censure; western nations were told what they wished to hear, making Israeli operations immune to external pressure.[75] This pattern has been perfected and institutionalized by the use of Special Forces and propaganda units, media spokespersons, and ministerial aides all participating in the creation of a series of complex false narratives, deflecting criticism and establishing Israeli impunity through exceptionalist arguments.

According to most historians, the war ended in early 1949, with a series of armistice agreements between Israel and its neighbors. For all the Arab states surrounding Palestine, the war with Israel, fought and lost so badly due to their limited and badly prepared intervention, has indeed ended then. Such states would soon find out that the Israelis only saw the agreements as provisional and transitory—more fighting would be required to satisfy even the minimum expectations of left Zionism, not to mention those of the right led by Begin.

Indeed, while the great powers deliberated on the exact lines for delineating Israeli control in November 1948, after fighting ended, Israel moved to take the Naqab, all the way to Eilat on the Red Sea.[76] Even hours before the final cease-fire was to start, after Israel was forced to remove its forces from Egyptian al-Arish, they attacked not just Egyptian forces, but shot down five British air force planes that were monitoring Israeli troop movements to make sure that Israel is indeed retreating as agreed. This almost led to military conflict with Britain when Ben-Gurion lied about the location of this conflagration and acted to tamper with the evidence.[77]

For Palestinians, the Nakba has never ended. Writing in 2008, Joseph Massad argues against containment of the Nakba to a single juncture:

> The Nakba is in fact much older than 60 years and is still with us, pulsating with life and coursing through history by piling up more calamities upon the Palestinian people. I hold that the Nakba is a historical epoch that is 127 years old and is ongoing.[78]

That Ben-Gurion himself was of the same opinion in 1937 is clear from a famous letter he wrote his son Amos.[79] He expressed similar views elsewhere, always in the privacy of a Zionist committee or a Mapai internal forum, not for publication. Such views, and the reliance on future military action to gain the rest of Palestine (and much more), were the guiding principle of left Zionism, and were enshrined and crystalized in military doctrine.

In that sense, the 1948 war is the initiator and progenitor of all following wars fought by Israel, setting the philosophy, doctrine, and methods that will become standard; more importantly, the war and the army that fought it have determined Israel as a polity—

controlled the social DNA forming the Israel of today. Actions, plans, and decisions enacted in 1948 have had long-term consequences, giving rise to the militarized Zionism of the last seven decades—a society in which "liberal" academics and historians find it palatable and rational to argue for ethnic cleansing as the solution to political realities, ones created by earlier injustices.

The human cost of the 1948 war is very difficult to determine. On the Israeli side almost 6,000 people, mostly combatants, lost their lives, out of some 650,000 Jews in Palestine—a very severe loss. On the Palestinian side, figures are more difficult to determine, but at least 15,000 died in war and ethnic cleansing operations—an even higher proportion of the population, mostly noncombatants. Most Palestinians who had lived in what became Israel lost their homes and human rights, and even those remaining in Israel lost their rights when they became a minority in their homeland.

To police and secure this militarized settler-society would require a very special type of army. The nature of the IDF that won the war and instigated the ethnic cleansing was in flux and bound to change, as Ben-Gurion sought to build a modern army that he could control at will. The methods employed are explored in the following chapters.

3

The 1956 War

In the second half of the twentieth century, we cannot hope to maintain our position in the Middle East by the methods of the last century. However little we like it, we must face that fact ... If we are to maintain our influence in the area, future policy must be designed to harness these [nationalist] movements rather than to struggle against them.

—Anthony Eden, 1952

Britain's Complex Role after 1948

The 1956 Suez War must be seen within the context of the Cold War. During this period Britain, France, and the United States were the main powerbrokers and armament suppliers of the Middle East, and the war made the Middle East into a sandbox where the superpowers competed for influence. Israel and Egypt found themselves at the heart of this conflict and had to choose sides.

Following the 1947–49 war, Israel had done all it could to develop its military strength. Indeed, many have argued it did so at the expense of urgent tasks of national importance, such as the resettlement of incoming Holocaust survivors and Jews who immigrated from Arab states. In 1954 Yisrael Ber, a senior official at the Ministry of Defense and a lieutenant colonel in the IDF, justified the military spending because it preserved and increased Israel's advantage over its neighbors.[1] In his analysis, Israel had the ability to build a strong and modern armament industry, which none of its neighbors managed to do.[2] It is an irony of history that Ber was unmasked in 1961 as an agent of the KGB and had passed to them highly classified documents on Israel's nuclear capability. This was top secret at the time, but like most of Israel's secrets, it had leaked to the public in fits and starts.

In articles published in IDF journals, Yisrael Ber argued that Western nations and especially the United States and the United Kingdom were either hostile to Israel and its defense interests or indifferent to the young state's eagerness to build links with the major powers. He argued that the West was an unreliable ally at best, if not an outright enemy. Such positions were not unique at the time and were held by people across a wide political spectrum. For example, the two co-leaders of the radical leftwing organization Matzpen, Akiva Orr and Moshe Machover, wrote a huge political treatise, *Shalom, Shalom Ve'en Shalom* (Peace, Peace, but No Peace), which was similar in outlook to that of Yisrael Ber.[3] The main villain of the article was Britain and its presumed one-sided support of the Arab states.

Many others of all political shades described both the United States and the United Kingdom as hostile to the new state and supporters of the "Arabs" because of their oil interests. Orr and Machover included hundreds of quotes of similar views by leading politicians and media outlets in Israel. Take for example this quote by Ben-Gurion:

> The [British] Government has broken the UN resolution and its own promise [to leave Palestine without causing difficulties], and it is no secret that this government is trying, by any means possible, to derail the UN decision on the setting up of a Jewish State.[4]

For two leftwing radicals to quote Ben-Gurion in support of their argument was certainly unusual, but it reflected a public mood, which has been abetted by the Zionist leadership—presenting the war in 1948 as a liberation struggle, not against the peasants of Palestine, who hardly played any role in the fighting, but as an anticolonial struggle against the British empire and its Arab client states. Although the war was fought against the Palestinians, the main enemy in Israeli propaganda and political argument was the foreign government from whose "colonial rule" the war has supposedly liberated Israelis. This folly has been disproved by a number of scholars, especially by Avi Shlaim, who looked at the Israeli claims that Britain had prevented Transjordan from making peace with Israel in 1949 and had supplied it with arms for fighting Israel.

According to Shlaim, such claims are untrue, as are the claims that Britain urged Transjordan to continue the fighting:

> Britain steadfastly refused to supply arms to the Arab Legion in contravention of the Security Council resolution of 29 May 1948 and declined Transjordan's requests to extend the application of the Anglo–Transjordan Treaty to the areas occupied by Transjordan in Central Palestine. The fact was that Britain did not have the capability on the ground to back Transjordan in a major military confrontation with Israel even if the political will had been there.[5]

As Shlaim notes, Britain "made no attempt to obstruct 'Abdallah's quest for a peace settlement with Israel."[6] But the claims presenting Britain as the colonizer and Israel as "postcolonial" continue to circulate.[7]

Reading about it now, one feels bemused by the success of this obscure narrative; after all, Britain was never a colonial force settling Palestine with English, Scottish, Welsh, or Irish people. It was an empire acting on behalf of an international Mandate, guided by its commitments expressed in the Balfour Declaration, and acting mainly on behalf of Jews of many countries, building a "national home" for them on the land of another people, the indigenous Palestinians. This they did because they were persuaded, up to a point, that a Jewish State would serve the interests of Britain; certainly Chaim Weizmann has tried to persuade them that it would. Without Britain, the Zionist project could not have been initiated, nor would it have succeeded, and Britain did play a role while denying the rights of the indigenous population. Even after 1946, when most resistance to Britain came from Zionist terror, the Mandate government stood aside while more and more of Palestine was taken from its inhabitants, immediately after November 1947.

Thus, in order to deflect criticism of the brutal treatment of Palestinians by the IDF during the 1947–49 war, Zionist leaders framed it as a liberation struggle against the British, despite everything

Britain did over three decades to enable and advance the Zionist project. Indeed, the leftwing Mapam Party went further and argued that the war was against "Anglo–American Imperialism" and that this struggle might also guarantee the

> completeness of Palestine, and we see the proposal of a pact between Israel and an Arab democratic state [in Palestine] as a way to ensure that Palestine is not divided, and this is understood not just by the Yishuv, the Jewish people and the Zionist movement, but also by the Arab democratic forces, who are our allies.[8]

That this could be thus expressed with a straight face, by representatives and leaders of a party that supplied many of the officers and men who have carried out the forced expulsion of more than 750,000 Palestinian Arabs in 1948, is a mark of the double-speak inherent in Zionism. This shadow-boxing with the British–American imperialist villain while dispossessing the people of Palestine, in which "Arab democrats" are described as fictive allies, is certainly one of the more disturbing examples of political delusion and propaganda taking over the public sphere in Israel in its early years. It would take decades for the shameful story of this war to be told.

The historical context is also emphasized by Benjamin Beit-Hallahmi's concept of the "periphery strategy," an Israeli doctrine "based on the idea that Israel should create alliances with the non-Arab nations on the periphery of the Arab Middle East, countries such as Turkey, Ethiopia, and Iran, in order to outflank the bordering Arab states."[9] This also included the Lebanese Phalangists, Yemeni royalists, rebels in South Sudan, and Kurds in Iraq. Such a regional "coalition of non-Arabs" was considered crucial to guarantee Israeli security. Such details are essential for a better understanding of the background in Israel before its attack on Egypt in 1956.

To believe Israeli claims that Britain was supporting an Arab coalition against Israel is nothing less than delusional. Even before the 1952 revolution in Egypt, the Egyptians were trying to liberate themselves from British rule; Egypt was tied to Britain by the 1936 Anglo–Egyptian Treaty of Alliance,[10] an agreement imposed upon Egypt and its docile king by the British empire, allowing Britain to keep 10,000 soldiers in a special military zone along the Suez Canal, an ex-territorial area under British control.

This agreement was signed, of all people, by the young Captain Anthony Eden. Despite resistance by the Egyptian government, the British maintained some 80,000 soldiers (down from 250,000 during World War II) in the Canal Military Zone, making it one of the largest garrisons on earth. This caused great consternation in Egypt and a wish to annul the 1936 treaty altogether. The pressure increased enormously after the 1952 revolution.

General Mohammad Naguib, the early ruler of Egypt installed by the Young Officers (maybe because he was not so young), spoke of "cleansing the Nile of British imperialist filth."[11] When he was replaced a year later by Colonel Gamal Abd al-Nasser, dealings with the Egyptian government became even more difficult. Naguib spoke toughly, mainly in order to endear himself to the Young Officers, but Nasser acted toughly; he was a committed nationalist with left leanings, soon to become one of the celebrated leaders of the nonaligned countries after the Bandung Conference of 1955. He would not do Britain's bidding. It was the death knell of Britain's power in Egypt, though it took a couple of years to become clear. Nasser became the main decolonizer of the Arab world.[12]

Depleted, postwar Britain watched developments without the ability to defend its interests in Egypt. By 1954, it had become too costly to keep the base at Suez,[13] and Egyptian opposition hastened its demise. In October 1954 an agreement to evacuate 80,000 soldiers was signed and the process of terminating the long presence of the empire in Egypt began. Britain, in Egypt since the 1799 invasion of the Nile delta by Napoleon, had now lost its prize possessions—India/Pakistan and Palestine. The importance of the Suez Canal could not be overstated, not just as a main shipping artery between East and West, but also as a strategic asset of the first order, as proved during both world wars. What helped Britain to swallow the bitter pill was the huge expense of the Canal Military Zone—the withering empire could no longer sustain the network of bases built at the height of its power. The sun was at last setting on the empire.

The Baghdad Pact

Despite its declining power, however, Britain continued to hold what looked like a firmer grip in Iraq, with Nuri Said in control and Britain's interests in mind; but this proved a short-term power base. Britain engineered a political coup against Nasser,[14] the Baghdad Pact of 1955, a treaty planned to move the center of gravity from Cairo to Baghdad. It included most of the Arab countries east of Egypt, together with Turkey and Iran, and later, India and Pakistan.

But British politicians found it gave them little of what they expected, and the power of Egypt under Nasser was becoming a counterforce to such machinations. Within a couple of years, it became clear that Nasser was a major player in the region, and Britain started to plan to get rid of him. After all, they had managed (with CIA help) to rid Iran of the left-liberal Mohammad Mossadegh in 1953 after he nationalized Iran's oil and install in his place a pliant, loyal Western agent, Shah Reza Pahlavi. This model had led the thinking before the Suez War.[15]

But Nasser proved to be a much tougher nut to crack. He was in total control of the army and enjoyed enormous popularity, not just in Egypt itself but across the Arab world, in the Maghrib and Mashriq alike. To remove Nasser through a coup was discussed but was out of the question and would mark any successor government as a collaborator regime installed by the West.[16] Nasser had international fame, commanding respect across the Third World. While various plots to remove him were hatched and ditched, he moved quickly to change Egypt's position in the Middle East—initiating the Aswan High Dam, the largest construction project ever planned, and getting a number of international donors to support the plan, including the United States and Britain.

At the same time, having failed to get Western arms, Nasser signed a secret arms deal with the Soviet Union, which supplied him with modern arms through Czechoslovakia at low cost. This shocked Britain's political elite; they persuaded Prime Minister Anthony Eden that the West must help Egypt to get the High Dam built to retain control. (No action was taken, however.)[17]

While support for the dam was being mulled over, Ben-Gurion, supposedly retired from politics and replaced by Pinhas Lavon as defense minister in February 1955, returned to his old position to replace the disgraced Lavon. He initiated military strikes against Gaza within weeks. Nasser reacted to this change by firming his stand against Israel and the West and started to make overtures to the Soviet Union, something he had rejected until then.

At the same time, the British government sought to balance its varying goals: to control Arabia's immense oil deposits, so crucial for the West; to subdue and humiliate the strongest country, Egypt, so as to "teach it a lesson"; and to support Israel, Iraq, and Jordan as the mainstays of UK policy in the region. Thus, the Israeli claim of British hostility was lurid propaganda. The agreement signed in July 1954 was presented in Israel as a betrayal by Britain. But Israeli actions against the revolutionary government in Cairo had started much earlier, designed to destabilize relations between Egypt, the United States, and Britain. Once signed, some in Britain, like Richard Crossman, believed it was far from certain that the agreement would support British interests, and were justifiably concerned about the behavior of the British government.[18]

The French Perspective

France, the other flagging European empire and also the other major holder of canal shares, was at the time struggling to keep control of its largest colony in Algeria and was justifiably suspicious about the links between Nasser and the National Liberation Front, so it had its own reasons to try and remove him from power. Discussions about this option were ongoing between the two governments since 1954. For obvious reasons, a joint operation was seen as preferable. France had also established strong links with Israel, both through supplying know-how and technological support for its budding nuclear project, as well as being its main arms supplier. This allowed France to argue for Israel to be included in Operation Musketeer, despite initial British reservations.

One element of Nasser's politics especially threatening to the three allies was his promotion of pan-Arabism. While sharing the anti-imperialist and anticolonialist views of his colleagues, he has gone further, seeking constructive solutions to the festering wounds of colonialism. In 1948, even as he languished in the Falouja enclave, surrounded by IDF forces, he meditated about the humiliation of the Arabs—the *umma* (Arab [Muslim] community) or *qawm* (Arab nation) rather than *watan* (single state or nation)—the great "Arab homeland," rather than the individual state:

> All our peoples seem, beyond our rear-line, the victims of a tightly woven conspiracy … I was convinced that what was happening in Palestine could have happened in any country in this region so long as it resigns itself to the dominance of these factors, elements and powers … And subsequent events have fully confirmed me in this belief … I began to believe … in common struggle.[19]

Such sentiments were not exceptional in the Arab world. Musa Alami, the Palestinian intellectual who found himself a refugee in Lebanon after 1948, reflected on the reasons for the defeat suffered by the Arab armies and concluded:

> The loss of Palestine was a great disaster with far-reaching results for the very existence of the Arab nation. If the Arabs hasten to face the danger before it overwhelms them, there is still time and opportunity…
>
> The first remedy lies in unity, so that we may become again a strong, cohesive body politic … (but) unity is not enough. There must be complete modernisation in every aspect of Arab life and thought.[20]

Alami's words seem prophetic in hindsight, accurately projecting what Nasser was to conclude only a few years later. This crucial ingredient of Nasser's political thought and influence in the Arab world was in direct opposition to both Britain's and France's policies in the Middle East. Early in his political rise, in 1953, Nasser voiced such views in an interview with the London *Observer,* clearly stating his political credo and providing an anticolonial challenge to prevailing British views.[21] From the moment his pan-Arab aspirations became known, Nasser was on borrowed time, as far as the old European empires were concerned.

Grabbing an Opportunity

In Israel, the attitude leading to the attack on Egypt in 1956 was even clearer than that in London and Paris. Ben-Gurion and Sharett, as well as the military leaders during the first decade of Israel's existence, had all made clear their strong hostility to Egypt and the Young Officers. In 1951, Ben-Gurion agreed to the establishment of special Unit 30, a death squad for carrying out *Peuloth Tagmul* (Retribution Operations) across Israel's ceasefire lines, against various Arab targets, mainly Egypt. Due to its inferior performance, this unit was disbanded in 1952 and replaced by infamous Unit 101, set up by Ariel Sharon, who had been recalled to the IDF for that purpose. This unit continued the earlier traditions established by the Special Night Squads and Palmach in the period 1939–47 of night-time attacks on villages to exact punishment for supporting the rebellion. Its many operations influenced the operational doctrine of the IDF for decades to come.

The Gaza Strip, home to numerous refugee camps and an Egyptian military base (as well as in Sinai), were frequent targets for Unit 101 attacks. These were often launched in "retaliation" for frequent intrusions by Palestinians crossing the lines in order to return to their abandoned villages, attempting to resettle or (more frequently) to reclaim some of their possessions or harvest fields and fruit during widespread hunger in the camps.[22] Israel treated the returning fellahin as mortal enemies; the return of the refugees, sanctioned by the UN, was seen as a sword of Damocles hanging over Israel. The IDF was therefore employed in hunting and killing these returnees (which it labeled infiltrators) or expelling them.

The tireless Yosef Weitz, director of the Land and Afforestation Department of the Jewish National Fund (Keren Kayemet Le'Israel), forever intent on emptying Palestine of its indigenous people, was quick to point out the danger of letting the infiltrations continue:

> Slowly but surely, abandoned villages are vanishing as they are resettled [by Arabs], partially or completely ... Many thousands of dunams [in original] formerly considered abandoned lands—are now being [cultivated] or claimed by their owners, and it is clear to me that, if the refugee problem is not shortly resolved

by resettlement [in Arab countries]—the day is not far when it will be solved by itself by the return to the villages of the [land's] owners.[23]

To halt this return, a Border Guard force was established in 1950, partly manned by non-Jewish Circassians, a pattern that was later repeated with the drafting of Bedouin and Druze men. This policy is still active, and the force is known for its brutality and disregard for Arab life. This use of small minority groups within Palestine to police the large majority of Palestinians (a divide and rule tactic) has proven generally successful, deflecting anger toward these groups and away from the IDF.

Israel created Jewish settlements in abandoned villages along the cease-fire line to block the return of refugees.[24] In many cases, the new immigrants were badly suited for the agricultural nature of such settlements; they lacked any knowledge of farming, as well as the tools and methods of tilling in a country with scarce water sources, where methods of sustainable farming were perfected over millennia and sparse resources had to be used wisely. Ben-Gurion had little faith in the refugee–settlers' ability to resist infiltrators[25] and thought they may abandon the "tripwire" settlements. Immigrants dumped in such villages were left to their own devices; unable to feed themselves, they experienced hunger and became a burden for the state, one it was unwilling to deal with.[26]

The Palestinian infiltrators were a long-term phenomenon, as pointed out by Morris; thousands were killed over a decade, and tens of thousands were expelled across the borders under false and illegal pretences[27] in order to get rid of Arab villages adjacent to armistice lines.[28] Morris claims that the number of infiltrators killed by the IDF and other Israeli forces is over 2,700 and may even be nearer 5,000.[29] What has taken place can only be termed ethnic cleansing—reducing the numbers of Palestinians so as to have a more "Jewish" state,[30] a quest for ethnic "purity."

Reality was different, as Morris notes:

> From 1954 the available statistics show a dramatic drop in the number of infiltrators killed by Israeli security forces, in some measure paralleling the drop in the number of infiltrators—some

4,500 or fewer annually compared with between 9,000 and 16,000 in 1952.[31]

However, by 1954, this number had started to drop. This is of great interest when reviewing the reasons given for the attack on Egypt in 1956. Zionist historians claim the main reason for that operation was the increasing number of infiltrators, but by 1956, the infiltration phenomenon was virtually over. Israel, projecting a racialized ideological dimension onto infiltration[32] and using this excuse, had expelled whole towns and villages after the war ended, such as the town of Majdal (Ashkelon) where "the bulk of the town's population fled to Gaza in 1948, and in the course of 1950 its remaining Arab inhabitants were transferred to the Strip."[33] The language used by Morris (including references to transfer as if it were a mode of urban transit) occludes war crimes—thousands were forcibly put on trucks and dumped in Gaza without any belongings. This was done at hundreds of sites across the country, in order to retain a Jewish majority. By 1956, this had been achieved, the process completed.

The Israel-France Relationship

The year 1954 was crucial in Israel's history; France started to supply Israel with modern arms, under favorable terms, and offered secret nuclear data and training. The IDF now received modern jet fighters, tanks, powerful mobile artillery, halftrack personnel carriers, large cargo airplanes and boats, new communication systems—the wherewithal of a modern army, enabling it to prepare to face the large Arab armies, whose equipment was obsolete.

At the same time Israel's leaders started to develop a policy whose fruits can be clearly appreciated today. Israel presented itself as an integral part of the West, a colonial partner in France's war against nationalists in the Maghrib—something Israel could and did help France with. The French also assisted Israel in laying the foundations for its military–industrial complex, which within decades would become its main export.

France's support for Israel's nuclear weapons development was a long-term commitment that spanned fourteen years (1953–67). After being barred from US nuclear secrets in 1947, France had searched for access to such data elsewhere. French nuclear researchers hoped that Israel could help them get classified information from the United States, through Jewish scientists at the helm of US nuclear programs. France started training young Israeli physicists at Saclay, their new nuclear facility, during 1953, laying the foundation for the nuclear programs of both countries. While Israel does not seem to have obtained major secrets through US Jewish scientists, it offered other benefits to the French. An Israeli scientist had developed a simpler and cheaper way of producing the heavy water required in nuclear reactors,[34] for example; Israelis had developed a method of refining uranium from low-grade ores,[35] which can be found both in the Negev and North Africa. There are various claims about Israel testing nuclear devices in the Negev, North Africa, and the South Atlantic, but most have not been proven.[36] While Israeli nuclear devices did not play any part in 1956—France itself tested its first bomb in 1960—this unique and highly secretive liaison evinces the closeness of the two states and France's crucial role in building up the IDF for the Suez War.

Those most responsible for the new turn and the building of a large offensive army were Shimon Peres, the undersecretary for defense, and Moshe Dayan, a career officer who became the IDF Chief of General Staff in 1953. Dayan completed the process of turning the IDF into a modern aggressive force used to risk-taking, incursions into enemy territory, and mobile surprise attacks to demoralize and destabilize the region. With the new armor France gave the IDF, this was a real option.

As opposed to earlier, more conservative IDF chiefs of general staff, Dayan was a veteran of the Wingate Special Night Squads units, trained in the tradition of night fighting, brutal preemptive strikes, commando raids, and retaliatory attacks on civilian populations. From very early in his career he was involved in politics; he proved to be masterful in maneuvering others into positions he wished them to occupy. Dayan's most palpable distinctive trait was his need for constant conflict and his use of war as an instrument of policy.

Since 1954, Dayan wanted war, and tried to prod the Arab states into military conflict so that Israel could conquer Sinai or the West Bank, not to mention destroy Egypt's army. This was what drove the IDF strikes in 1955 at Kuntilla and Sabha and against Syria in December 1955.[37]

Indeed, Ben-Gurion and Dayan had even more far-reaching plans for the neighboring countries. During 1955, Foreign Minister Moshe Sharett thwarted Ben Gurion's and Dayan's unhinged plan to "buy" a Maronite officer as a way of intervening in Lebanon and taking over its southern region.[38] Only Sharett's intervention saved Israel from getting "bogged down in a mad adventure that will only bring us disgrace,"[39] foretelling the disastrous war Ariel Sharon started in Lebanon in 1982.

Israeli Offensive Plans and the "Lavon Affair"

By operating in this vein, Dayan did not invent something novel; he was simply following an established tradition of elite units in the IDF. His contribution was in making adventurism, offensive planning, and baiting the enemy and using force to achieve political goals the modus operandi for the IDF and for the political layers above him. This became the norm of Israel's young polity. While this strain in Ben-Gurion's thinking during the 1947–49 war was toned down by previous chiefs of general staff, Dayan was prepared to take unnecessary risks in a small country without strategic depth. In his opinion, it must initiate all and every military engagement with its neighbors, never waiting for them to take action, precisely because it did not have strategic depth.

This practice was not limited to military risk-taking. In 1954 Pinhas Lavon was appointed defense minister, following the resignation of Ben-Gurion, who retired to Kibbutz Sde Boker. The new minister's outlook was similar to Moshe Dayan's, despite coming from the left (in Israeli terms) of the political spectrum. He was keen on risky operations beyond enemy lines, and initiated or agreed to many raids, baiting Israel's neighbors into conflict during his short term of thirteen months.

Now Israel had three swashbuckling politicians at the defense helm —Dayan, Peres, and Lavon—and the inexorable countdown toward the next war began. Lavon supported sting operations in the Arab countries and was involved (exactly to what degree is still under debate) in authorizing the infamous Operation Susannah in Egypt in the summer of 1954, activating an Israeli cell in Egypt set up in 1953.[40] This operation consisted of IDF Military Intelligence recruiting Egyptian Jews—supplying and training them to carry out a terror bombing campaign against Egyptian, American, and British targets, and planting false evidence implicating the Muslim Brotherhood and Communists. The operation was conducted with staggering amateurism, with people hurt in the blasts, as well as one of the operatives who had a bomb explode prematurely in his trouser pocket, leading to his arrest and the exposure of the network, which had not used even basic security measures.

The operation was designed to scupper the agreement to close the Suez Canal Military Zone between Britain and Egypt, but it failed miserably.[41] IDF Military Intelligence wanted to delay the British withdrawal from the Canal Zone through acts of sabotage in Western countries that it would attribute to Egyptian nationalist groups, with the goal of souring relations between Egypt and those countries. The amateurish saboteurs bungled the operation and were caught red-handed; they admitted the plot. The results of their trial were the execution of two members, and the suicide of two others. It also came out that the Israeli officer activating the group betrayed the network to the Egyptians.[42]

Israeli reporting of the Lavon affair is highly selective; even Schiff does not mention that many bombs were planted in Cairo and Alexandria, causing casualties. Yet even such accounts make it clear that the ruling clique in Israel had taken decisions that were not only illegal but also irrational, without recourse to mitigating powers. The head of Mossad at the time, Issar Harel, later described the intention: "to undermine the confidence of the West in the existing regime by creating public disorder."[43]

What was planned as an action to drive a wedge between Egypt's leaders and Western powers turned out to be a great embarrassment for Israel. The operatives were all caught, disclosing their failed operations at the trial and most were executed or committed suicide. While the person responsible for this dark chapter of criminal ineptness was never publicly named, both Dayan and Lavon shared responsibility as Chief of General Staff and defense minister, respectively, with Ben-Gurion as the likely puppet master.

But the question of personal responsibility is irrelevant; what enabled such actions were the attitudes of the IDF and the Israeli leadership and a doctrine of using armed force as a political extension for achieving illicit aims and as an instrument of aggression rather than defense. Apparently, not only the high command believed in revenge and baiting strikes; so did the other ranks:

> From time to time, the troops—socially the middle-echelon officers in the elite units—had to be thrown a bone: they were forever (especially in Unit 101 and the Paratrooper battalion/Brigade) straining at the leash. It was not only a matter of vengeance, punishment, and deterrence; the officers wanted to demonstrate their own and their units' mettle.[44]

Dayan went even further—he saw the reprisal and preemptory strikes as "the 'life-giving drug' for the nation as a whole: Without them we will not have a fighting people and without the regimen of a fighting people we are lost."[45] By that time, it was the IDF running the government, rather than the other way around. Prime Minister Sharett was hardly in control and was frequently lied to.

While people abroad knew of the Lavon affair, Israelis were unable to learn about it, because all media channels were subject to military censorship. And while Israel buzzed with rumors and bizarre theories, no one knew for certain what transpired, unless they listened to foreign broadcasts in Arabic, English, or French. For almost a decade the details of this shameful episode were not released, and most Jewish Israelis assumed that the "Cairo Martyrs" (as they became known) were heroes acting for their country, rather than a bungling bunch on a black-propaganda mission. Instead of this episode becoming an object lesson for the Israeli public, IDF, and politicians, it was seen as yet another episode of the imaginary global antisemitism that lurked everywhere, according to the government.

While this type of subterfuge worked in Israel, where most of the population believed the lies and obfuscations, it was less effective abroad. Even before this bungled operation, Israel's image abroad was already tarnished, especially after the murderous attack on the village of Qibya[46] in 1953, part of the "educational" strikes Dayan so valued. In the attack, led by the already infamous Ariel Sharon, commander of Unit 101, more than seventy villagers were murdered and fifty houses were flattened, some over their inhabitants, as well as the school and the village mosque.

In Israel, the detailed releases were typically made up of lies, and the public supported the attack when they heard Ben-Gurion on the radio giving his version of what happened, claiming this was not an IDF action, but the act of angry civilians from border settlements:

> The [Jewish] border settlers in Israel, mostly refugees, people from Arab countries and survivors from the Nazi concentration camps, have, for years, been the target of ... murderous attacks and had shown a great restraint. Rightfully, they have demanded that their government protect their lives and the Israeli government gave them weapons and trained them to protect themselves. But the armed forces from Transjordan did not stop their criminal acts, until [the people in] some of the border settlements lost their patience and after the murder of a mother and her two children in Yahud, they attacked, last week, the village of Kibya across the border, that was one of the main centers of the murderers' gangs. Every one of us regrets and suffers when blood is shed anywhere and nobody regrets more than the Israeli government the fact that innocent people were killed in the retaliation act in Kibya. But all the responsibility rests with the government of Transjordan that for many years tolerated and thus encouraged attacks of murder and robbery by armed powers in its country against the citizens of Israel.[47]

Even in Israel, many people doubted this version of events, which included not a shred of truth. Outside Israel, the attack was widely condemned, including by the US State Department, UN Security Council, and even many Jewish communities abroad.[48]

Benny Morris has devoted three whole chapters of his book (including a special chapter dealing with the Qibya strike) to the retaliatory and punitive strikes by Israel in the period 1949–56, and it is an impressive (though partial) portrayal of the many, varied, and brutal IDF attacks of Arab targets in the period, so there is no need to repeat the list here; suffice it to say that Dayan fully implemented his (and Ben-Gurion's) policy of using strikes on Arab villages and army outposts as the main tool for training an offensive army for bigger, more important wars.[49]

Israel's leaders were also extremely worried about highly secret US efforts to bring about a regional rapprochement through Operation ALPHA,[50] which they saw as dangerous, demanding Israeli concessions they strongly opposed. The 1956 war was the sure-fire way of terminating ALPHA.

By the early 1950s, lying about its own activities had become systemic in Israel. One need not take a moralistic line about such lies —most states lie about their real intentions, as well as the exact details of covert actions. In the case of Israel during the 1950s, lying became so normalized as to form a pattern, and lies were used to placate not just the overseas audience, but also the Israeli population. It really started biting when members of the cabinet started lying to one another, and army commanders lied to ministers and to one another. A web of lies replaced normal discourse, which became polluted by half-truths, outright lies, and preposterous inventions. It will not be possible to enumerate such lies here; a few examples suffice. No one in the Israeli leadership seems to have been immune to this double-speak.

On the return from the Qibya operation, Sharon, the commander of the attack, reported that around "ten or twelve Arabs" died in the operation, while knowing the actual number was close to a hundred.[51] Sharett, the acting prime minister and only "moderate" in the cabinet, suspected this was probably inaccurate and demanded to see the operational order for Unit 101. Lavon, the acting defense minister in Ben-Gurion's absence, gave Sharett a doctored copy of the order, from which the following line describing the nature of the strike was removed: "to attack and temporarily to occupy the village, carry out destruction and maximum killing, in order to drive out the inhabitants of the village from their homes."[52]

As the full horror of the murders became known across the world through UN reports, Ben-Gurion saw fit, with the full agreement of his cabinet, and with Sharett's careful eye for detail, editing the statement above, to read what was a tissue of lies over Israeli Radio. This further exacerbated the horror of the raid and death toll by blaming it on the most vulnerable sectors of the Jewish population of Israel, Holocaust survivors and Arab Jews, two groups that have suffered a combination of dismissive neglect and active disdain from the ruling elite.[53]

On an earlier occasion, in the summer of 1953, IDF attacked the Bureij refugee camp in the Gaza Strip, murdering more than twenty refugees before retreating. The operation was a disaster by any measure, and Sharon, the commanding officer, was almost killed by a crowd of unarmed angry refugees surrounding his small group of soldiers. After the raid, Dayan lied to Lavon, insisting that the refugee camp was not the target but that the unit was derailed from attacking the military target by some happenstance and found itself in the middle of the Bureij refugee camp, where "they had been forced to kill civilians."[54] Lavon obviously knew this was a lie, but it was the lie he wished to hear. He would himself lie later on about Operation Susannah, not only to Sharett, but to the various official inquiries conducted after the debacle. Sharett had himself lied in a rather grand manner, claiming that the Israeli government's four-day silence after Qibya is itself proof that the action was not by the IDF. This was one lie Sharett knew no one would believe.[55] He says later, thinking of the long list of officially sanctioned lies: "Who will believe that we are telling the truth?"[56]

Thus, deceit had become the modus operandi of the Israeli government; it created a space for illicit IDF operations, as the various players knew that it is both necessary to lie and that they could get away with it, even at the highest echelons. A pattern was set, then, in this formative period of the Israeli polis, one that continues to be followed in the twenty-first century.

Nasser and Arab Nationalism

Early in the life of the revolutionary regime, which had shaken Egypt in 1952 and turned it into a republic in 1953, the Israeli government led by Ben-Gurion and Sharett had taken a position of animosity toward the modernizing regime of the Young Officers. This was hardly surprising. The activating of the 1953 cell mentioned above was only one instance of Israeli efforts to destabilize Egypt's leadership. Once Israel decided that Nasser must be challenged and removed, there was no way back.

In the relationships between Egypt and Israel, there seemed to be a point of no return—one side was carrying off one raid too far, finally persuading the other side of a premeditated pattern. At some point—February 28, 1955, to be specific, after the Gaza Raid—it was impossible for Nasser to pretend that the IDF attacks were isolated incidents.

Following a series of pinprick attacks (more annoying than efficacious) by Egyptian forces, Ben-Gurion returned as defense minister, replacing the disgraced Lavon and giving war-mongering Dayan a powerful boost in trying to start a war with Egypt. Ben-Gurion was extremely keen on such an offensive and supported Dayan in his various attempts to provoke Egypt.[57] The two managed to press Prime Minister Sharett—increasingly under pressure from his activist cabinet—into sanctioning what was described as a limited operation against an Egyptian army base in Gaza. (Later he would recognize this as a trap and greatly regret his decision.)

The IDF action went wrong in more than one way. Targets were misidentified and heavy losses to the Egyptians defending their base were matched by heavy loses to Israeli units, to the shock of Dayan who was awaiting the returning troops.[58] The operation was the opening gambit that would lead to the Suez War and was seen as such by Israel's military leaders.[59]

The attack caused shock waves in Cairo, alarming Nasser and his ministers. It was now clear to them that Dayan would stop at nothing to get his war. The Egyptian leadership therefore decided on a series of measures, all of which indicated that they recognized the severity of the situation.[60] The first one was a sharp turn toward the Soviet Union (until then blocked by Nasser). With Western nations cold-shouldering Nasser on his request for assistance with the largest project in the history of modern Egypt—the High Dam at Aswan—and their refusal to supply Egypt with modern armaments, the options before Nasser were limited. Yet his move, and its immediate success, threw the Israeli as well as other intelligence services off balance—none had foreseen this astute maneuver. Israeli military analysis, undermined by the Soviet spy Yisrael Bar, believed that the likelihood of the Soviet Union arming Nasser was nil.

The second action taken by Nasser was, as Sharett admitted many times, a direct reaction to the humiliation suffered in the Gaza raid.

Nasser had moved within days to set up a commando army unit, the fedayeen, made up of mainly Palestinian volunteers, aimed at harming Israeli morale by using methods similar to the IDF—deep penetration and sabotage behind the lines, against military and civilian targets alike. This guerrilla fighting against Israel was a development of tactics used by the Egyptians between 1952 and the end of 1954, when similar sabotage harmed the Suez Canal Military Zone, before Britain finally agreed to vacate it.

Some of Israel's leaders recognized the repercussions of their own actions: "Sharett understood that the Egyptian harassment, which amounted to a systematic campaign, was a 'direct result' of the Gaza Raid, and proved the 'inefficacy' of the retaliatory policy."[61] Dayan, however, saw this differently. His continued efforts at baiting the Egyptians were at last producing some long-term results. Nasser was about to give Israel a casus belli for a preemptive strike, as Dayan and Ben-Gurion saw it.[62] The long-held intention of using war to improve Israel's strategic depth was about to materialize.

Nonetheless, Israel was in a tough spot. Not only was it faced with a guerrilla campaign, for which its army was badly prepared and unsuited, but it also was doomed to face a much better equipped army in Egypt, backed by a superpower. The need for modernizing the IDF became urgent, almost desperate, as the United States and the United Kingdom refused to arm Israel without it agreeing to a peace process with its Arab neighbors, something Israel refused point-blank. France saved the day.

The two arms deals, the Egyptian one with the Soviet bloc and Israel's with France, materialized in mid-1956 and made war inevitable. Now both armies were poised to attack one another, but the readiness and intention to go to war was decisively more real and urgent in Israel, with Nasser being intent mainly on noisy posturing. Thus, Dayan carefully plotted his next move while the IDF trained with the bounty of modern arms speedily pouring across the sea from France.[63]

Operation Musketeer and Its Background

By the spring of 1956, France had decided that due to Nasser's support of the Algerian resistance movement, the National Liberation

Front, he would have to be toppled and replaced by a "more friendly regime." French planners conceived it either as a common war with Israel as ally, or by "using the IDF operated by remote control," as Morris puts it.[64] Mossad supplied detailed information about links between Egypt and the Algerian resistance, and France started to plan the war against Egypt in collusion with Israel. For both countries, the task of removing President Nasser from power became urgent. Britain, which had hatched unsuccessful plots against Nasser for a number of years, would soon come to the same conclusion. Nasser was not likely to be toppled by a simple Western-financed coup; a major military operation would be required. The combined interests of Britain, France, and Israel, and to a certain degree also the United States, were all pointing toward military action, albeit one built on "creative" projections and wishful thinking, as Nasser was about to prove.[65]

Israel moved toward war in early 1955, after the failed Gaza strike. A special tax called *Yahav Magen* (defense tax) and a "voluntary" fund called *Keren Magen* (defense fund) were launched, and all social institutions were recruited to support the war effort. Companies, schools, unions, and political and cultural organizations were pressed to hold special events, at which speeches were made about the dangers facing Israel and the urgent need for rearmament. Even schoolchildren (including this author) were successfully harangued into donating their meager pennies. IDF troupes, especially the new (and very popular) Lahakat Ha'Nachal, were performing new patriotic songs, urging the public to donate toward rearmament.

The campaign was very successful. While the government did not really need the savings of children and the poor to buy weaponry, it cherished the hysteria and the nationalistic fervor that spread throughout Israel, muzzling any opposition on the road to war. Israeli society was recruited to the war a long time before the new tanks started rolling toward the Suez Canal on October 29, 1956.

It was during the summer of 1956, with both sides quickly getting familiarized with the new weaponry, that Nasser learned that both the United States and the United Kingdom finally declined his requests for loans toward the cost of the High Dam, the make-or-break element of his new economic and social plan. One week later, on Egypt's Independence Day, Nasser announced immediate nationalization of the Suez Canal Company, a move he was urged toward by many Egyptians but avoided taking until his other options vanished. In doing so, Nasser was taking on some of the strongest powers on earth.

The countdown for war had started, not just in Israel, but also in London and Paris. Most historians argue that Nasser misjudged the reaction of the colonial powers to his move. After all, Britain had agreed to remove its forces from the Canal Zone, completing the evacuation a short time before the crisis. With new arms for his military and international support for his popular move, Nasser was sure that Britain and France might indeed be angered but would not act militarily.

But Nasser did not foresee the tripartite plan of attack, with Israel providing the technical excuse for Operation Musketeer (dubbed in Israel Operation Kadesh) by attacking Egypt and thus "threatening" the Canal Zone. David Tal described Nasser's surprise on learning of Israel's actions: "Could it be that Israel really wanted war? If so, he could not see why ... What is it all about?"[66]

In order to satisfy US demands that all nonmilitary options would be tried, Britain convened a Canal Users Forum in London during August 1956. Eighteen nations took part and Prime Minister Robert Menzies of Australia was mandated to present proposals to Nasser. Thus, while preparing their forces for the attack, the allies covered up their intentions by feigning diplomacy. It was clear from the outset that Menzies's proposal for placing the canal under international supervision would not be acceptable to Egypt, and Nasser duly rejected it.[67] In the meantime, black-propaganda stations set up by the allies started airing anti-Nasser programs in several languages.[68]

While Britain and France had planned an attack on Egypt even before the nationalization of the Suez Canal Company, the recent arming of the IDF and the meeting of interests between France and Israel created a new and dangerous reality. France had been persuaded by Israeli politicians and commanders to include Israel in the attack and use its incursion into Sinai as the casus belli.[69]

The planning by the three allies was based on a draft presented by Moshe Dayan to the French team of planners during the beginning of October 1956. But having been adopted, it was thrown into disarray by a stern warning by US President Dwight Eisenhower, relayed to the Prime Minister Ben-Gurion, on October 28, as Israel was preparing for its entry into the Sinai. All this was taking place as the Soviet Union was involved in a brutal suppression of the Hungarian uprising started a few days earlier, on October 23, drawing media and public attention away from the Middle East.

Ben-Gurion took a snap decision to attack, worried that the opportunity for attacking Egypt might disappear. In great haste and without prior notification to their allies, the IDF entered Sinai the following day, surprising even those in the know. The careful and secretive preparations, whereby the French and British forces covertly made their way to Malta and Cyprus to await the Israeli attack, had all gone awry. Struggling to adjust to this serious hitch, the two allies issued stern instructions for "both forces to get away from the Canal to a distance of ten miles."[70] Israel moved its forces as prearranged, while Egypt refused to do so, and hence exposed its army to the full force of the bombing by the tripartite forces, as well as to the occupation of the Canal Zone by a combination of naval and parachute units. The change in the timetable at the last moment has caused much of the chaos, which typified the bungled military occupation of the Canal Zone by both European allies.

Israel clearly could not have taken Sinai without the aerial and naval umbrella of its allies. In a sensationalist account of the war, the French Bromberger brothers claim that to facilitate the Israeli attack, France needed to act in support:

> As the Israeli army of General Moshe Dayan went into the attack, in came the French squadrons to guard the sky behind him against the calamitous retaliation raids that everyone expected to follow—at once.
>
> It was the guns of the cruiser Georges-Leygues, too, that blew to bits the Rafa [*sic*] resistance, the strong fortifications behind the Gaza strip ... It was the French fighter-bombers that halted the big Egyptian reinforcements column coming up through Sinai. They worked independently of the Israeli command, under the orders of General Brohon in Cyprus, but sometimes carried out missions at the request of General Dayan.[71]

Indeed, there were no Egyptian fighter planes or bombers in the air[72]—the combined air forces of the Allies had heavily bombed all Egyptian airfields, destroying MIG and Ilyushin planes recently supplied by the Soviet bloc.

It took more than a decade for British and Israeli sources to corroborate this account. The enormous force collected at the ports in Malta and Cyprus was the largest such operation since the D-Day attack. The French military command, dismissive of the British obsession with size, dubbed it the Hundred Ships Armada; Paul Johnson called it "a war machine of such monstrous size"[73] but noted that despite its size, it could not "be exerted against a small, weak and recalcitrant nation."[74]

The enormous force had hardly served its purpose. Nasser had ordered the canal blocked by sinking prepared tankers into the waterway, as his ultimate measure against the three armies,[75] but more was to come: "With the Canal unusable, the flow of oil to the West was disrupted when Syrian Army Engineers blew up the pumping-stations to the Iraqi oil pipeline."[76] This was not going well.

What the tripartite partners were not prepared for was the vociferous reaction by both the Soviet Union and the United States (as well as other nations), who—despite the ongoing Soviet incursion and repression of the wayward Hungarian reformist government—both agreed that the tripartite forces must immediately and unconditionally vacate Egypt and the Canal Zone. This unique concord of opposing sides during the Cold War was not something the three aggressors could have predicted; indeed, it never happened again. This apparent consensus was based on a complex set of assumptions made by the leaders of the two superpowers.

The United States accepted the brutal suppression of Hungary as an "internal" Soviet issue; it protested and used it in its propaganda but never considered intervention. The Middle East was different. The spent forces of Britain and France were, in US planners' minds, part of a tainted history of imperialism, now at an end. The deflated empires were fighting to retain an outdated mode of control, while being replaced in the Middle East by the emerging interests of the two superpowers. For the United States, there was an urgent need to stop Soviet influence in Egypt, and the tripartite attack was the wrong way of going about it and complicated US objectives in the region; the attack indeed cemented both Nasser's hold over Egypt and the Soviet Union's role in the country. Both the superpowers temporarily agreed on the need for a swift end to the war.[77] There was nothing the three allies could do against the combined force of the superpowers.

What also influenced the British and French decision to retreat was the stiff defense put up not just by the Egyptian army but by thousands of armed volunteers, operating as classic guerrillas and causing painful losses to the occupation forces. In both countries large demonstrations against the war were taking place, demanding retreat, and political upheavals developed in nonaligned countries. In Britain the economy started suffering immediately,[78] and the government requested urgent assistance from the United States, so resisting demands for withdrawal was impractical.

Results of the War

For both European empires of the past, the Suez War was the last gasp of imperialist gunboat diplomacy. The politicians involved were disgraced and most retired or resigned shortly after the Suez fiasco. Decades would pass before these (and other) countries would again be tempted into an even more destructive Middle East war. In Egypt, the political defeat of the Tripartite Alliance was crucial to the following success of the Nasser regime, which, despite its military defeat, emerged the political victor of the conflict.

For Israel, which had stayed longer in Sinai than its two partners did in the Canal Zone, this war was of great historical importance. A number of earlier patterns were reinforced as a result of the war.

Israel was visibly placed on the power map of the Middle East through its imperial alliances, outdated though they were. This deceptive parity with world powers has greatly influenced Israeli political and military aspirations and the ways Israel was perceived by others. The close relationship with the United States on a variety of military and political issues dates back to 1956, achieving its current modality in 1967. Israel always wished to build itself as a crucial foundation for US policy, and this was the starting point of such changes, presaging a move from French to US sponsorship and financing.

The war also established proof of Israel's value as an agent and ally of Western interests in the region. Ever since Herzl's journeys to the various centers of power, Zionism has done all it could to persuade various imperial powers of its strategic value in the Middle East. While Israel fought on the wrong side in 1956, with partners that history had indicted for colonial and imperial crimes, and while the operation was a debacle, it emerged out of it stronger than before, while the other two partners suffered a deep decline. Furthermore, Israel proved its value as a regional sheriff for US interests, and this role would mushroom over the coming decades, making Israel the main receiver of US military and civilian aid.

The conflict also consolidated the position of the IDF within Israel. It confirmed that war was politics by other means and that the offensive doctrine of the IDF was optimal. This idea that attack was the best form of defense was crystallized in the development of a dynamic, "moving battle" concept—a version of the German Blitzkrieg—started in the late 1930s, under the tutelage of Orde Wingate; the preemptive war would become a standard strategy of the IDF and Israeli leaders, as would the vengeance/retaliatory raids perfected during the 1950s.

But war was only made possible by the conversion of the state itself to a permanent security state, both in terms of politics and the economy. The emergence of its military–industrial complex, fuelled by the war, became the cornerstone of Israel's economy, identity, and raison d'être. The building of that vast, lucrative industry of death and destruction started with the difficulties of rearmament during the mid-1950s, in the period leading to the Suez War.

Thus, despite the painful gap between the swift victory over the Egyptian army, enabled by tripartite air power, and the humiliating retreat from the newly occupied territory, the war of 1956 resulted in long-term benefits for Israel, ones that would be built upon and improved in future conflicts. The Suez War clarified and intensified tendencies established in 1948 and, in turn, led inexorably to the 1967 war.

After 1956, there could be no doubt about Israel's long-term commitment to the Western neocolonial project in the Middle East and its positioning against regional modernization and democratization. Israel's commitment to the aims of Western power brokers such as the United States, the United Kingdom, and France had been established, as was its opposition to Arab progressive governments. The foundations had been laid for both Israeli nuclear arms and its military–industrial complex. This positioning has remained intact for six decades, and it is not likely to change anytime soon. In that sense, the 1956 war was the foundational, formative event in the lifecycle of militarized Zionism.

But this tale will not be complete without relating an incident from the evening of October 29, at Kafr Qassem, in central Israel, as Israeli forces moved into Sinai. The village is located in what is called in Israel the Triangle of Arab villages southeast of Haifa. Kafr Qassem had been placed under a curfew that afternoon that had come into force at 5 p.m. that evening. The many villagers who worked in the fields did not know of the curfew and the commander of the force guarding the village was aware of this fact. He asked his commanding officer, Colonel Yissachar Shadmi, how to treat the villagers returning home from work. He was told to shoot all who were outside their homes:

> At his trial, Shadmi denied ordering the killing of curfew violators. Whatever the case, the result was a disaster. Between 5 pm and 6 pm, 47 Arabs, returning to their homes were shot to death by Border Guard troops. An additional victim, who was elderly, had a heart attack after he learned that his grandchild had been killed. In the end, according to the villagers, the total number of victims was 51.[79]

This event remained unknown to most Israelis for a number of months, as the military censor prohibited its reporting in all Israeli media. It took a prolonged legal campaign over several months, led by Arab lawyers and members of the Knesset, to make some of the facts available on Israeli media, but it has taken more than sixty years for the real story to emerge.

Under public pressure, Shadmi was tried by court-martial two years later, in 1958; he was found not guilty, fined 1 agora [Israeli cent], and resumed his military career unscathed. It has recently emerged that it formed part of a top-secret Israeli plan, unearthed by the historian Adam Raz, who believes the IDF tried to hide the existence of a secret program called Operation Mole, intended for the expulsion of Arabs from the Triangle across the Jordan.[80] This plan, a virtual continuation of the ethnic cleansing of 1948–49, was based on "emerging opportunities" and could be immediately triggered.

There is little to suggest that such a plan is not active now, but it is no longer secret; it is discussed on the pages of daily papers. In an opinion piece in Hebrew in *Haaretz*—not translated in the English edition—Kobi Niv exhorts:

> At the end, we will do it. What alternative is left? What did we not try? We besieged, barred and blockaded, starved and darkened—this is it, we have reached the end of the scale. We ran out of ideas and they are still there. We won, but what kind of victory is this? A victory is not a "regularisation." A victory is not real if there are no heads rolling on sidewalks, no blood in the streets. Therefore, Gaza needs destroying, leaving not a single stone, expelling, then killing anyone left.[81]

Israel is nothing if not methodical and consistent. The two objectives—expanding territorially and ethnically cleansing Palestine—have gone hand in hand ever since 1948, if not before.

One aspect of the 1956 war that remained relatively unknown was the clear intention to annex the Sinai and Gaza Strip to Israel, just as the battles were ending in Sinai. This fascinating aspect was revealed through the gradual release of archive materials, though much remains locked in the archives and might never come to light. In December 2016 an article in *Haaretz* revealed that Ben-Gurion had secret far-reaching objectives:

> Then-Prime Minister David Ben-Gurion was in euphoria on November 6, 1956, immediately after the Sinai Campaign in which Israel quadrupled the territory under its control. In a letter to then-IDF Chief of Staff Moshe Dayan, Ben-Gurion proclaimed the beginning of the third kingdom of Israel.[82]

That Israel had, as early as 1956, planned to defy international law for the second time on this issue and hold on to territories won through an act of unprovoked aggression, as it did in 1948, casts light on its behavior in 1967, when the territories were put under military law and prepared for future annexation, with East Jerusalem annexed before the guns had a chance to cool.

> Classified documents from that era that were viewed by *Haaretz* show how close Israel was to applying Israeli law to the territories of Sinai and the Gaza Strip, which would have amounted to Israel's *de facto* annexation of those territories, according to senior public figures at the time ... The archived documents ... indicate a clear intent to take over territories captured in the 1956 war, known as the *Operation Kadesh* ... According to the papers, the state attorney general at the time, Haim Cohen, compiled a draft declaration applying Israeli law to occupied territories for Ben-Gurion.[83]

That the intended action was contrary to international law worried at least one official in Ben-Gurion's administration: "The Foreign Ministry's legal adviser at the time, Shabtai Rosen, cautioned that what these documents amounted to in effect was the annexation of the Gaza Strip and Sinai to Israel, in violation of international law." He wrote:

> This morning at the Justice Minister's office I saw the proclamation regarding IDF rule in the Sinai Peninsula and emergency regulations regarding occupied territory prepared yesterday by the attorney general and transferred for the defense minister's signature. I explained to the Justice Minister and the state attorney general that in my opinion these documents are in violation of international law ...
>
> I fear that if we don't follow international law, we shall create complicated political problems for ourselves.

The next day, November 6, 1956, Rosen sent a more detailed opinion, designated classified, in which he cautioned that signing the documents compiled by Cohen would lead in effect to annexation of the occupied territories. He stressed that annexing Gaza would be legal, since it was a part of the Land of Israel in Mandatory times, but that annexing Sinai would lead to severe legal and international problems.[84]

This episode of Israel trying to legalize its illicit intention to annex the Sinai and the Gaza Strip demonstrates, yet again, that the 1948 Nakba was but the first iteration of a systemic policy of annexing any territory that Israeli forces may occupy and hold. During the decade between the two wars, Israel had carefully prepared to annex the rest of Palestine through a military campaign, which meant that the 1967 conquests could be retained for many decades through a complex set of measures. These included illegal occupation and control, denial of all rights for the residents of the West Bank and Gaza, and the imposition of a military government of occupation.

A total rejection of all UN resolutions that demanded Israeli withdrawal from the Occupied Territories was a logical corollary of such actions, and the UN has never acted to enforce its resolutions, in effect enabling Israeli infractions. The resulting five decades of Israeli control are concrete proof that such methods, despite their illegality, have been effective in detecting the deep flaws and inconsistencies within the UN constitutive bodies. As far as Israel is concerned, its control of the territories, including the Golan Heights, is "forever."[85]

Considering the history of Israeli militarization, the 1956 war played a crucial role—it proved the efficacy of advancing political aims by military action, as well as pointing out the need and potential for closer links with the United States; it also showed the potential of the military–industrial complex, which would be further developed over the next decade. The period preceding the war was crucial as a building block of Israeli militarism, supplying proof that for politicians in Israel, war was the safest option for establishing a long-term career. On the day Israel was forced to withdraw from Sinai, it began to prepare for the 1967 war, an event that would have devastating effects on the whole region.

4

The 1967 War

Can anyone seriously believe that another defeat will make all the Arabs stop bothering Israel and go away? Yes, people seriously believe that, even a whole nation believes that. As if the effort was no greater and no more difficult than ridding a small area of a nest of rodents.

—Edward Said, "Arabs and Jews"

It would be difficult to justify Israel's subsequent foreign policy of deception whose aim was to preserve the territorial status quo of 10 June at the expense of a peace settlement.

—Avi Raz, *The Bride and the Dowry*

Unfinished Business

If the 1956 war in Sinai had all the hallmarks of international intrigue and neocolonial folly, the conflict of 1967 bore a very different parentage. The confidence that Israel acquired in the earlier campaign—both militarily and politically—transformed it from a small and insignificant state into a regional and even a global player. The close cooperation between Israel and France on nuclear capability transformed the future of the Middle East, making Israel much more inclined to military adventurism, as it moved closer to developing operational nuclear weapons. Indeed, by the 1967 war, Israel possessed two nuclear devices, which undoubtedly influenced its decision to start the war.[1] While this was not public knowledge, security services of various countries had the information, and its deterrent factor was real.

The introduction of modern, devastating armaments into the Middle East permanently changed the conflict's dynamics. In 1949–50, Israel and the Arab states left the armistice talks at Rhodes unreconciled to the resulting cease-fire; the conflict was hardly resolved by these arrangements, leaving both the Israeli occupation of most of Palestine and the refugee problem as festering wounds. Arab societies and the Palestinians were deeply traumatized by the results of the 1948–49 war and vowed to secure the return of Palestine's refugees. This became a ritual announcement for domestic purposes, quelling the anger within the Arab countries; such anger was fueled by the shortcomings of Arab regimes and galvanized by the combined humiliation of losing the war and capitulating during the Rhodes negotiations.

Similarly, Israel under Ben-Gurion was far from satisfied with the results of the war and the 78 percent of Palestine it had gained in 1948. This still did not include the Old City of Jerusalem or the historical parts of the West Bank—the heartland of biblical Judaism. Ironically, Israel now controlled the coastal areas of Palestine, originally inhabited by Philistines and Canaanites, which had not been part of the Hebrew biblical latifundium. Palestinians remained in the traditional Hebrew heartland, now under Transjordan's reign, as well as in strategic Gaza, under Egyptian control.

Ben-Gurion was never reconciled to this outcome, and during the war repeatedly tried to force his cabinet into taking Gaza and the West Bank, without success.[2] That was not only because it would have broken the agreement with King Abdullah, but because of the high cost of the war in Israeli lives and the clear understanding that fighting the Arab Legion for the West Bank would be bloodier than any other battles fought in the war. The cabinet also realized that after such a conquest Israel would be left with many Palestinians in a state covering the whole of Palestine. This was not a prospect they were prepared for, being committed to a Jewish State and hence to a substantial Jewish majority. But the plan to take the rest of Palestine was never dropped—it was merely shelved, awaiting a suitable juncture.

The situation after 1956—when Israel had flexed its muscle and took Sinai in a week, causing thousands of Egyptian fatalities—was anything but stable. Israel was forced to withdraw from its new conquests by the United States and Soviet Union, but the knowledge that the conquest was so easy stayed in the forefront of Israeli leaders' minds. The question remained: how to correct the mistakes of 1948, when not all of Palestine was taken and large numbers of Palestinians stayed behind?

What was not possible to do in 1948 would become possible later, with Israel both stronger and more populous.

It was clear that the circumstances of such an eventuality would have to be portentous, allowing an occupation to pass without international sanctions. Since 1956, Israel has developed its military–industrial complex with French military assistance; it was dedicated to quality weapons production, adapted to the needs of the IDF and tested in battle. This, combined with the secret development of nuclear weapons completed in 1966, gave Israel a strong lead over the Arab forces, despite their larger size and Soviet support. The IDF became a modern, dynamic military force, capable of mobilizing its large reserves within a few hours and launching war at short notice. Based on the earlier model of the Special Night Squads, Palmach, and Unit 101, as well as the lessons of 1956, the IDF was mobile and trained for offensive action; it followed the doctrine of "taking the battle into enemy territory" and launched preemptive strikes at a time of its choosing.[3]

Western intellectuals, especially on the left, found it difficult to decode Israeli society and were invariably inclined to support it, in part because of the Holocaust. Even the neoimperialist war of 1956, in which Israel joined the old imperial powers against Egypt, did not weaken this popular sentiment; the Palestinians were invisible and the clash was mistakenly conceived as an Arab–Israeli conflict. One of the reasons for this was the kinship felt toward an entity conceived as Western and European, consistent with Herzl's argument that the Jewish State would serve as a European bulwark to face Asia.

Even such an astute intellectual as the late Tony Judt perceived Israeli society as European.[4] However, Israelis of European origin were already a minority in Israel by the 1960s, and while such Israelis (or their parents, more accurately) may have stuck to some traits of the old home country, Israel itself vehemently rejected this identity. The historic Diasporic Jewry was viewed as tainted, an amalgam of the antisemitic stereotype of the Ghetto Jew and the helpless victim of the Holocaust. The New Jew was a negation of European Jewry—its traditions of humanitarian and cosmopolitan universalism were anathema. From such positions grew not only cultural attitudes, but also the new militarism that characterized Israel.

The Rise of Arab Nationalism

The appearance in 1964 of a new kind of Palestinian resistance, the Palestine Liberation Organization (PLO), added complexity to the equation. Whatever expectations Israelis had about the Arabs abandoning Palestine after the defeats in 1948 and 1956, the formation of a nationalist liberation movement on their front door was not the response they anticipated. The Arab League summit in Cairo, in May 1964, had decided to establish the PLO and the Palestine National Council and to build an armed resistance movement, led by Ahmad al-Shuqairi.

Furthermore, instead of waiting for the Arab armies to liberate Palestine, the PLO presented a strategy of guerrilla warfare, which Israel was unprepared for. Nuclear arms, tanks, and fighter jets cannot stop committed and trained guerrillas, as had already been proven by the fedayeen in the 1950s. While the full military potential of the PLO did not get developed until after the 1967 war, it was clear to Israel's leaders that it was but a question of time before its long and unprotected border would attract guerrilla attacks, some of which started before the 1967 war.

An additional aspect of regional politics contributing to Israel's decision to start the war were the changes in the Arab world since the military defeat of 1956. While the Egyptian army was indeed vanquished in 1956 by the combined forces of two European empires and their client state, the political victory was Nasser's. Not only did Operation Musketeer fail to achieve its military aims, it had also failed to replace Nasser with a collaborationist regime. Indeed, the very opposite occurred: Nasser grew in stature and influence, well beyond the vast boundaries of his country. In addition, the war taught Nasser and other Arab leaders an important lesson—if the colonial aggressors can combine forces, so can Arab societies wishing to defend themselves against such adventurist politics.

The result was the United Arab Republic (UAR)—the 1958 political and military union of Egypt and Syria, with the possibility that Iraq would eventually join. This emerged naturally out of Nasser's pan-Arabism doctrine: the argument that divisions within the Arab world only benefit the West and act to the detriment of the Arab masses. A united Arab political entity, on the other hand, made much sense. The ruler of the most populous and militarily strongest Arab country could not fail to notice that his large population was almost devoid of natural resources, in contrast to tiny societies in the Gulf controlling much of the planet's petroleum. A political coming-together would create an influential, strong entity, with hundreds of millions of Arabs controlling massive resources over a large area in Asia and Africa, as well as some of the planet's most strategic zones. From marginal, backward, and conservative societies ruled by regional potentates, the new pan-Arabia would become a world player, a force to be reckoned with. Nasser even spoke about returning to the glory of the lost empires emerging in the wake of the Prophet Mohammad, covering vast tracts of land, from India to France.

This argument was easy to understand and appreciate, and Arab masses everywhere were greatly attracted to Nasser's rhetoric and ideas. A leader of the Arab world had at last emerged, and this was noted by the West and by nonaligned countries (soon to be labeled the Third World), sharing a strong anti-imperialist sentiment personified by the young and charismatic Nasser. The 1956 Egyptian Constitution promised social welfare for all Egyptians, a unique progressive feature of social engineering in the Arab world of the time (and one still missing from most Arab countries even today).[5] The social changes started affecting the country's millions almost immediately. A minimum wage was enforced, subsidized housing for the poor and basic food prices became a reality, and modern education and health services were built up.[6] Nasser made a positive change in the life of most Egyptians, probably for the first time in millennia. In the education system, for example, enrollments in primary schools had tripled within a few years, creating a sound foundation for modernizing society.[7] The High Dam at Aswan promised to regulate agricultural production for the first time since the Pharaonic era, removing the insecurity of fluctuating inundations of the Nile. Egypt was on the move, and its citizens supported the changes.

Such dramatic transformation did not escape notice in other Arab countries, ruled by a motley crew of tribal patriarchs or violent and mercurial officers, very much resembling the Latin American pattern. One of the new investments by Egypt's leaders was the building of a powerful propaganda machine, based on radio and the press. This brought the voice of Egypt to the whole Arab world. To see such fast and thorough changes taking place in the largest and oldest polity of the Arab world was a mesmerizing spectacle; millions of people, from Iraq to Morocco, conceived of themselves as Nasserists, expecting similar changes in their own societies. All of a sudden, nothing was impossible—the energy radiating from Nasser's Arab socialism attracted the masses like a magnet. Nasser invoked the Palestine cause as an icon of Arab unity on many occasions: "We can never relinquish the rights of the Palestinian people, because, as I said in the past, the honour of the Palestine people is the honour of the Arab nation."[8] Such statements endeared him to Palestinians, who believed that he actually meant what he said, unlike many Arab potentates voicing similar sentiments.

This, more than anything else, was seen as the greatest danger to Israel. Such changes could transform the balance of power in the Middle East against Zionism. As long as Israel was presented as the main modernizing force in the Middle East, its proclivities and aggression were sanctioned by the West, but the new regime in Cairo spelled a likely end to this isolated position of power. Nasser told a Western journalist, somewhat disingenuously: "I do not think of myself as leader of the Arab world, but the Arab peoples feel that what we do in Egypt reflects their collective hopes and aspirations."[9] This was an accurate depiction of the sentiments of Arabs everywhere, even as the leader played down his own aspirations, one suspects, in reaction to growing Western fears of his growing influence.

While social reorganization and especially land reforms were at the heart of Nasser's great popularity, his investment in the military did not go unnoticed. Nasser vowed to reverse the defeats of 1948 and 1956 and to assist the return of the Palestinians to their homeland. This, he said, was part and parcel of the pan-Arab responsibility and communal solidarity. This created a huge following in Palestine, on both sides of the cease-fire lines: Palestinians with Israeli citizenship, who were closely monitored and politically suppressed under a military government that ended only in 1966, and those in the rest of

Palestine (Gaza under Egypt, the West Bank under Jordan). Nasser was seen as the Great White Hope of the entire Arab umma. Even more than in 1956, Israel was totally opposed to such modernizing and unifying potential for obvious reasons; the fear of an all-Arab union under Nasser terrified them.

The short-lived political union between Egypt and Syria, lopsided and biased with Nasser as the president of the UAR, lasted only until 1961, when a Syrian coup ended it. But the concept of pan-Arabism did not die.[10] The early end to the UAR was not altogether surprising: Egypt controlled the political scene in Syria through repressive means, including the abolition of political parties and the removal of some Syrian politicians and replacing them with Egyptian bureaucrats, and this could not last. Also, both the socialist features of the Syrian polity and its capitalist economy were more advanced than those of Egypt, contributing to deep resentment in Syria.[11] The UAR was doomed because it was not a union of equals. But despite the break, both militaries continued working closely together.

Elsewhere, Nasser got entangled in the Yemen quagmire, sending military forces to back up the leaders of the military coup in 1962. This involvement cost Nasser dearly: by 1963, a third of the Egyptian Army (some 70,000 troops) had been trapped in Yemen fighting a hopeless war.[12] This bled Egypt in military and political terms; the Egyptian forces were not trained to fight guerrilla forces. It also posed an acute financial burden for Egypt. More than 10,000 Egyptian soldiers were killed in Yemen, and many more were wounded, bringing about a reawakening of opposition to Nasser in Egypt. Army morale was badly hit. All this greatly increased Israeli appetite for another war with the enfeebled Egyptian army, trapped in Yemen with no way out, like the US army trapped in Vietnam at the same time. (Parallels were drawn in the Arab world as well as in Egypt itself.) Nasser deeply regretted his decision to intervene but found no way of extracting his forces from Yemen.[13]

The Israeli Recession

Israel faced its own problems when, in early 1966, a deep recession hit the country. Some claim it was caused by government decisions, which seems likely.[14] As the economy started to contract, the government of Levy Eshkol proscribed shock treatment, but no miracle took place; a dark mood of resignation and pessimism set in. The satirical book *All Eshkol's Jokes* contained the blackest of Israeli humor, presenting the prime minister as a modern *shlemiel* (Yiddish term for a provincial bumpkin). There was nothing Eshkol could do to stop the tide, and even the IDF was affected. Yitzhak Rabin, the chief of staff, referred to Eshkol and his ministers as "the Jews" during general staff meetings; this was a hard insult in 1967 in a society distancing itself from the ghetto Jew in favor of the Israeli *sabra*. The insult suggested a cowardly, reticent character, one not ready for hard decisions, such as initiating a war.

To his critics in the IDF, Eshkol and his ministers seemed too cautious, unnecessarily attuned to messages of foreign diplomats, rather than the aggressive military elite built by Ben-Gurion and Moshe Dayan. The economy was in free fall; unemployment was rising fast. Population growth had stalled: from a 4 percent annual growth in 1964, it went to zero in 1966. Israelis were leaving the country in large numbers.[15] Many businesses suffered a 10–15 percent contraction in revenues, and many young people could not find a job after leaving the army, in the rapidly shrinking economy. For the first time, Israeli society perceived itself as a polity with no future. This mood was reflected accurately by graffiti at the entry to Lydda airport, "The last one to leave, please switch off the light." In two decades of statehood, an earlier energetic Zionism was replaced by a jaded, negativist, individualistic hedonism. Unless something was urgently done, Israel stood to lose a whole generation of young people who did not believe it had a future.

By historical fluke, the results of the establishment of Fatah—the Arafat faction of the PLO—became evident during the same period. Many more guerrilla raids were carried out within Israel, with qualified success, mostly causing irritation but occasionally leaving behind victims. During October 1966, some high-profile acts of sabotage by the PLO were impossible to ignore. On October 25, the main railroad between Jerusalem and Tel Aviv was mined, resulting in the derailment of a large goods train. The saboteurs narrowly missed the passenger train passing a couple of hours earlier. Although there were no casualties, this incident caused deep shock in Israel. A short time later, on November 11, three paratroopers were killed when their vehicle hit a mine near Arad in the Naqab (Negev). As usual, the killing of soldiers was perceived as an especially heinous crime that must be avenged. The IDF pulled out its prepared scenarios, deciding to attack the village of Samu'a, south of Hebron.

Operation Shredder, which the cabinet was told by Rabin had limited scope, went badly haywire, and in the end included even an air battle between Israel and Jordan, in which a Jordanian jet was shot down and its pilot killed. Military casualties include fourteen Jordanian soldiers and officers killed and some forty wounded; an Israeli battalion commander was also killed and ten Israeli soldiers were wounded. About one hundred houses were destroyed by dynamite, many on top of their inhabitants. It was mindless carnage, disproportionate to the event initiating it. Nevertheless, on returning from the scene of bloodshed, the paratroopers held a victory parade through the streets of Beersheba.[16]

The rest of the world found less to celebrate. The international anger was all the more poignant, as it was quite clear that Jordan and the Arab Legion had no part in setting up the attacks the operation was supposed to avenge, and Israeli intelligence sources knew that Syria had assisted the saboteurs. But the IDF found it easier and simpler to attack a defenseless village in the West Bank than face Syrian fortifications in the Golan Heights. This argument did not cut mustard with the State Department in Washington, which wasted no time in telling the Israeli ambassador that, were such an attack to be repeated, the United States would "re-examine" its arms supply policy to Israel.[17]

Dayan and Rabin knew what they were doing, however. The Israeli public, fed for years on propaganda and fearful of further acts of sabotage by Fatah, supported the raids, even if the cabinet members were secretly worried and others were shocked by the results of their reluctant agreement to the raid. The raid's function was less to punish the Jordanians than to steel the cabinet to approve an all-out war planned by Dayan, Rabin, Yigal Allon, and others since 1948.

Some of these veteran politicians in the cabinet, like Israel Gallili and Allon, belonged to the highly nationalistic (and activist) Achdut Ha'Avoda Party and started to press the hesitant Eshkol to agree to further raids. Eshkol—isolated, despised, ridiculed, and unfavorably compared to Ben-Gurion who always gave the army a free hand—dared not refuse when the public mood was evidently so bellicose. This was cynically used by IDF command and activist ministers, with Eshkol their permanent prisoner.

The IDF had no proper plan for defending the country against guerrilla attacks; arguably, such a plan could never be made effective, and it certainly was not the type of fighting that the IDF had prepared for or preferred. In the words of Dayan, "The IDF is a decidedly aggressive assault army in the way that it thinks, the way it plans, the way it implements. Aggression is in its bones and its spirit."[18] With such an outlook, defensive strategies were out of the question. By 1966, the IDF was, for the first time, recruiting Israeli youth who were born in the Jewish State and habituated to the kind of threats and violence that the birth of Israel had presented. The older politicians such as Eshkol, preferring to resolve differences by negotiation, were swiftly becoming unrepresentative of the new Israeli society.[19] The sabra was poised to replace the European Ghetto Jew at the top echelon of the Zionist leadership, and they had different attitudes and methods.

War as a Solution to Social Unrest

Arguably the IDF replaced in Israelis' mind the all-powerful biblical deity—God of the Hosts. Large billboards in the cities declared: "Israel, have faith in Zahal" (IDF), swapping God with Zahal in this famous Hassidic expression. Zionism has crafted a secular religion, where the IDF with its military might stood for the Almighty. Indeed, in the general atmosphere of desperation and despondency, with deep mistrust of the political class, the IDF retained its relatively untarnished image as the protector of Jewish Israel. That this was usually done with extreme brutality and disproportionate use of force was not a difficulty for most Israeli Jews, but further proof of its power and fitness for purpose. Israelis felt they could not trust politicians, businessmen, the banks, employers, or unions, but the IDF was on the transcendental pedestal of "just force."

It was the only institution that offered all sectors of Jewish society a common goal and arena of action, while all other avenues of social action were segregated. Israel's social groups—kibbutzniks, urban types, Mizrahim, new immigrants, rightwing Revisionists, communists, the ultra-Orthodox, Palestinian citizens—were isolated from one another. In the life of a young Israeli the only substantially shared space was the obligatory service in the IDF, from which only the ultra-Orthodox and the Palestinian Arabs were excluded. Ben-Gurion's dream of the People under Arms and the IDF as the vehicle of forming and uniting Israeli (Jewish) society was at last realized.

So, if the economy could not be fixed, if the emigration of young people could not be stemmed, if the despondency and depression could not be wished away by the Eshkol government, the social cohesion of Israel could be renewed through the old fix that always worked: the IDF carrying out an aggressive action against "enemies." Urgent calls by IDF generals for military action were carefully listened to by the Eshkol government and the "Jews" in his cabinet. The media—mainly the press, before Israel had a television service—was part of the social machinery building the militarist polity and rarely criticized the army. Segev described the militaristic musings of Yosef Lapid, a leading correspondent of *Ma'ariv*, one of Israel's main evening papers.[20] "Israelis worshipped the IDF, wrote Lapid, and that was as it should be. But the dangers of projecting such a perfect image was that the army itself might start to believe in it: "An officer who reads in the papers every day that the government appears powerless and the Knesset is helpless, and at the same time reads only praise for the IDF, will sooner or later start to believe that he is made of better stuff."[21]

During the same time, Gadna had more than 70,000 members (the highest membership ever recorded), and a special arms-purchasing "charity" collected great sums of money from the public. Preparations for military action were palpable, and even the children and youth weeklies brimmed with militaristic gibberish. Even the most critical of Israel's weeklies, *Ha'olam Haze*, edited by Uri Avnery, cultivated admiration of the IDF as the essential national institution.[22] Avnery was a nationalist and past member of the Stern Gang during the 1940s, no wallflower in this militarist ballroom.

The plans to take over the West Bank were not made in the summer of 1967. As early as June 1963, a short while after becoming Israel's prime minister, Levy Eshkol met the IDF chief of staff, Zvi Zur, and his deputy, Yizhak Rabin, to inquire about their ideal borders for Israel. Their view was simple and clear: The borders should become the Jordan River in the east, the Suez Canal in the southwest, and the Litani River in the north.[23] A military plan named Whip followed, outlining the planned takeover of the whole West Bank, establishing the Jordan River as the new boundary. While this was not as expansive as the boundaries wished by the Begin-led Herut (Liberty) Party, which coveted all of Jordan as well, it meant that Israel would control the whole of Palestine. This was in contravention of UN resolutions and the declared policy of most states.

But grabbing territory was not sufficient; most crucial was the detailed planning for the aftermath of the war, or it could end with a forced retreat, like the 1956 war. To hold on to the territories while displacing as many Palestinians from the land demanded a complex procedure. The behavior of the occupying forces had to be carefully calibrated to make life in the Occupied Palestinian Territories very difficult, leading to a gradual emptying of the land—a continuation of the Nakba by other means.

The long-term plan was worked out in 1963—according to documentation unearthed by Ilan Pappe—at the Hebrew University campus of Givat Ram.[24] A large group of lawyers and legal officers had secretly congregated in order to plan Israel's future occupation of the rest of Palestine and work out the legal codex for such an eventuality. The resulting work was shared by the CoGS [the chief of general staff] with senior officers on May 1, 1963, preparing the IDF for the task of controlling the West Bank and its population through a specially developed network of legal and administrative systems.[25]

> By May 1967, the plan became operative and the actual appointment of military governors and military judges to the West Bank and the Gaza Strip moved to a more detailed stage (it included also preparation for installing a regime in what the army called "Syria").[26]

With such meticulous planning, the ground was prepared for the war, and what remained was to find or produce a casus belli. The situation that supplied the trigger was a spate of tense confrontations on the border with Syria. The first issue was Israel's one-sided diversion of the Jordan River, near the point where it flows into the Sea of Galilee. This project, dubbed the National Water Carrier, pumped huge amounts of water from the lake up the eastern reaches of the Galilee hills, sending it down a large-caliber conduit all the way to the northern Naqab (Negev), the driest part of Israel. As the Jordan River originates in Syrian, Lebanese, and Israeli territories, passing along the cease-fire lines between Israel and Syria, and then Israel and Jordan, before ending in the Dead Sea (which is divided between Israel and Jordan), four countries had to agree on the use of the water of this crucial, rather small river. This agreement never took place, and in a sense, could not happen: Syria and Jordan would not agree to Israel's diversion of the water, as long as Israel refused to return the Palestine refugees and resolve the many illegalities of the 1947–48 war. Israel has thus gone ahead and built the National Water Carrier despite its neighbors' protestations.

When Syria tried to construct a similar diversion project using some of the Jordan's water to feed its own agricultural lands east of the river, Israel bombed it, rendering it useless: "In the battle over the water, the IDF gained an easy and cost-free victory, managing to disrupt the Arab diversion plan by simple, cheap, and effective means, striking locally without risking full-scale military conflict," wrote the Israeli historian Ami Gluska.[27] This was done by "staging" the conflict (bombing the Syrian heavy machinery after dubbing it "military equipment"), a practice confirmed by Rabin.

The second issue causing friction was the agricultural land that ended up in the demilitarized zones between both countries. Neither side was allowed to place soldiers or armaments in these areas, but this did not stop Israel from using threats and the IDF to evacuate two Syrian villages in the vicinity and send their inhabitants packing all the way across the cease-fire lines.[28] The continued presence of soldiers caused friction; the IDF posed as "farmers" in the demilitarized zone, working in armored tractors, in order to lure the Syrians into reacting so that they could attack with force. This was common practice throughout the 1960s.

Syrian-supported operations by Fatah (Arafat's organization, part of the PLO), which often took place when active cells left from Jordanian territory on their missions across the lines, were a third issue. Fatah operations started in 1965, becoming more numerous and audacious during 1966. This aggression was used by Israel as a justification for attacks against Syrian forces and villages, because the IDF had no effective method of countering such intrusions into home territory. Indeed, sometimes the reactions against such acts of sabotage were not only disproportionate, but in many cases caused larger loss of Israeli life than the original saboteurs, not to mention the greater loss of life among the Palestinians in villages under attack.

But all this was par for the course. Israeli political and military leaders were mainly concerned about increasing the size of the territory under their control, wishing to gain the Jordan River as a natural border with Jordan, and such friction and the resulting carnage were needed as an excuse for war. Two of the leading proponents for such a war, apart from the chief of staff, Rabin, were General Uzi Narkiss, commander of the central command bordering Jordan, and General Rechavam Ze'evi, of the Operations Branch of the IDF headquarters. Both wanted a war against Syria and Jordan, and one plan (among many) named Axe outlined the conquest of much of southern Syria, including Damascus.[29]

Preparing the War

From the summer of 1966 until April 1967, the tit-for-tat operations in Syrian and Jordanian territory became a constant feature of IDF routine, as the stopwatch ticked down toward war. In April the skies saw an aerial battle between Israeli and Syrian fighter jets after the Syrians were lured into a trap by a series of Israeli provocations, during which seven of their jets were downed over Damascus in as many minutes. This show of force ignited military tensions across the region, as did the decision to hold the May military parade in Jerusalem for the first time—a blatant departure from normal practice and a move that was hotly criticized by the Arab countries, and especially by Jordan and Egypt, who saw it as a provocation, which of course it was. Further destabilizing was the fact that Israel was developing nuclear weapons and the United States was failing to stop it.[30]

Israel poured oil on the flames, announcing that compulsory military service of Israeli men was to be extended by a further four months, thus enlarging the army at a pen stroke. At this point in early May, Soviet intelligence sources had informed Syria and Egypt that Israel had moved large forces into the north of the country, facing the Syrian border.[31] The Syrians acted on this inflated information and demanded that Egypt should support them against the danger of an Israeli attack. Threat of an attack on Damascus was clearly mentioned in a meeting of Aharon Yariv, head of Military Intelligence in the IDF, when he briefed foreign correspondents on May 12.[32] Because Israel also refused to cancel the Jerusalem military parade or move it elsewhere, Egypt, spurred on by Syria, openly moved forces into Sinai. This maneuver was not considered a threat by the IDF; the intelligence services considered an Egyptian attack totally unlikely at least until 1970. They were right; Nasser moved the troops into Sinai not as preparation for an offensive, which would require surprise and a covert reinforcement, but as a demonstrative act warning Israel against crossing red lines. Nonetheless these events together started a snowballing of action and counteraction that led to the Israeli attack against its three Arab neighbors: Egypt, Syria, and Jordan.

A day after the parade in Jerusalem, Egypt upped the ante and asked the UN to remove its emergency force, which was positioned to separate the foes from the Sinai desert to Gaza. The general secretary of the UN, U Thant, lost his nerve. Most historians conclude that he badly misread Nasser's intention. U Thant was not as experienced as his predecessor, Dag Hammerskjöld, who probably would have moved immediately to resolve the conflict by direct diplomacy in regional capitals.[33] U Thant demanded that the emergency force, which Nasser wanted moved to Gaza, was either removed totally or stayed put, directly confronting the Egyptian president. In such circumstances, Nasser had no choice but to ask that the force be totally removed. The mistake by U Thant had thus compounded Nasser's own misstep—after all, he could not go back on his word—and the force was removed, surprising both sides. This gave the events a twist of fait accompli—none of the sides was able to retreat from their positions of braggadocio and they moved units into position along the cease-fire lines.

Not only foreign diplomats were concerned about the moves by Israel and Egypt causing tension in the region, but also Israeli politicians, especially Moshe Dayan, retired from the army but serving as

Rafi Party Knesset member together with Ben-Gurion. Dayan claimed Nasser did this to protect Syria; he had no plans to attack Israel and needed a symbolic action to retain credibility. Dayan thought Egypt was likely to bomb the Dimona nuclear facility or close the Straits of Tiran.[34]

Two MiG-25s were indeed chased from Israeli airspace after they circled the Dimona nuclear reactor without bombing it. Nasser was warning Israel not to mess with Syria and thought such a maneuver would suffice, avoiding aggressive action. Israel was unable to either prevent or intercept two such incursions on May 17 and May 26; the cabinet, which was informed about the moves in real time, was deeply shocked.[35] Isabella Ginor and Gideon Remez, journalists with close Mossad and IDF ties, introduced a counterintuitive thesis that the Soviet Union was behind the event, using its newest jet fighter, the MiG-25, to propel Israel into attacking Egypt, even though their own evidence clearly shows that the Soviet Union was continuously curtailing Egyptian reactions to Israeli taunts, so as to avoid the blame for a war started by an ally. This attempt at rewriting the history of the conflict by placing responsibility at the door of the Soviet Union fits well into the Zionist narrative but failed to persuade historians.[36]

Dayan had in the past demanded the removal of the UN Emergency Force, to facilitate an Israeli attack. Eshkol was furious with Dayan for expressing his views, accurate as they were, as his intervention was likely to increase the likelihood of war. Gradually, though, the mood of the leaders and the public changed, and the realization that a war was likely took over. If Nasser was aiming to confuse the Israeli leadership, he succeeded. At the end of the first week of the crisis, on May 23, Israel learned that Nasser had sealed the Straits of Tiran, as foreseen by Dayan. This was cause for war, according to some Israeli politicians who were eager for it.

With the straits closed, Israel lost the single achievement of the 1956 war—open straits for Israeli shipping. The IDF (with Rabin at its helm) now strongly pressed for a surprise attack. The command structure of the IDF promised Eshkol that the air attack on both the Syrian and Egyptian air forces would take a few hours, after which Israel would rule the sky, so the war was a "safe risk." Eshkol faced constant IDF pressure, and a thrust toward war in the cabinet; at that stage, Eshkol was still hoping diplomacy might resolve the tension, as he may not have trusted the information given him by the military leaders.[37]

Eshkol also knew that despite the blockage of the straits to Israeli ships, Egypt had decided to allow vessels flying the American flag, thus the blockage could be circumvented by the symbolic act of using US-flagged ships. Because the straits had been closed to Israeli ships between 1948 and 1956, some Israeli cabinet ministers suggested that claiming the closing was a casus belli was rather tendentious; they tried hard to stay Eshkol's hand, while his Achdut Ha'Avoda colleagues—mainly Gallili and Allon—pressed for an attack. Eshkol was besieged and the pressure was telling on him; as a way of shedding some of the responsibility and spreading it around, he agreed to invite the rightwing Herut Party to share in a wider National Emergency Government, a throwback to Churchill's move to include the Labour opposition in his wartime government.

In hindsight, the inclusion in government of rightwing parties, and the return of Moshe Dayan as minister of defense, made the route back to the status quo ante impossible.[38] From then on, and until the initiation of hostilities on June 5, Israel inexorably moved toward war despite earlier resistance, dithering, and the strong international pressure from major blocs toward a negotiated mutual climb-down and peaceful resolution. The hardliners in Israel had waited for exactly such an opportunity since the end of the 1948 war and especially since 1956. They were not about to give up the chance for conquest when it so clearly beckoned.

The way they saw it, war was a great chance to enlarge Israel; it became a forgone conclusion, despite the exhortations by Charles De Gaulle, who threatened that France would blame whoever shot the first bullet. As Israel's armaments supplier, this was not an idle threat. But the temptation to act in the manner preferred by Dayan and Rabin was too strong. The truth was that Israel felt safer initiating an offensive war than waiting for an attack that may or may not materialize. With the whole IDF mobilized at great cost to Israel, a delay was not possible.

What followed was a series of bluffs and counterbluffs, each upping the ante and making restraint more difficult and less likely. It is impossible to outline here the many small moves that sealed the warpath for Israel, though it knew well that Nasser was not going to attack, not to mention Syria, paralyzed as it was by fear. Now, buoyed by its new weaponry and the timely completion of its first nuclear devices, the IDF was eager to go to war.

The Aftermath of the 1967 War

What made this war noteworthy was not the fast military victory against the three Arab armies or the acquisition of vast areas; this had happened before. What distinguished it were its lasting results, in contrast to the 1956 retreat, forced by the combined efforts of the United States and Soviet Union. Such a combined effort was not likely in 1967 for a variety of reasons.

If in 1956 both the United States and the Soviet Union had been vying for control of the region, and both saw Britain and France as spent forces that stood in the way of the new superpowers, the 1967 situation was different. The Soviet Union had made huge strides with the progressive part of the Arab world, having signed contracts for both military and civil aid with Egypt and Syria. The United States was having limited success with the most conservative regimes in the Gulf.

By then Israel saw itself (and was perceived) as part of an anti-Soviet dynamic in the Middle East and offered the United States a unique opportunity for regional influence. It stood as a complementary force to that other important pole of influence and power, Iran under the Shah. Both regimes had collaborated closely; Israel depended on Iranian oil and on Iran's anti-Arab, conservative politics. Both countries were becoming more dependent on the United States. The intelligence and security services in both countries, Mossad and Savak, continuously collaborated; so did their armies.

The United States had created a system of client states that could be described as interdependent—such states supported one another, complementing the basic support from the United States. Such regional interdependence was an important plank of the US foreign policy in the region, and this extended to other interdependent client states, such as apartheid South Africa. The three countries exchanged resources and assisted one another; for example, Israel had assisted South Africa in developing nuclear weapons,[39] and apparently the two states combined in testing a nuclear device in 1979, in the Indian

Ocean[40] (the Vela Incident).[41] Iran and Israel had also assisted South Africa in defying the international antiapartheid boycott, and Iran buffered Israel against the Arab boycott.

The United States was offering Israel important military and political aid, well beyond the size of its population or territory, because it was seen to offer a stable and dependable underpinning for US interests in the region. There was an additional benefit to supporting Israel: Jews from the Soviet Union, many of whom had held important positions in the Soviet science and military industries, started arriving in Israel (at first a thin trickle) so that Israel became a source of crucial information on the capabilities of Soviet arms, captured both in 1956 and 1967. This accorded it special value for the Pentagon and the US military–industrial complex. The IDF was involved in fighting Soviet hardware using US-made weaponry, testing it to the limit in real battle conditions. Israel had become the largest arms-testing laboratory and would continue to develop this role in decades to come. For such a crucial function, the United States was prepared to offer not just the largest aid budget of any country, but also unlimited and unconditional political support in the UN and beyond. Israel has found its new imperial master and its specialized role. It has never looked back.

If in 1956 the United States was one of the forces to evict Israel from Sinai, beginning in 1967 it became Israel's vigilant protector, guaranteeing its impunity from any international sanctions. The UN General Assembly may indeed pass resolutions against Israeli actions on numerous occasions, but such resolutions were symbolic; the United States was there to veto any serious move against Israel in the Security Council, using its veto powers hundreds of times. It was the power that made Israel, in turn, militarily and financially powerful. As long as the United States was safe as the world's leading power, Israel was secure in its continued occupation and oppression of the Palestinians.

To qualify for such a special spot in the affections of the United States, Israel had to serve US interests in the region. Israel's power, and its readiness to punish Arab states that stepped out of line, were the most useful qualities it could offer its patron.

Through its access to Israeli intelligence, the CIA and its military intelligence became privy to specialist knowledge of the Middle East, which they badly needed. Israel was, in fact, a large American aircraft carrier beached on the eastern Mediterranean coast, exactly where the United States would need to build a base. Israel's leaders were following Herzl in tying Zionism to the lead empire, but in 1967 they were much more successful than he was.

Israel's success rested on its Spartan, bellicose posture, rather than its ability to resolve conflict—indeed, conflict made it richer and stronger. This pattern was now well-established and would not change in the next five decades; if anything, it became more ingrained. Israel has chosen the Dayan option—to live on (and off) its sword.

The War's Phases

While the war lasted six days, its outcome was decided in the first three hours, during which a lightning attack of Egyptian and Syrian airfields brought about the destruction of both air forces, leaving the IDF ruling the sky. Not only were the planes destroyed, but the whole infrastructure was damaged beyond recognition—the radar units, the ground control, the antiaircraft defenses, and the actual runways.[42] Very few Egyptian planes were able to operate during the raid or afterward.[43] In the first day of fighting, 410 Arab aircraft were destroyed—a military feat without historical parallel.[44] During the first two days of fighting, the Israeli Air Force planes spent more than 80 percent of the time in the air—an unprecedented record, which enabled the wiping out of their foes' airfields.[45] At the same time, IDF units started the offensive in the West Bank, concentrating on the Jerusalem region. King Hussein has agreed a week before the war to sign a defense treaty with Egypt and Syria, which also meant that his forces were to be commanded by an Egyptian officer, General Abd al-Munem Riad. Riad was a gifted officer, but he had no knowledge of the West Bank theater, and this impaired his decisions.[46]

The fighting in Sinai was no less surprising. The Egyptians expected the IDF to move most forces toward the strategic Sharm al-Sheikh and deployed the bulk of their forces accordingly.[47] The IDF decided to avoid this coastal route and concentrate on the least defended one —the northern coastal region of Sinai. Early in the morning of June 5, three divisions started clearing the three main roads toward the canal. The northern armored column defeated a large force at the corner of the Gaza Strip, opening the coastal highway to al-Arish, the main military base in Northern Sinai.[48] It took less than eight hours to occupy al-Arish by the force commanded by Brigadier General Israel Tal, IDF's commander of the armored forces and its chief strategist.

The second column, commanded by Brigadier General Ariel Sharon, targeted the fortified base at Abu Ageila, south of al-Arish and on another main route crossing Sinai. In between the two columns operated the third force, commanded by Brigadier General Avraham Yoffe; it surrounded the Egyptian forces under attack by the armored division of Brigadier General Tal, blocking the retreat of the Egyptian forces and deciding the outcome. Despite the long period of preparing for the war, the Egyptian forces were not ready for this blitzkrieg; deprived of air support, their forces were doomed—the Israeli air support helped to defeat them. But resistance in the Gaza Strip itself, with the Egyptian forces trapped and unable to retreat, was especially stiff, and the area well-defended by mines and a system of trenches with antitank guns. The Gaza Strip was only taken after a massive attack on Gaza by a large force aided by massive artillery force and air bombing, and it fell to Israel on the night of June 6. The bombing was indiscriminate and included the UN base at Gaza, where fifteen UN soldiers died and many were wounded.[49] By that time, the IDF had a major problem of dealing with thousands of Egyptian prisoners of war, which included the military governor of the Gaza Strip, General Abdul Monam Husseini. Once the Gaza battle was over, the IDF ground forces had the upper hand in Sinai, despite difficult resistance faced by the Sharon-led force in Abu Ageila—a well-defended fortification controlling the road to Suez. It took a very tough battle lasting twenty hours for this stronghold to fall.

Another force headed south toward Kuntilla, another stronghold on the route to Sharm al-Sheikh, and then changed direction toward the crucial Mitla Pass, which defended the approach to Suez. Within thirty-six hours the force was at the Mitla, with the main Egyptian airbase in Sinai, Bir Gafgafa, mortally disabled on the first morning, hence unable to offer any support. On the third day of fighting, the IDF controlled the three main routes in North Sinai and were in a position to attack the canal bases once they had taken the Mitla Pass.

The Mitla battle raged for twenty-four hours, with over a thousand tanks involved on both sides—one of the largest armored battles in such a limited war theater. The Mitla Pass has become a death trap for the retreating Egyptian forces, fiercely attacked by the Israeli air force; the heaviest casualties of the war occurred there.[50] The following day, the IDF engaged in wide-ranging mop-up operations in northern and western Sinai, where the remains of seven Egyptian divisions were scattered, unable to retreat or move forward, constantly hammered by Israeli air support. Lacking support from what was left of their air force (which was unable to operate from the destroyed airfields), the Egyptian units fought valiantly, but they could hardly win against the overwhelming advantage of the IDF's unchallenged airpower. On the same day, another IDF force had taken Sharm al-Sheikh without a battle, lifting the naval blockade of the Gulf of Aqaba, the original casus belli.

By June 8, IDF tank columns were headed toward the canal bases, Ismailia and Suez. By that time, both Israel and Egypt accepted a cease-fire called for by the UN Security Council. It seemed at that point that the war might be over, and this message was broadcast to the people of Egypt on the morning of June 9.

However, this was not the case. During the same period, fighting occurred in and around Jerusalem, where some of the fiercest battles took place, with the Arab Legion fighting for every yard. Jordanian positions in the stronghold of Latrun fired at Tel Aviv, where civilians were killed and wounded. Within twenty-four hours, the IDF was surrounding the Old City, which was cut off from the Legion's main forces; they were encircled by Israeli forces on the hills surrounding it. On June 7 a force of paratroopers broke into the Old City through the Gate of St. Stephen and headed for al-Aqsa Mosque and the Wailing Wall, while Jordanian forces continued to target Israeli West Jerusalem. On the same evening, the Legion forces in the Old City officially surrendered to Brigadier General Uzi Narkiss; the Holy Places were intact and undamaged. By that time, the Legion forces were defeated in the north of the West Bank by a pincer movement of two Israeli forces, and Jenin officially surrendered. The Arab Legion was decimated across the region. By the evening of June 7, all Legion units in the West Bank had surrendered to the IDF. The frequent and unnecessary moves of the Legion's armored units, dictated by misleading information broadcasted from Cairo about great successes against the IDF, had been decisive in the rout of the Arab Legion.[51]

While the forces of Egypt and Jordan fought the IDF and were overwhelmed by its air superiority, the Syrian forces in the Golan were also hampered by lack of air support; their air force had been destroyed on the morning of June 5. While no actual fighting took place, the IDF and the Syrian army exchanged artillery battles intermittently before June 8, but without major damage on either side. With both Jordan and Egypt out of the war, Syria was on its own, and the battle to take the Golan started with a three-pronged attack on the heavy fortification along the borderline, on the afternoon of June 9, while international efforts were trying to prevent the onslaught. The IDF knew that its window of opportunity to destroy the Syrian forces in the Golan was very limited, and some of the toughest fighting in the war was taking place in the deep trenches of the Golan strongholds. All Syrian border posts were taken during the night, after bloody hand-to-hand fighting, with heavy casualties on both sides.

On the following day, the last day of the war, Israeli forces moved deeper into Syrian territory, past the main city of Kuneitra, and toward Damascus. The Israeli air force had inflicted heavy losses on Damascus and the forces defending it, but the IDF made no attempt to take the city. The third cease-fire agreement of the war, brokered by the UN, was signed at Kuneitra on June 11. This brought the short and fierce war to an end.

The Arab world was in a state of shock. Early false Egyptian radio reports about great achievements against the IDF had been reassuring; the reality of total defeat had begun to sink in, and difficult questions were directed at all Arab leaders, and not just in Egypt, Syria, and Jordan. While Arab societies were in mourning and shock, Israel had started celebrating in ways never seen in its short history, as well as inflicting the first postwar damages against civilians in the area now under its control.

In mid-June, the IDF cleared the large Moroccan Quarter (*Haret al-Maghariba*), which bordered the Wailing Wall. Given fifteen minutes notice, a few thousand Palestinians were driven away from their homes, which were destroyed immediately by waiting bulldozers to make a large empty piazza, where thousands of religious Jews pray daily. At the same time as the Moroccan Quarter was being ethnically cleansed, and using the chaos and confusion of the war, more than 250,000 Palestinians from various locations in the West Bank were expelled into Jordan. None of this was arbitrary or accidental.

While the guns were roaring on the third day of the war, June 7, Defense Minister Moshe Dayan told Lieutenant General Yitzhak Rabin, the chief of the general staff, that the aim was to empty the West Bank of its inhabitants. When the hostilities were over, Prime Minister Levy Eshkol coined a metaphor, which adequately encapsulated the Israeli ambition. In the metaphor, Israel's territorial conquests were a "dowry" and the Arab population a "bride." "The trouble is that the dowry is followed by a bride whom we don't want," Eshkol repeatedly said.[52]

While saying repeatedly that Israel is not interested in conquests and territorial gains,[53] the task of separating the bride from the dowry had begun in earnest and continues to this day.

Israel's New Empire and the Settlements

The realities of the new Israeli empire were stark. "Israel—whose population in 1967 was 2.77 million (2.38 million Jews and 393,000 non-Jews, mostly Palestinian Arabs)—now controlled more than 1.4 million additional Palestinians."[54] The bride was indeed substantial, as was the dowry. Within a week, Israel had changed from a threatened community of Jews surrounded by hundreds of millions of hostile neighbors to a mini-empire devoted to enlarging its real estate.

In hindsight, the five weeks before the 1967 war were some of the oddest Israel has ever endured. On every media outlet, politicians, public figures, and even army officers were telling Israel (and the rest of the world) that Israeli Jews faced a new Holocaust. Idit Zertal claims that the hysteria that possessed the Israeli polity dates back to the Eichmann trial in 1961. In this imaginary formulation, Nasser was the new Hitler, and the war a Holocaust facing Israel. This set in stone the structure for the most consistent argument of the Israeli propaganda machine—that antisemitism is an eternal, timeless, ahistorical phenomenon, that each generation faces its Hitler and its Holocaust. The Eichmann trial, a carefully orchestrated media event "curated" by Ben-Gurion, was the launchpad of this historical (hysterical?) thesis and bequeathed it the emotional power required for overwriting historical detail and reality. This crucial asset of Israeli propaganda was seemingly never exhausted, despite frequent abuse. Between the Eichmann trial and the start of the 1967 war, Ben-Gurion and many other public, cultural, and political figures in Israel managed to turn this false thesis into a tenet of Israeli identity: whoever criticizes or opposes Israel is an antisemite and continues the work of Hitler.

By the time the state was preparing to attack Egypt in June 1967, this was so deeply ingrained into the Israeli psyche that there could be no doubt in Israelis' minds that Nasser ("the new Hitler") was preparing a second Holocaust. Such crude claims hit a raw nerve, both in Israel and the West; media images of Egyptian commandoes goose-stepping energetically juxtaposed with Israelis digging trenches in Tel Aviv were especially conducive to the myth of a tiny Israel, surrounded by enemies, isolated and outnumbered.

This hysteria served the construction of the "second Holocaust" mythistory (to use Shlomo Sand's term), spawned by the press and radio. Ben-Gurion, out of office but not far from the political arena, was agitating for the prime minister, Levy Eshkol, to be sacked and to replace him as the defense minister, partly as a personal vendetta and partly because he worried that Eshkol might resist the pressure to start the war. In a press conference he called on May 29, he said: "None of us can forget the Holocaust inflicted on us by the Nazis."[55] Besides the report on Ben-Gurion's exhortations, *Haaretz* published an editorial titled "The Danger of Hitler Is Returning."[56] Invoking the political formula Nasser = Hitler ("Hitler is always with us") was never more persuasive than during the five weeks of nervous energy that preceded the war. An integral part of this toxic myth was the perceived need to get rid of the metaphorical bride—the non-Jews who now made up 37 percent of the people living under Israeli rule, between the Jordan River and the sea. Something had to be done, and quickly.

~

The day after the war, Eshkol invited his life-long confidant, Yosef Weitz, for a consultation about the newly conquered regions. Weitz brought his son Ra'anan, the former assistant of Eshkol, now chief of the Settlement Department, like his father before him. This was not a social meeting for idle talk; it would dictate the future of the occupied territories of Palestine/Jordan, Syria, and Egypt.[57]

Eshkol was already being pressured by his cabinet to reclaim the settlements of Gush Etzion, south of Jerusalem, abandoned during the 1948 war, as the start for a larger settlement drive. Eshkol needed Weitz and Son to aid his decisions; the discussion was arduous.

Weitz, no shrinking violet even at his advanced age, suggested that repopulating the old settlements in Gush Etzion was not a good idea, because it might lead to parallel claims of Palestinian return to the many hundreds of cities, towns, and villages depopulated in 1948. Settlement activity should be part of a wider policy, he argued; if territory was to be annexed, then its inhabitants would need to be defined as citizens. The government must first decide on the fate of the territories and the population in them, rather than launch a settlement process without resolving such issues, he suggested.

For Eshkol, who could see the rationale of Weitz's position, there was one problem he did not wish to postpone: Jerusalem. The three agreed that Jerusalem should be annexed immediately, and the debate focussed on the extent of annexation. Eshkol agreed to father and son's suggestion that there was an urgent need for "a detailed study of the territories and their inhabitants"[58] and asked Dayan to conduct it.

If the Gaza Strip was to be annexed to Israel, as suggested by Ra'anan Weitz, what would happen to its one million inhabitants, asked Eshkol. Ra'anan had a ready solution: part of Sinai will be annexed with the Gaza Strip, and the whole population could be transferred there. While Eshkol did not accept this proposal, such logic became normalized by the surprise victory, dictating the moves that followed. For Weitz, the Green Line of 1949 was a temporary feature, and with Israel in control of Palestine for the first time, it was time to reshape its borders.

The first legal move was the annexation of Jerusalem, or more accurately, of Jerusalem and 652 square kilometers around it, making it one of the largest small cities on earth, with a population (at the time) of less than 250,000. This was voted upon by the Knesset on

June 29, some three weeks after the end of the war, and all parties agreed to the move, except for the Communist Party. The whole legal move was kept secret, and its various stages were completed in a single day. Eshkol made sure that it was completed before midnight. Thus started the great land theft.

In this way the first and most meaningful step was taken to control the territories occupied by Israel, to avoid returning them to their legal owners. The message of the annexation was clearly understood by the public, and a long and painful debate emerged, between supporters of annexation and others wishing to use the territories as a bargaining chip toward negotiated peace with the Arab world. Only a tiny minority in Israel believed at the time that the territories should be returned. Neither the Arab regimes nor Israel were keen to discuss the end of the conflict. As a result, the territories were to remain under its control, and only Sinai was negotiated back to Egypt in 1979, when a peace agreement was signed by President Anwar Sadat and Prime Minister Menachem Begin.

Crucially, members of the cabinet immediately started to arrange for Jewish settlement building in the occupied territories. Both agricultural and urban settlements were built in the Sinai, the Golan Heights, around Gaza, in the West Bank, and around Jerusalem, in what was dubbed the Iron Ring—a circle of fortresslike concrete settlements surrounding the city, controlling all the routes into it. The Arabs of East Jerusalem became (noncitizen) residents of Israel, while the rest of the Palestinians in the West Bank and Gaza became prisoners in their own country, nonpersons lacking human, civil, legal, property, or political rights. In effect, they started on the painful road to becoming examples of Giorgio Agamben's "bare life existence" —life denied and opposed by the controlling state, akin to that of concentration camps' inmates, devoid of any rights.[59]

Israel started a process of using the land originally controlled by the Jordanian authorities as "state land" and building settlements on it, as well as a process of confiscating privately owned land without due legal process or compensation. This incarcerated the population of the West Bank in less than 10 percent of the area; they had no way of building legally, because Israel controlled the permit system. The only people who had no right to live in Palestine were its indigenous people; they remain without any rights five decades after that war.

The 1967 war posed an existential problem for Israel; it now controlled the whole of historical Palestine, and no one was prepared to force it out, despite UN resolutions and conceited speeches by Western diplomats. But what of the Zionist project, requiring a state without non-Jews, or at most, a tiny minority of them? It was clear that the two are not compatible—holding on to the Occupied Palestinian Territories with millions of Palestinians in it makes a Jewish State impossible. The option of further ethnic cleansing presented itself to the Israeli politicians. More than 250,000 Palestinians were indeed led to the Jordan River and helped to the other side by the IDF in the days following the war, while the world looked elsewhere. But this process of forced expulsion had to be stopped, because it was attracting attention. So, what could have become the continuation of the Nakba had to be deferred, and Israel found itself with more than a million Palestinians in the West Bank and another million in Gaza. These residents lost their Jordanian/Egyptian civic status, becoming both stateless and status-less.

Israel had decided to square the circle: to hold on to the territories, making the inhabitants of the land permanently stateless, in effect nonpersons. They would live under military law and Emergency Regulations, a leftover from the British Mandate legislation and in force ever since May 1948. Israel is the only country in which Emergency Regulations have been in force for every minute of its existence.

The decision to maintain the status quo was a systematic undermining of basic existence for millions of Palestinians. The reasoning behind this was simple: "Until we can push all of them out," Israel's politicians seemed to say, "we can make their life so miserable that many will leave of their own accord—which we will facilitate." They believed the Christian middle class and the educated would emigrate over the years; those remaining would either have to get used to living in an apartheid state or be pushed out when another opportunity for ethnic cleansing presented itself. The decision was to cleanse Palestine by other means, gradually, making it practically Arabrein. The model was the herrenvolk democracy of South Africa, which physically removed its black population to the Bantustans.

Thus far, this project has failed, even though many thousands of Palestinians have left the country since 1967. Those who have left were mainly Christians with a chance of finding a home in Canada, Australia, Latin America, and even the United States, but this was not enough to satisfy the Zionist dream of a land without goys. This problem is still with Israel today, looming over the Jewish State like a racialized scepter. There are now (in 2020) more Palestinian Arabs than Israeli Jews between the Jordan River and the Mediterranean Sea.

Another result of the war was the boundless adulation of the IDF, placing the army at the center of a semi-religious system. If Israel entered the war as a deeply divided, ailing society, with young people emigrating to find a life abroad, it quickly found solace in military success and the new sense of power emanating from its victory. IDF generals, never low in public estimation, were the new elite, exactly as Ben-Gurion had wished. The IDF was never just an army but the vanguard of the nation, the bearer of identity, healing the wounds of the mid-1960s.

A new Israel has emerged from the war—larger, richer, more militarized, and more coherent than at any point in its short history. The adulation of the IDF found its clearest expression in a deluge of the so-called "victory albums,"[60] the partaking of IDF officers in every conceivable cultural and social function, a great wave of victory memorials erected across the country—monstrous steel and concrete specters celebrating various IDF units. This quickly transformed Israel from its depressing prewar existence into a confident mini-empire, with thousands of Jewish and non-Jewish doe-eyed foreign volunteers working in kibbutzim, "learning about Israel." The additional two million Palestinian consumers forcibly added to the Israeli market, with the shekel enforced as legal tender, also helped Israel's economy boom quickly, with full employment. The Palestinian workers—a reserve army of nonunion labor—were a captive market as workers, as noncitizens lacking rights, as prisoners in territories whose legal status is unclear. Nonpeople in a never-never-land, their country was pulled from under their feet.

The economic boom revived Israel's economy, now more than ever focussed on defense and security. Israel had found its vocation. Jews, who lived in small and crowded European ghettoes for many hundreds of years, only to leave them in the nineteenth century, have now constructed the largest of ghettoes in the Middle East—the Jewish

State. Now, instead of being surrounded by non-Jews who ruled their lives and periodically terrorized them, this new ghetto was modern, included the latest amenities, and was surrounded by non-Jews who were under the Israeli-Jewish jackboot. The boot was on the other foot, so to speak—Israel controlled millions of subjects who had no legal standing whatsoever, their lives, land, homes, and families prey to Israel's merest whim.

This mini-empire not only gave rise to an even more virulent form of racism than ruled Israel before, but also to a new economy that would flourish and boom under the occupation: the economy of war, of unending violence, of conflict as a mode of survival. The words of Moshe Dayan, spoken in April 1956 when he served as the IDF chief of staff while standing over the grave of Roy Rotenberg, a twenty-one-year-old soldier who was killed by Arab infiltrators near the border with Gaza, became the beacon for the new, brutal, militaristic Israel. His eulogy merits quotation because the ideas expressed became the main tenet of the Zionist settler-colonial state and its innate militarism:

> What can we say against their terrible hatred of us? For eight years now, they have sat in the refugee camps of Gaza and have watched how, before their very eyes, we have turned their land and villages, where they and their forefathers previously dwelled, into our home ... Let us take stock with ourselves. We are a generation of settlement and without the steel helmet and the gun's muzzle we will not be able to plant a tree and build a house. Let us not fear to look squarely at the hatred that consumes and fills the lives of hundreds of Arabs who live around us. Let us not drop our gaze, lest our arms weaken. That is the fate of our generation. That is our choice—to be ready and armed, tough and hard—or else the sword shall fall from our hands and our lives will be cut short.[61]

Dayan was candid, speaking of the refugees' justified and understandable hatred toward those who had robbed their land and destroyed their homes and lives. The rationale of such thinking served the building of a war machine like no other; the "people of the book" became the people of the gun. Since 1967, "security" has filled every last nook and crevice of Israeli existence—guards in cinemas, clubs, shopping malls, and schools, involving the development of intricate procedures and machinery. Following Dayan's exhortation, Israel built the largest laboratory of conflict, with the Palestinians forced to perform the role of the enemy on which weapons, technologies, and tactics are tested, in order to get the stamp of approval: "tried in action."

The other foundation of the occupation was settlement of the land. The powers that allowed Israel to rule the OPT since June 1967 were already detailed in the Shacham Plan of 1963. During May 1967, while preparing to attack Egypt, Jordan, and Syria, each army unit received a large pack of documents that included the Shacham Plan[62] as well as copies of relevant international treaties, so as to equip it to rule in the occupied areas.[63] Pappe claims that the Shacham Plan, based on British Mandate Emergency Regulations of 1935–36, was the basis for all government decisions:

> The resolutions taken in that short period of three months, between June and August 1967, charted clearly the principles to which future governments in Israel would religiously adhere and from which they would not diverge, even during the most dramatic events that followed in years to come: be it the first or second Intifada or the Oslo peace process and the Camp David summit of 2000.[64]

A short while after the decision was taken by the Israeli cabinet, some of its members—Begin and Allon, and later Peres—initiated the settlement drive that, by 2019, placed 700,000 Jewish Israeli illegal settlers in the West Bank alone.

At the time, ethnic cleansing was also discussed, but it was not feasible to start this form of genocidal expulsion. However, cabinet members shared an understanding that Israel enjoyed impunity in the matter of land expansion; the international community was unwilling to confront Israel over it. Hence, the annexation had to remain only de facto. While areas Israel occupied in 1948 were given the status of Israeli territories, the 1967 holdings were seen as occupied territories. Thus, the population could not be expelled, but Israel would not consider offering them full citizenship. This was a difficult circle to square.[65]

The Dilemma: Territories versus People

Control of the territories was decided upon with the expectation of permanence. This could not take place without populating the areas with Jewish settlers, thus continuing the age-old Zionist credo—Israel stretches as far as Jewish settlement does. While full annexation was impossible unless Israel was prepared to give civil rights to the population under occupation, it chose to annex the land while leaving the people living on it in a juridical limbo. Israel would populate the territories with Jewish settler-colonists while imposing a brutal military regime over the indigenous population. The Israeli government believed that most Palestinians would be forced out by such conditions, presumably into the surrounding Arab states, as happened in 1948, and the world would come to live with full Israeli control over the whole of Palestine.

This preplanned and carefully implemented project was also paired with a "peace-making process" (or, as Moshe Dayan more accurately described it, the "piss-making process"). While this never offered Israeli withdrawal from the 22 percent of Palestine that Israel conquered in 1967, it mollified Western hankering for meaningless peaceful noises. A resolution, Israel claimed, depended on the Arabs, who could be relied upon to reject it, removing any danger of real peace and a need to vacate the OPT. This worked after 1948, so why would it not work after 1967? But what if the plan would not work? What if the Palestinians would not leave as planned?

Pappe points out the Israeli dilemma: the wish to hold on to the territories while expelling the Palestinians. Because there was no way that a forced expulsion could be successful after June 1967, Israel was faced with an impossible conundrum—how could it achieve both aims?

The minutes of the meetings display the incompatibility of these aims. The wish for more land is complicated by the Palestinian population living there. The committee develops the solution that was adopted—annexing territories without their inhabitants.[66]

The 1967 war changed the Middle East beyond recognition. Through a complex transformative process, Israel turned adversity into a successful business, building a whole new class of workers—academic researchers, technology experts, military tacticians, security and policing specialists, security "trainers," industrial workers building and testing the thousands of products, and so on. Israel's most burgeoning financial and industrial activity is its military–industrial complex, whose strength is the result of the 1967 war. What started as a tool crucial for retaining the occupied territories became a way of life. The other substantial vehicle for growth was the new territories and the great potential for Israel's economy in exploiting a captive market, powered by captured workers—a subproletariat lacking rights and unable to defend itself. On these foundations the New (Zionist) Jerusalem was built.

5

The 1973 War

There are those in Israel who call the Yom Kippur War "the sobering war." This adjective will be justified, if and only if it will become clear that it helped to terminate the nationalistic mysticism which flourished between the two wars; if and only if it will increase realism and assist the peace dynamics.

—Amnon Kapeliuk, *Lo Mehdal: The Politics Which Led to War*

Israel's military elite had been hoist on its own illusions and rigidities.

—Patrick Tyler, *Fortress Israel*

The Post-1967 Hiatus

The 1973 war, a direct progeny of 1967, was the only war in Israel's history not initiated by the IDF but forced upon it. Most of the war required that Israel adopt a defensive stance, not the offensive mode the IDF was trained for or excelled in. The war marked a serious failure in Israel's armed response and a disastrous breakdown of the extravagant defensive system along the Suez Canal. That Israel was taken by complete surprise is a testimony to intelligence failures that presaged future debacles, especially in Lebanon. The war also marked the last of the interstate wars between Israel and its neighbors; it led to the Oslo Peace Accords, an attempt to remove them from the circle of "conflict states."

After the 1967 war, it did not take long for the world to divine the direction of Israeli intentions. Two weeks after the guns fell silent, Israel annexed Jerusalem, despite repeatedly stating that it did not fight to gain territory. In the immediate aftermath, some 250,000 Palestinians were expelled across the Jordan River, some of them for the second or third time in their lives. A military government was imposed on the occupied territories, political activities of any sort were criminalized, and within months illegal settlements began in the West Bank and the Golan Heights, spurred on by such "doves" as Yigal Allon and Shimon Peres.

If anyone still harbored doubts, Israel clarified its plans with illegal expulsions, military settlements, and abrogation of civil and political rights in the Occupied Palestinian Territories. In response, Israeli "doves" and "hawks" started another round of shadow boxing. On the one side were those who would use the recent territorial gains to achieve lasting peace and, on the other, those supporting the original tenets of Herzlian Zionism—ejecting the indigenous population by "transferring" them, to build a Jewish State without non-Jews. Both camps had a common understanding: this debate was an intra-Jewish affair, excluding the very people whose fate was in question. The Palestinians' view of their own destiny did not matter. Avi Raz tells the tale of the many urgent meetings of Palestinian notables and activists with operatives of the security services, whom they tried to persuade that the time for a peaceful and sustainable resolution of the colonial conflict has arrived and that Israel held in its hands the key to this desirable future that could transform lives for all in this small country. Solutions ranged from an independent Palestinian state beside Israel to a single, secular democratic state of all in Palestine.[1] A minority suggested a return to the status quo ante—Israel returning the West Bank to Jordan. Israel's political leaders were regularly apprised of such ideas, starting within days of the end of the war.

But Moshe Dayan, Yitzhak Rabin, and Golda Meir were in no hurry to find a resolution: each had his or her own version of the same sentiment—"we are waiting for the Arabs to call us." A popular song of the period was "Nasser Awaits Rabin's Call." While speaking of peace, they concentrated on building more settlements, denying rights to the occupied, and banning Palestinian political activity.

The most decisive and unguarded of Israel's politicians was Defense Minister Dayan, considered the architect of victory in 1967.[2] He offered a political vista of the future:

> We have to see ourselves also in the Occupied Territories as the permanent government, planning and carrying out everything possible and not wait for peace, which may be a long time coming ... That is why we must use every opportunity and build Jewish settlements, rural and urban, in the areas conquered in 1967, as many settlements as possible, as this is *our homeland*.[3]

Golda Meir famously opined, "There is no Palestinian people," in the *New York Times* in January 1976, having used the sentence for years before on Israeli media.[4] Another pearl of Golda's wisdom: "A Palestinian entity is the invention of Jews with twisted minds."[5]

Nasser, who had resigned after the 1967 debacle, returned to government, millions of Egyptians having refused to see him leave power. The defeat was a deep shock to the Soviet Politburo, as recounted by Petro Shelest, leader of the Ukrainian Communist Party:

> Everyone was in a depressed mood. After Nasser's boastful declarations, we had not expected the Arab army to be routed so quickly ... Everything had been staked on [Nasser] as the leader of the "progressive Arab world," and this "leader" was now on the brink of an abyss. Political influence was lost ... Most of the military equipment was captured by Israel.[6]

Having tried and failed to get Israel to retreat from the territories it had occupied, Nasser realized another war was required. Even before the war was over, he acknowledged the new realities: "What was taken by force will be returned by force."[7] Despite deep misgivings, the Soviet Union decided to back Nasser fully, attempting to recoup lost territory and reinforce its naval base in Alexandria.[8]

To put pressure on Israel, Nasser started the War of Attrition in March 1969, causing hundreds of Israeli casualties and spreading despondency among Israelis, who had believed that the 1967 victory would bring a lasting respite from war. In retaliation, Israel bombed Cairo's industrial base, using the latest US fighter bombers, causing many civilian casualties. The escalation was spurred on by the Soviet Union supplying Egypt with modern and capable weapons, especially SAM missiles, with Soviet military experts offering advice and training.

With both sides rearming and intensifying the conflict, the countdown to the next round started. The first indication of "a Soviet plan to participate in a limited Arab offensive against Israel" was noted by the CIA in 1968;[9] it is likely that this bulletin was passed to Israel.

After heavy Egyptian bombing raids in October 1968, Israel started building massive defense fortifications along the canal. The debate on defensive options within the IDF was intense and involved Sharon, commander of the Southern Command; Israel Tal, commander of the Armored Forces, who supported a dynamic mobile system based on armored forces; and Chief of General Staff Haim Barlev, who preferred a static system of reinforced strongholds along the canal. Barlev won the argument and Israel spared no expense to build a defense line of awesome size and complexity.[10]

The "Conception" of the IDF

The Barlev Line consisted of some thirty strongholds separated by seven miles and with tank emplacements in the rear creating a second line of defense. It was considered invincible, like its historical precedents such as the Maginot Line. The great expense, innovative bunkers, electronic and other gadgetry, the elevated dirt mounds along the whole canal to provide a tank barrier, and a secret system of flooding the canal with burning petroleum oil—a floating firewall against crossings—all contributed to the Israeli conviction that the line could not be breached.

The confidence of the IDF planners in their incontestable superiority was based on the "lessons" of 1967, specifically, that Arab soldiers generally and Egyptian soldiers specifically were inferior. Israel placed its trust in the formidable Israeli air force; the IDF's edge in armored battle; the "strategic depth" offered by the Suez Canal, "the best antitank obstacle in the world"; and especially, the immense abilities of the Israeli Mossad and IDF Military Intelligence service, "which never make any mistakes."[11] As Israeli planners had spent years analyzing the reasons for the "weakness of the Arab fighter," they felt confident. Ezer Weizman penned the line "war is not for the Arabs,"[12] and this misconception (tinged by racism) became a truism in the IDF after the 1967 victory, the third in a row. Some, like the Arabist Ezra Danin, gave it a pseudoscientific sheen by arguing that the Arab soldier is devoid of patriotism, never ready to sacrifice his life for the collective, without which wars cannot be won.[13] Others, like the head of Military Intelligence, Aharon Yariv, "talked about an inherent "fundamental flaw in the Arab person's character." The Arabs, he said, were *drek*—Yiddish for rubbish or shit—and would remain that way.[14]

The disdain expressed by the IDF "analysts" was shared and popularized by the Israeli media, spoon-fed by the IDF. The press and media accepted the need to play their part in raising morale by deceiving audiences. Military correspondents of all channels were intimates of IDF commanders, assisting "their generals" in their moves into powerful positions in politics, finance, and industry. In so doing, they inculcated the racialized underestimation of Arab military and society among the wider public and IDF soldiery; a false conception of innate superiority infected Israel's polity. The toxic internal discourse of the IDF had become common sense among the Israeli public.

The translation of such prejudices into active policies was not long in coming. The strategy was known as the Galili Plan (September 1973), even though it was worked out by a trio consisting of Golda Meir, Israel Galili, and Moshe Dayan, the cabinet's right flank. It encompassed a new city and "Jewish port" in Sinai (Yamit), intensification of the settlement program, an industrial-base conurbation in the Golan, law amendments to allow Israelis to purchase land in the Occupied Palestinian Territories, and substantial investment in the "territories." For Dayan, the need for settlements was political—he never pretended that they support Israeli security, a lie intended only for public consumption.[15] This Mapai policy was sending a clear message; Israel was telling the Palestinians and Arabs that "we are here to stay."[16] A mere month after this great political "victory" by the trio of political wizards, they would find themselves stunned by Sadat and Assad.

~

Global Context

The 1973 war needs to be situated not just within local politics but also with reference to the power struggle between the United States and the Soviet Union over the wider Middle East. This contest started properly in 1956, with the Soviets offering to help with the High Dam and the Czechoslovak arms deal. Since then, the Soviet Union had established contacts with Egypt, Syria, Iraq, and (to a lesser degree) Libya through arms sales. By 1972, however, the canal had been closed for six years, with no income accruing to Egypt. The canal cities had been destroyed by Israel, their million and a half residents turned into refugees.[17] The Soviet Union was committed to reversing the situation.

In contrast, US powerbases in the region could be found throughout the Gulf states, led by Saudi Arabia, as well as Israel, Iran under the Shah, and Turkey. The animosity between both superpowers was closely reflected on the ground, turning the Middle East into a three-dimensional political chessboard of immense complexity. Nevertheless, the superpowers were invested in tension, not wars. Controlling their client states through arms purchases, they turned the Middle East into a testing ground for military technologies and doctrines. The problem arose when trying to control the pawns—the superpowers soon realized that they had limited control over client states' moves.

Within this chaotic scenario, therefore, it was difficult to tell who initiated the moves—superpower or client. Israel was always trying to ensnare the United States into complex traps, while the United States was not "anxious to come over there and fight the Soviets for you," as Dean Rusk advised Abba Eban in 1967.[18] The game became a complicated balancing act, each side liable to be wrong-footed or maneuvered into a military–political trap.

Ginor and Remez have suggested that the Soviet Union caused the 1967 war through intelligence manipulations leaked to Israel, combined with MiG-25 flights over Dimona in May 1967.[19] This far-fetched thesis is unsupported for the 1967 war, but seems applicable to the 1973 war. The Egyptian attack on Israel, the most successful in the history of the conflict, would have been impossible without a massive deception of Israel and its imperial master, unique in modern military history in its multifarious intricacy. This could not have happened without Soviet Union collaboration throughout the process of preparing for war, which lasted more than three years.

The United States was susceptible to hoodwinking because of its entanglement in Vietnam. Both superpowers were looking for ways to end that conflict, with Nixon and Kissinger making it clear that they might agree to a quid pro quo. However, the Soviets wanted the payback in the Middle East and expected the United States to push Israel into negotiating a resolution to the conflict. This was hardly promising, but the United States (meaning Kissinger) was willing to make an effort; this meant accepting Soviet domination over Syria, Egypt, and Iraq.[20] These actions brought about détente, a gradual lessening of tensions between the Cold War foes; they attempted to defuse flash points through frequent meetings of leaders and establishing a hotline for urgent developments. In the United States, Watergate was greatly limiting what Nixon and Kissinger could do. After Nixon's 1972 electoral victory, he became hopelessly ensnared in the illicit web he had himself woven, his authority eroding with every passing week. The Soviet leadership watched the president turn into a lame duck after his reelection and knew they could press the administration for political compromise: withdrawal from Vietnam and concessions in the Middle East. Nixon's decline and his Watergate difficulties gave Secretary of State Kissinger a free hand in building a power network through military coups, repression, and militarized regimes. His apparent power persuaded him that nothing was beyond his reach, and this played a crucial role in the great deception by the Soviets and Sadat.

The Great Deception

The subterfuge planned by Egypt and Syria with Soviet support depended on a sophisticated, long-term, and multilayered analysis of IDF philosophy to fool Israel through playing to its prejudices. The laying of the trap was a masterstroke of military and political planning, with the Soviet Union playing a crucial role. Without total surprise, an attack on Israel was doomed to immediate failure. The network of false leads took years to design, build, and perfect—

continuously waving "threats" of peace at Israel, a peace predicated on Israeli retreat, as well as a resolution of the outstanding Palestine issues, a prospect Israel was neither ready to discuss nor to consider seriously.

Sadat was not naïve enough to think that he might convert Israeli attitudes and knew that the United States was not about to assist him, so only the military option was left. Hence, his frequent use of the "peace option" in discussions with the United States and the Soviet Union was a sophisticated ruse. Either way, he could not fail: if the Israelis listened, he would win the peace; if they did not, it assisted the grand deception needed to ensnare and lull Israel. The more he combined periodic threats of armed conflict with demands for negotiations, the more Israel and the United States were calmed, convinced that his were empty threats.

It worked perfectly. By alternating warlike noises with peaceful pleas, a pattern was established in the minds of the Israeli leaders—Sadat spoke of war, but really wanted peace, presumably because he knew he could not fight Israel militarily. This underestimation, shared by the CIA, perfectly suited the Egyptian plans. Sadat also established that without certain equipment, such as medium-range missiles like the SCUD and top-of-the-range fighter-bomber jets, he was unable to attack Israel successfully.

Sadat's assessment of his military needs was leaked to Israeli intelligence through double agents, especially the infamous Ashraf Marwan, Nasser's son-in-law. As far as we know, this highly placed official, with responsibilities including armaments' manufacture, had contacted Mossad in early 1970 or possibly late 1969, offering his services for a hefty fee.[21] While it is quite possible that the ruse might have worked without the services of Marwan, his involvement added the authority of a highly placed source, by feeding Israel much accurate information, helping to build the misconceptions crucial for the success of Operation Ramadan. This helped to secure the plan; Mossad leadership bought into Marwan's services, helping to create what in Israel was termed "the conception" during the postwar debate about the debacle.

A typical example of the procedure used by Egypt to confound the IDF was the use of dummy surface-to-air-missile (SAM) batteries along the canal. These dummy batteries were introduced during the

War of Attrition and ignored by the IDF as targets. In 1970, after the cease-fire that ended the War of Attrition, Egypt covertly replaced the dummy batteries with real ones.[22] It took some time for the IDF to discover the move, by which time it was unable to move against the SAM batteries; such batteries had not been used before October 6.[23] Zeira claims that the IDF did not fully realize that this Egyptian move was a quantum leap, pointing toward the future attack; this, he believed, was one of the main reasons for the debacle.[24] Many such small steps were planned and executed over years on the way to the 1973 attack, but the IDF had not connected the dots. Learning from the meticulous planning of the IDF in 1967, the Egyptians and Syrians seem to have improved on the Israeli record.

A crucial part of the deception build-up was the "crisis" with the Soviets staged by Sadat and the supposed "expulsion" of Soviet experts. Marwan was used to leak "information" about this Egyptian "loss of confidence" in Soviet assistance, first to the British services, then to others, including the Mossad.[25] Indeed, some of the experts left but, it later became clear that these were lower level staff. The strategic and command grades stayed behind, with more joining in late 1972. This manufactured crisis helped to reinforce the view in Israel that Sadat was unable to prepare for war, especially without the Soviets, as his own forces were incapable of planning or fighting.

The Israeli assessment of the situation was based on a complex set of interconnected assumptions by IDF Military Intelligence and Mossad leaders about the options facing Sadat and Assad; most were based on wishful thinking.[26] The IDF stopped challenging its own assumptions and doctrine. So confident were Israel's political masters that they were prepared to publicly speak on the improbability of a war, just weeks before it started.[27] "The most common expression was 'we never had it so good.'"[28] On Massada, Moshe Dayan called for

> a new Israel, mighty, with wider borders, better than what we were forced into after the 1948 War. We are facing conditions that never existed before, and importantly, the might of the IDF, and Israeli control from the Jordan and unto the Suez Canal. When in the past could we settle the Golan, build airfields in Sinai and expand Jerusalem?[29]

Speaking before high-school students, Dayan noted: "We are turning an Arab country into a Jewish one."[30] The sky seemed to be the limit. Assumptions of superiority and impunity led Israel to deeply unlawful actions. In February 1973, the Israeli Air Force intercepted a Libyan passenger jet above Sinai, after it lost its way in a sandstorm, and shot it down without warning, killing everyone on board (103 civilian passengers and crew). In August of the same year, Israel forced a Lebanese passenger jet on a routine flight from Beirut to land at an Israeli military airfield; the Mossad believed that George Habash, the leader of the Popular Front for the Liberation of Palestine, was on board. There, the plane was surrounded by hundreds of commandos, but Habash was nowhere to be found, and Israel allowed the flight to return to Beirut. Both instances were not only illegal but qualified as piracy. Despite muted mutterings abroad, Israel remained immune from sanctions. There were no UN resolutions, no court hearings, no compensation paid. With good reason, Israel saw itself as above the constraints of international law.

In this atmosphere, Sharon and Zeira offered crucial interviews to the *Armed Forces Journal* in April 1973, arguing that with Egypt unable to cross the canal a war was unlikely.[31] As the war grew nearer, Dayan and the IDF grew ever more confident. Speaking in an interview with *Time*, Dayan noted: "The current boundaries of Israel will stay as they are in the next decade, and there shall be no major war."[32] That he said this shortly before the war came to haunt him afterward and led to his resignation. But Dayan was not posing; like all Israel's leaders, he genuinely believed in the boundless power of the IDF and the unerring efficacy of Israel intelligence organizations.

At the time that Dayan was expressing such inanities, the preparations by Egypt and Syria for Operation Ramadan, the complex training for crossing the canal, had been completed and the technical systems developed for the operation were ready and waiting. The Egyptians had covertly progressed the SAM 2, SAM 3, and SAM 6 batteries, replacing dummy batteries sited in 1971 without the IDF noticing. Special equipment, including large water cannons and diggers, was placed close to the canal, ready to open breaches in the antitank mounds built along the waterline.

The preparations were on a scale that Israeli Military Intelligence and the CIA could not altogether disregard. But Dayan "knew" what this was about—every year, the Egyptian army had held the Fall Maneuvers, with three armies and almost 100,000 soldiers. This annual "routine" had conditioned the IDF to see the 1973 movements as part of the normal exercise of the army. As a consequence the IDF and the political leadership stayed put, confident that war was impossible. As these exercises escalated, Military Intelligence advised Dayan and Golda Meir that there are indeed large concentrations of forces by the canal and also in Syria but they were purely defensive and posed no threat.

Nobody knew the date set for the attack, and most of the officers and troops found out less than an hour before when the war was about to start. King Hussein, who met Sadat some ten days before the launch, was told that the attack was near but was given no further details. Sadat and Assad chose the date October 6 barely one week earlier.

At the time, Ariel Sharon was interviewed almost daily, broadcasting the army's "conception," as it became known after the war: "In the next war the retreat-line of the Egyptian will be Cairo. They have no other option and it will bring about terrible destruction of Egypt. Total destruction."[33] To put no finer point on it, Sharon announced:

> "Israel is now equal to such powers as Britain and France. There is not a target between Baghdad and Khartoum, including Libya, which the IDF cannot occupy," in his view. No wonder that General Sharon believed that "in the current borders we have, actually, no security problems."[34]

Every hollow statement proved to Sadat, Assad, and the Soviets that the ruse had worked. They fixed the operation deadline for 6 p.m. on October 6.

One of the very last confusing signals was sent by Sadat on September 28, the anniversary of Nasser's death. On that day, a very large and well-advertised canal-crossing exercise took place, with rubber boats and bridging equipment. Sadat, in a broadcast to the nation, reminded Egyptians that he had not given up on the need to free Sinai, which he promised to do "with God's help." Meanwhile Israel's Military Intelligence saw this as another example of ritual threats, exactly as intended by Sadat. The same reading was articulated in CIA advice to Kissinger.[35]

On October 4, Mossad reported to a special emergency cabinet meeting that Soviet families were leaving Egypt and Syria in waves of military jets, regular flights, and military boats. The CIA also shared this information with Israel.[36] The news was seen as another falling-out with the Soviets. After a long discussion, the decision was made to marginally reinforce units in Sinai and the Golan, but the IDF saw no real danger of a surprise attack. A morning meeting was scheduled for Saturday, October 6, to review developments.

The next day Ashraf Marwan requested an urgent London meeting with Mossad head Zvi Zamir and told him just after midnight on October 5 that an attack was planned for the following day at 6 p.m. What he did not know was that the timing had been moved forward to 2 p.m. Zamir phoned back this message to Dayan in the early hours of October 6. This gave the IDF almost no time to react, because it took at least seventy-two hours to mobilize the reserves.

Israel Caught Unawares

It also happened to be Yom Kippur, the Day of Atonement, on which no services operated, media broadcasts were off, no cars were on the roads, and many Israelis were fasting and praying in synagogues. Very early in the morning, Dayan and Chief of Staff David Elazar went to Golda Meir's official residence to report Marwan's revelations. As the information came from the most important spy in the service of the Mossad, with excellent access to Sadat and the army leadership, they had to consider it, even if it did not fit into their "conception." The reserves were quickly mobilized to reinforce the depleted units facing Egypt and Syria. Meir sent an urgent preemptive message to Kissinger—"Tell the Soviets urgently that we have no intention of attacking Egypt or Syria, they should not fret"—which Kissinger forwarded to the Soviet ambassador at 1:00 p.m. Israel time. The reaction was shocking: The Soviet ambassador told Kissinger that Israel had attacked Egypt and Syria and that they were fighting to stop Israeli advances. There were no such attacks.

In fact, the Egyptians and Syrians attacked at 1:55 p.m.

~

This war was different from any Israel had ever fought before or since. The fact that it was forced upon Israel at a time it least expected was crucial. The Israeli cabinet was recalled in great haste at mid-day on October 6, as reserves recruiters were scouting synagogues for officers and soldiers. Meanwhile, just as Golda Meir was telling her colleagues that an attack would take place at 6 p.m., her assistant burst into the room, telling them that the war had started.

Head of Military Intelligence of IDF, General Eli Zeira, was removed from his post after the war and the inquiry that followed. Years later he wrote a book defending his actions before and during the war.[37] In it, he lists thirteen military campaigns that started with a surprise attack. In none of the cases, he argued, did the defenders succeed in foiling the surprise offensive. Although the full attack plan of the Egyptian army was discovered by IDF Military Intelligence in 1972 and extensively discussed by the IDF General Staff, there was no way of working out with certainty when and if it might take place.[38] Hence, there was no way for the IDF to stop the combined attack, especially in the initial period.

Despite its name, the IDF was never good at defense. But the situation in 1973 was not normal, because the IDF did not believe it needed to properly defend the lines. This was the reason, as pointed out by the Agranat Commission, that the IDF had "no detailed plan for the eventuality of a surprise total enemy attack."[39] In 1973, the defending forces were not only tiny, but badly prepared, assured by their commanders that nothing would happen. This brought about a total collapse before the attacking armies on both fronts within hours.

Within twenty-four hours, the defense lines along the canal and in the Golan were overpowered and routed. Even after reinforcements on October 4 and 6, they could not stop the large forces advancing on them. And quickly it became clear that the IDF was fighting the wrong war. Both tank forces relied on promised air force support, but such support was not forthcoming, with SAM missile batteries inflicting heavy losses on Israeli jets. Not only was the Israeli Air Force unable to destroy Egyptian and Syrian jets on the ground as it did in 1967, it could offer little support for IDF ground forces. Indeed, during the war, most SAM batteries attacked by the IDF were destroyed by armored forces moving to the rear of the Egyptian army, rather than by Israeli fighter planes.[40] The armored forces were assisting the air force rather than the reverse, as planned by the IDF. The Israeli army also faced a new factor for which they were unprepared: infantry-carried shoulder antitank guided missiles, which required enormous courage from operators—staying in the path of advancing tanks to target them. In this way, hundreds of Israeli tanks, attempting to support the canal strongholds, were destroyed or disabled in the first hours of fighting, two-thirds of them on the first day.[41] Meanwhile, the main body of Israeli armor was still being mobilized and transported from hundreds of miles away to the front. When the main body of Israeli armor finally arrived in Sinai on October 13 and 14, they were met with more than a thousand Egyptian tanks. Military historians have noted that this was one of the largest tank battles in history, involving more tanks than used by the Germans, Italians, and British combined at Al Alamein.[42] Losses on both sides were staggering. The bombing raids and artillery volleys were no less horrific—at least one raid used more bombs than were used in the entire 1967 war, and the Israelis used more ammunition than in all previous wars combined.[43] Similar volumes were used by the Egyptians and Syrians.

The initial shock was further aggravated by the inept and confused reaction of the IDF command on both fronts. The strongholds and tanks protecting them soon ran out of ammunition and were unable to defend themselves. Meanwhile, supply lines were long and vulnerable to air strikes. Of the thirty-two strongholds along the canal, only half were manned, and the others were locked up. This allowed the Egyptians to bridge the canal by a variety of means, in the long stretches between strongholds, quickly routing the whole defense line. The Israelis faced similar problems on the Golan front.

During the first week of the war, Israel lost half its tank force and almost half of its jet fighters, without seriously denting the Arab offensive.[44] At the same time, the superpowers started emergency operations to replenish both sides. Tens of thousands of tons of ammunition and materiel as well as personnel were flown in convoys from Soviet and US bases, to keep both sides fully supplied.[45] Sadat was so incensed about US support given to Israel that he spoke of "fighting the US."[46] While the Egyptians had at their disposal medium-range missiles with conventional heads capable of reaching all Israeli cities, they carefully avoided using them, despite civilian casualties caused by the IDF in Egypt.

It took eleven days before an effective Israeli counteroffensive was launched, by which time the two superpowers were pulling out all the stops, trying to achieve a cease-fire through the UN. The Israeli counterattack created a bizarre situation—the two Egyptian armies (corps) that had crossed the canal were parked a few miles inside the Israeli-held side, while a smaller Israeli force had crossed the canal, in between the two Egyptian forces, and was positioned on the western bank, cutting them off. In effect, both armies were cutting each other off—an untenable situation. It would take a few more battles over the coming days and a number of false starts for the cease-fire to take hold, as the IDF tried to continue its advance toward Cairo, presumably to make good on Sharon's promise of "total destruction."

On October 24, IDF forces surrounded the city of Suez, causing civilian casualties and cutting off the Egyptian Third Army. This was in defiance of UN Security Council Resolution 338, accepted by Egypt and Syria. At the same time, the Soviet Union gave an ultimatum to the United States: either join in sending a force to the Middle East to enforce the resolution or allow them to do it. The Soviet Union was worried about

> the eventuality of an Israeli air strike on the Aswan Dam ... which might cause a nuclear war ... It was rumoured in the highest corridors of power in Moscow that in response to such a development, our air force would have to land a nuclear blow on Israel.[47]

The United States was badly positioned to force a confrontation with the Soviet Union. A few days earlier President Nixon had fired the legal team investigating his illicit activities in the Watergate affair, and moves to impeach him were afoot. Kissinger, who wished to salvage the Vietnam negotiations,[48] had no choice but to quickly rein in the IDF, forcing Israel to negotiate. The series of agreements signed between 1973 and 1975 eventually led to Sadat's visit to Jerusalem in 1977 and the signing of peace agreements with Israel, which returned Sinai to Egypt, but not the Gaza Strip.

The Results of the War

While Sadat achieved all his goals, at high human and military cost, the Israeli leadership had foundered. Despite clear information about the concentration of forces on the boundaries, they had not acted to prevent the attack. While the war was not a total defeat of the IDF, it could hardly be presented as a success for Israel. The IDF had failed to predict or defend against the attack, through forswearing such a possibility. Israel lost more than 2,500 soldiers, with some 10,000 wounded.[49] Hundreds of Israelis were captured by both attacking armies and paraded on television, further humiliating the IDF and causing disenchantment. Both the political and military leadership had manifestly failed.[50]

To give an idea of Israel's financial burden after the war, one needs to look at the percentage of the defense budget as part of the GDP. In 1973–74 (before the war), the defense budget was 18 percent of GDP, —second only to North Korea. However, the actual defense spending that year was 43 percent of the GDP, due to the war, and it only dropped in the 1974–75 budget to an astonishing high of 33 percent. No country can survive financially with such levels of security spending, as all other categories of spending shrink. To have caused Israeli society this level of infrastructural damage was one of the most enduring failures of the leadership.

But the most serious failure was not "technical" but political. Israel's determination to control the conquered territories at whatever cost failed. Israel's leaders brought about the conditions for the war and the defeat that followed. This also meant that the political solution to the Palestine problem, so systematically avoided by Israel, had now become a contradiction in terms. As a result, the Israeli public was angry, blaming its leaders for a series of serious debacles, expecting heads to roll, even if they did not call for a change of policy.

Public anger was further fueled by the denial of responsibility on the part of the political leadership and their refusal to face the seriousness of the situation, especially the loss of so many lives. A young demobilized officer, Moti Ashkenazi, has pitched a tent by the official residence of Golda Meir, demanding that the government admit responsibility and that it immediately sack Moshe Dayan. Ashkenazi had been the commander of the only stronghold not overrun by the

Egyptians. For a number of weeks before the war he tried to get his commanding officers to realize that the strongholds were badly neglected and underprovisioned, all to no avail. Before the war, he wrote an article for an IDF military journal, pointing out the flaws in the military doctrine and asking for an urgent rethink, without success.

Despite attempts to remove him, Ashkenazi stayed put—"If the Egyptian army could not move me, the Israeli police will not succeed either," he quipped—and after weeks of campaigning many started joining him. A protest movement was born. But its stance was not politicized, its demands nebulous, moralistic, unfocussed. The protest was about ways and means, not substance. It never questioned the military doctrine, only the performance of the army. This was noted by Edward Said, who quotes from the letter sent by a large group of "dovish" Israeli academics to the *New York Times*, on October 17, 1973, noting their typical rightwing positions, and by Nakhleh.[51] A similar point is made by Elias Shoufani, who examined Israeli opinion polls on the war.[52] The social mask of the elite had been shattered.

Similarly, the 1973 elections, delayed by the war, brought little change in the political direction of Israel. The ruling Ma'arach (a later apparition of Mapai) lost six seats in the Knesset, and the Likud gained eight more seats than it had, not enough to unseat the government. All in all, the public, like the protest movements, leaned heavily to the right. As the protest continued, the government had to appoint a commission of inquiry chaired by Supreme Court Justice Shimon Agranat, with powers to call witnesses and make recommendations.

The hearings of the commission were used to settle personal accounts, mainly against the commanding officer of Military Intelligence and the chief of general staff, who became the scapegoats of the political class. The commission's members guaranteed an unfocussed inquiry; the findings were published fourteen months after the war ended. And, in the end, the political leadership was let off the hook, with some of the IDF leaders sacked, including the chief of general staff, the commanding officer of the Southern Command, and the two most senior officers leading Military Intelligence. But despite this rehabilitation of the political leadership, both Golda Meir and Moshe Dayan had been forced to resign even before the final report was published, due to great public resentment, especially on the right.

Meir was replaced by Rabin, the "blameless one," who had spent the period as ambassador in Washington.

At last, the first generation of Israeli leaders had vacated the political stage, with the "1948 generation" taking the helm, promising the interesting combination of "continuity and change," delivering more of the first and less of the second.[53] The door was opened for the right to take power, and Begin's Likud won in a landslide in 1977.

The cost of the war in 1973, for both Israel and the Arab states, could have brought about real movement toward a just peace and the defusing of military tensions in the Middle East. While the Arab countries were more prepared than at any time in the past for such a move, which would include a just solution for the Palestinians, Israel became trenchant and extreme, more interventionist and less circumspect.

The peace accord signed in 1979 was an exclusivist peace, omitting all the other issues, especially Palestine, refugees, the occupied Golan Heights, and the so-called final status issues. Israel would use the ending of military conflict with Egypt to increase the settlements, further limiting and suppressing the Palestinians. Indeed, in the Occupied Palestinian Territories, the war had set off a process of political autonomy: "With the October war the West Bank Palestinians began to establish themselves as a recognisable political force. The Palestine National Front was formally established in August 1973 to coordinate activities and national forces within the occupied territories."[54] This sociopolitical process would lead to the first intifada of 1987.

The IDF had spent years studying the 1973 war. But, impressively, it had failed to divine the deep reasons for the defeat it suffered. This became clear when it initiated the next round of armed conflict, in 1982. Despite this, most Israelis continued to blame the politicians and to consider IDF a "body of perfection and efficiency, a body capable of any task, even ones which seem impossible, a body immune to societal shortcomings, and untainted by politics."[55] Such sentiments are unsurprising in a society that more than any other had promoted army officers to political leadership. Almost counterintuitively, the war reinforced this tendency in Israel.

The interdependence between Israel and the United States and the generous support afforded to the IDF only increased after the war.

Even before the war ended, on October 19, Nixon requested from Congress the allocation of an interim "emergency security assistance" for Israel of $2.2 billion.[56] This proved to be only a down payment. Israel afforded the United States access to the Soviet weapon systems that remained on the battlefield. This in turn helped Israel to develop one of the most advanced military–industrial complexes anywhere. After this war, it became the main partner of the United States, and this primacy has not been questioned or undermined since.

The relationship between the superpowers had survived a very dangerous juncture, which like the Cuban missile crisis, could have led to nuclear confrontation. This strengthened détente, both blocs wishing to avoid direct conflict and move toward "normalization." In hindsight, this was an early sign of the decline of the Soviet Union as a world power, to be further eroded in the next decade. In the Middle East this process started almost immediately.

An epilogue to the 1973 war must include the total frustration of Soviet ambitions in Egypt, after decades of massive investment—financial, military, and political. In 1974, a cooling of the relations starts with Sadat's launch of infitah, a policy of "openness" to the West, converting Egypt to a market economy. The rift was to intensify with the signing of the 1975 US-negotiated agreement between Israel and Egypt. By 1977, the honeymoon was over, just as Sadat was heading to Jerusalem. The Soviet empire, itself on borrowed time, was left with Syria as a main base in the Middle East. By then, secrets of Soviet military technology were shared with the United States, through Israel and Egypt, both of which had no diplomatic relations with the Soviet Union.[57] This resulted from Sadat's belief that the Soviet Union was involved in planning a coup against him. He may have had a point—he would be murdered by an assassin from the Muslim Brotherhood, on October 6, 1981, the eighth anniversary of the October War, a little irony of history. Ginor and Remez find reason, in 2017, to criticize as pure propaganda the Soviet Union's media denunciations of the 1979 peace treaty as an American ploy to gain "deeper military and political influence."[58] In hindsight, one wonders what is inaccurate in this portrayal.

After 1979, it emerged that Sadat had damaged his former backers in an especially harmful deal—Egypt had sold its surplus of Soviet arms to the United States, which supplied most of it to the mujahideen in Afghanistan.[59] Sadat would be replaced by another military officer—Hosni Mubarak, who, in 1984 renewed diplomatic relations with the Soviet Union. By then, the Soviet Union itself was on the way out.

While Sadat achieved most of what he planned since assuming the presidency in 1970, his later policies were hardly a success; he ended the conflict with Israel but made little reference to Palestine and the other Arab nations. He was isolated and excoriated before his death, and Egypt fell victim to a toxic combination of untrammeled market forces and deep-seated corruption located within the armed forces. This wiped out all the social advances that Nasser had introduced in exchange for promises for a modern, richer, and stronger Egypt that were illusory.

For the Palestinians, the peace treaty between Israel and Egypt was a double blow: From the widely admired Nasser, revered though unable to reverse the defeats, history moved to Sadat, who, despite dealing Israel the most painful blow in the state's history, ended up betraying the Palestinian cause and cutting a separate peace deal with Israel. This never became a real peace, but brought about the end of a crucial phase in the conflict's history—the interstate war between Israel and the Arabs was at an end, at least for some decades. The next armed conflicts would all be with nonstate players—Fatah, Hamas, Islamic Jihad, and Hezbollah. Even the conflict with Iran would not, for the time being, become a war between states, but war against a state proxy. It was not at all clear that this phase of the conflict would prove easier for Israel, either militarily or politically.

6

Israel's Longest War: Lebanon War, 1982–2000

From the outset, the war was an exercise in national deception. Seldom had a military enterprise so large and so ambitious been launched with such a modest description of its purpose.

—Patrick Tyler, *Fortress Israel*

Come to us, dear plane
Take us high into the sky
We shall fight for Sharon
And return in a coffin.

—Soldiers' song during the 1982 war,
based on a popular nursery rhyme

The War Lab of the IDF

Here is a paradox: each new Israeli war introduces radically new elements into the Arab–Israeli conflict, while, in contrast, each is a direct continuation of the one before. This results in part from the Israeli norm of ratcheting up the violence in the hope of making the enemy hesitate before taking any action. Because this never appeared to deter the opposition, Israeli military planners and politicians drew the conclusion that not enough force was being used. Thus, by definition, there was a continuous and unrelenting escalation in Israel's military brutality.

It is just possible that such planners were aware that escalation was a planned excess, designed to initiate a level of aggression that might demand a final solution to the "Palestinian problem." If so, Israel had been very successful in managing world reactions while it maintained its aggression. What would have been unacceptable in 1948 passed almost unnoticed in 1956 and, later, in 1967. Such gradual intensification of the callousness of the Zionist project normalized what would be unacceptable if done at a single stroke.

This approach hardened after the 1973 war. The rationale of the debacle seemed to suggest that Israel would never again allow itself to be surprised by anyone. The United States had chosen Israel as its local power broker in the Middle East, and as it replaced the Ottoman and British empires as the real power in the region, it did not have to place large forces on the ground: "Instead of having to maintain foreign garrisons, the US was able instead to rely on the power of deterrence—the availability of rapid deployment forces to intervene from a distance—and on Israel."[1] This also meant that Israel was totally dependent on the United States for ammunition and supplies, which was less than ideal.[2]

The deep shock suffered in 1973 left its mark on the whole social order, facilitating even more brutal attitudes toward Arabs in general and Palestinians in particular. This failure had severely dented Israel's self-image but had not damaged its inclination for furthering the political agenda by military means; if anything, this tendency became even more pronounced.

Ironically, the results of the 1973 war were complex. This was to be the last war the IDF fought against an Arab army; later battles were fought against resistance fighters and civilian populations. In this way, the 1979 peace agreement with Egypt enabled Israel to concentrate its wrath toward "imperial" ventures—the redesign of the Middle East. There was no clearer signal of this change in intensity than the growing focus of wrath toward the PLO.

Lebanon and Its Complex Background

For nearly a decade, Israel had been fighting the PLO and its constituent organizations wherever it found them: in Jordan, Lebanon, Syria, and beyond. Carrying out antiresistance operations in countries surrounding Israel had harmed many innocent civilians, but this was regarded by Israel as the necessary price of security. This was clearly demonstrated in September 1970, otherwise known as Black September, when Jordan's armed forces managed to eject the PLO from Jordan. The majority of the surviving PLO fighters and leadership found a refuge in Beirut, with some small groups relocating to Damascus. This led to the creation of a Palestinian power base in Lebanon, as the PLO started taking active part in the complex politics of Lebanon—a fragmented society, divided between four main groups: Maronite Christians, Sunni Muslims, Shia Muslims, and Druze. The arrival of a large body of armed fighters undoubtedly complicated an already unstable situation.

The Lebanese constitution, adopted in 1943, was confessional; power was divided roughly between Christians and Muslims in the Lebanese parliament, as were the offices of state; by law, the president would always be a Maronite Christian and the prime minister a Sunni Muslim. This created severe tensions that in 1958 led to a civil war, sparked by the great power imbalances between the communities. The Christians controlled the economy, media, banking, and tourism, while Muslims were left behind and Lebanon's economy raced ahead. By the 1960s Beirut became the hub of Arab culture and politics, the meeting place of East and West. The city of two million, around half of Lebanon's population, was home to many newspapers, radio and television stations, and inter-Arab publishers.

Muslims felt deprived of a place at the table in this successful city, underrepresented politically by the iron girdle of the inequitable constitution. The conflict was resolved by brute force: the Lebanese president appealed in 1958 to the United States to send troops to support the "constitution" (in other words, Maronite dominance); the marines arrived, and the crisis abated, but the deep resentments were merely suppressed; they would rise again as the success of Beirut continued to exclude the mass of Lebanese.

The Maronite community had never united behind a single figure for most of the period since 1943. It had a number of military organizations controlled by leading clans: the largest was the Phalange of the Gemayel clan, whose founding father, Pierre Gemayel, was a fascist sympathizer who had set it up in 1936 as the Kataeb Party; it became the strongest and most influential political and military entity in the country. The Phalange borrowed more than its name from fascism; modelled on fascist organizations under Franco and Mussolini, it shared their extreme agenda and methods. In the 1970s, it joined forces with other militias such as the Guardians of the Cedars (Lebanese Renewal Party) set up by Etienne Saqr, al-Tanzim, the Marada Brigade, and the Tyous Team of Commandos. These militias opposed another broad coalition, the leftist Lebanese National Movement led by the Druze leader Kamal Jumblatt. In addition to the Druze forces of the Progressive Socialist Party, the coalition included the PLO; the Rejectionist Front of Palestine guerrillas; Amal, the militia of Shia villages of South Lebanon; and a few other, smaller groups. The two main military/political blocs represented broadly the right and left sectors of Lebanese politics, which, due to the constraining constitution, never made allowances for each other, causing a bitter and prolonged civil war, in which Syria, and later Israel, would come to play an especially destructive role.

The actual percentage of Maronites (not to mention the wider Christian community) in Lebanon is not known. The last time Lebanon held a census was in 1932. This is a subject of great sensitivity, as the confessional constitution is based on the assumption that Christians made up half of the population, though this was patently untrue. The most up-to-date figures are approximate, collected by the United Nations High Commissioner for Refugees and published in 2008. According to those figures, Christians make up around 33 percent of the population, with Maronites (the largest Christian denomination) at 22 percent. Even this figure may be inflated; in 1982, the forces of Bashir Gemayel, the leader of Lebanon's Maronites, initiated a population census, the results of which remained secret. The reason: it revealed that the real percentage of Christians in Lebanon was 30 percent, lower than anybody had suspected, especially the Israelis.[3] The rest of Lebanon is mainly Muslim: 28 percent are Sunni, 28 percent are Shia, and 6 percent Druze, making up at least two thirds of the population. This raises a serious question about the legitimacy of the constitutional arrangements guaranteeing the presidency to the Christians—if such an arrangement was ever valid.

None of the political alliances in Lebanon were trouble-free. On the Christian-Maronite side, long-term rivalries between the

Gemayel, Chamoun, Chehab, Frangieh, Sarkis, and other clans generated obstacles for collaboration, with most involved in murders and assassinations of other clans. The 1970 arrival of the PLO exacerbated the tension.

In this febrile atmosphere, full of intrigue, double- and triple-dealing, betrayal and murder, the Maronites who once controlled the country were by 1970 losing their advantage through internecine conflict. Leftist parties were boosted by the arrival of the Palestinian fighters, with comparatively trouble-free collaboration, as well as a larger population base supporting them—most with little to lose, as things stood. A war for survival was just a matter of time.

The two sides were preparing for conflict. The Maronites considered themselves the only true Lebanese, descendants of the Phoenicians and followers of the Crusaders. They were a "branch of Europe" in the Middle East, while the Arab Muslims were considered by the Maronites invading barbarian hordes. This all made the Christians attractive partners for Israel. The United States was a quiet partner of this relationship, maneuvered into this corner by the misconception that the PLO was part of the Soviet array of forces in the region—for which error Kissinger was responsible.[4] Based on this solid support from the United States, Israel announced in 1979 a new policy of preemptive attack in Lebanon. This was the last brick in the shaky foundation on which the 1982 war was constructed.[5]

For Maronite leaders, the existence of Palestinians in Lebanon was a blight and a danger to Lebanese society. There had been no love lost between the Lebanese elite and Palestinians ever since 1948. The Palestinian refugees had been treated by the Lebanese government and society in a despicable manner; they were denied basic rights, including the freedom to leave the camps or to work, and they lacked regulated civic status. Sharing a deeply Islamophobic outlook with Israel, most Maronites saw Palestinians as unwanted infiltrators causing trouble by irritating Israel. This is why both Sharon and Begin, obsessed with fighting the PLO, were so taken in by Bashir Gemayel, a swashbuckling, strutting young warrior of the Maronite tribe, who was only too ready to bring an early end to the Palestinian presence in Lebanon. Only later would they learn that his plan was to get the IDF to do this on his behalf.[6]

The Gemayel brothers, Amin and Bashir, commanded the Phalange militia and looked to Israel for assistance in chasing the PLO out of Beirut—and if possible, out of Lebanon altogether. They saw an opportunity for claiming common cause against the Palestinians and acted on it. Fighting began on April 13, 1975, when much of the downtown area was destroyed by running battles. This led to a demarcation line that cut the city in two. However, the Lebanese army itself was as divided as the society it nominally protected, and many soldiers deserted, joining militias on both sides. The fighting led to massacres, mainly of Palestinian refugees in Karantina and Tel al-Zaatar, but also of Maronites in Damour. The Phalange carried out such massacres to force the Palestinians to leave Lebanon; they were inspired by massacres committed by the IDF in 1948, which had caused large-scale expulsions of Palestinians.

IDF as Part of the Lebanese Miasma

Throughout this period, Israel acted as the main weapons suppliers for the Phalange, which it supported and financed[7]; this collaboration, kept secret by the Maronites, was well known in Israeli elite circles. This trade proved insufficient for the Maronites to gain a conclusive victory. In March 1976, with the war going badly, the Phalange secretly dispatched an emissary to Israel to request support.[8] This mission was successful; two senior IDF officers were sent to meet Bashir Gemayel on a gunboat off the coast. They were unimpressed by his casual and carefree manner, describing him as an unreliable partner. Nonetheless, this meeting led to others with higher ranking IDF officers, including Colonel Benjamin Ben-Eliezer (aka Fuad), who had a somewhat more favorable impression of Bashir.

Despite this seeming progress, negotiations were scotched by Prime Minister Rabin, who warned against any dealings with Gemayel. The turning point came after the surprise election of rightwing populist Menachem Begin in 1977. Begin was eager to enter Lebanon and eject the PLO from Beirut,[9] and Gemayel offered him a chance to do so. But it all had to be delayed due to the secret peace talks between Begin and Sadat, the highest priority of Israel.

Some of Begin's actions after the signing of the Camp David agreement were rather difficult for Israel's traditional backers to stomach, but they were crucial milestones on the road to the war against the PLO. The first was the bombing of the Iraqi nuclear facility in June 1981; the second was the illegal annexation of the Golan Heights, in December of that year. Both events, beyond their obvious functional role in Israeli politics, were designed to normalize the "long arm" tactics based on Israel's putative right to strike anywhere, under the pretext of self-defense.

Israel, as opposed to other countries, seemed now to have an exclusive right to act illegally wherever it chose, confident that it would not be stopped. Much was made of the Holocaust as a "kosher certification" for Israeli iniquities. To question Israel was to be antisemitic, Prime Minister Begin often stated. His daily diatribes made use of the Warsaw Ghetto uprising (a symbol he grafted onto Israeli "self-defense"), of "Jewish blood" spilled by the PLO, and of Hitler's bunker in Berlin (to which he compared Arafat's Beirut hideout). During a hiatus in the fighting in July 1982, Begin, in his parallel reality, extemporized in a memo to Reagan: "In a war whose purpose is to annihilate the leader of the terrorists in West Beirut, I feel as though I have sent an army to Berlin to wipe out Hitler in the bunker."[10] The US president was not impressed by the metaphor.

While some Israelis could see through this manipulative verbiage, most were caught in the trap, as were politicians abroad, making them unwilling to oppose Israel's extreme actions. This reversal of perpetrator and victim achieved miracles in 1982 and has done so ever since. By describing Israel as the victim and the weakened, entrapped PLO as the Nazi menace, propaganda gold was produced from the base metal of daily realities. This was a crucial stage in preparing the world for Israel's coming act of "liberation": the destruction of an Arab capital and the killing of some 25,000 people, mostly civilians, in the Israeli effort to expel the PLO from Lebanon.

The IDF War on the PLO

Gradually, preparations came into focus. Following reelection in 1981, Begin appointed Sharon to the Ministry of Defense, where he was given a free hand to start "to eliminate the traditional mechanisms which mitigated or blocked the government's natural propensity towards extremism."[11] After a couple of visits to Washington to meet Secretary of State Alexander Haig, hoping to get tacit approval for the attack, Sharon took an incredible next step. In January 1982, he was secretly flown to Jouniye, the large Maronite enclave north of Beirut, to meet Bashir Gemayel. Arriving with some of his advisors, he was taken that night into Beirut and saw the theater of conflict for himself from a tall building on the Maronite side, the divided city laid out below him. This was clearly a government "now looking for opportunities to make war."[12]

This was quite an unprecedented act; an Israeli minister of defense covertly visiting a hostile Arab capital was a clear sign of Sharon's commitment to the planned war. Sharon, who lied unashamedly to his cabinet colleagues, on May 10, 1982, promised them a limited excursion, lasting some "24 hours."[13] He was more candid with Gemayel and his aides:

> I do not wish to go into details of our plans, as they are still being finalized. Too early to speak of that. But one thing is clear and will form the foundation of IDF's future action in Lebanon: when the time comes, we shall not halt at the Litani, or the Zaharani [two rivers dissecting the roads north, toward Beirut] but continue due north, and will get to the environs of Beirut on the coastal road. Then, as I told you, you shall have the historical opportunity to take the whole city ... I see no problem for you to control this area and cut off the Beirut–Damascus route at an early stage of the fighting; thus we shall have a situation whereby Beirut will be surrounded and you can start cleansing the city of the terrorists [meaning the PLO and the Druze leftist forces] and those who assist them.[14]

Gemayel agreed with Sharon at that meeting that they would need the IDF to stay in Lebanon for at least three months after that.[15]

Now came the task of persuading the cabinet: "Sharon herded the Israeli cabinet in and through a war it did not want and had not approved, treating it 'like a kindergarten,' in the words of one of its members."[16] During the whole period between Sharon's appointment and the beginning of the war, the coming conflict was widely and openly discussed in the Israeli media, the Israeli public being prepared for the coming bloodshed. This was doubly confusing, as the PLO faithfully stuck to the US-negotiated agreement reached with Israel in August 1981, which forbade it to launch offensive operations from Lebanon. Sharon was hoping to incite it into action, initiating bombing raids on PLO positions, but the United States has advised the PLO against reacting, so as not to play into his hands. The PLO had listened and avoided any action, frustrating Sharon's plan.

But PLO restraint was of little interest for Begin, Sharon, Eitan, and other IDF leaders. Begin approved Sharon's and Eitan's plan for a takeover of most of Lebanon including Beirut, ousting the Syrian army, which had been there since 1975 as a "guarantor of stability" and expelling not just the PLO, but most Palestinians,[17] to bring about "the establishment of a 'new political order' in Lebanon."[18] This aim was shared with Gemayel, their chosen candidate for ruling Lebanon as a dependency of Israel.[19] Needless to say, both sides used one another in this complex political game, but it seems that Gemayel was better at pushing the Israelis—especially Sharon, who admired him—into difficult corners.

The Israeli Web of Lies

During the months of preparations, the PLO and Arafat, who initially had put his faith in the United States, soon came to realize that this is an exercise in self-deception and that the president and his secretary of state would support Israel in any conflict. In truth, no force existed that would stop the Israeli buildup for war, even if the PLO carried out the agreement to the letter. After the bombardment of April 1982, the countdown to the invasion started in earnest. The first launch in May was deferred due to weather conditions but was rescheduled for June 6. The cabinet was still under the spell of Sharon's promise that the battle, limited to a range of forty kilometers,[20] would last twenty-four hours—just as the tanks and armored personnel carriers (APCs) rolled over the border that morning.[21] By chance, Arafat was vising the Gulf states to raise support for the PLO that very day and was unable to get back to his command post in

Beirut.[22] One of the bloodiest military plots in the Middle East was about to unfold.

The IDF plan called for confronting the Syrians and encircling Beirut from the outset. The number of tanks used was staggering—of the 3,700 modern battle tanks held by Israel "up to 1,600 tanks were used in the invasion, consisting of M-60s, M48s, Centurions, and Merkavas."[23] This huge force was joined by 1,600 troop-carrying APCs, together a force of more than 120,000 heavily-armed soldiers.[24] This was the greatest number of armored vehicles ever to be used in a (small) theater of war. Ironically, this massive force was creating enormous difficulties for the IDF—supplying it with fuel and ammunition proved almost impossible, especially on steep, narrow mountain roads. This force was deployed against 4,000 PLO fighters.[25]

There were further obstacles once they had crossed into the city. The narrow, steep roads the force was traversing in the Shouf allowed Syrian and Palestinian ambushes to easily target tank convoys and destroy them. Retrieving the dead and wounded was very complicated, and helicopters were used as many of the main roads were now blocked by disabled tanks. This explains the high number of IDF casualties despite their enormous numerical advantage. Such huge formations were of little use in most areas:

> The tanks showed very little capability or versatility when fighting in urban areas. ... Beirut showed that tanks fought blindly and could not advance if the way was not cleared by infantry. ... The conclusion is that despite the brief experiences of street fighting in Suez and Qantara in 1973, the IDF had not learned how vulnerable and unwieldy tanks were in urban settings.[26]

Again, the IDF was fighting the last war.

Apart from this mass of armor, Israel used some 130 helicopters[27] and 670 fighter jets, and bombers made "100 to 200 sorties a day at the peak of the fighting involving up to 400 formations."[28] Such huge numbers of fighting units, with the navy and artillery added, were deployed for hours and even days before infantry or tanks were sent in, creating expanses of scorched earth where most remaining civilians were killed or wounded.

At the same time, the PLO was in the process of morphing into a traditional army—which negatively affected its fighting ability.

> By June 1982, the Palestinian military had not evolved from guerrilla units into regular forces using classical modes of operation, despite the considerable development of its armament and structure in that direction ... The Palestinian forces had lost the guerrilla's advantage of mobility, flexibility, and relative invisibility, without gaining the advantages of a regular army.[29]

Arafat soon realized that the transformation of the PLO into a regular army was a mistake, making it stationary and visible for the Israeli war machine, no match for it in size, equipment, or training. When infantry supported by tanks broke into the fortified camps in Tyre, Sidon, and Beirut, it became clear that urban guerrillas had many advantages over conventional forces. In the view of military historian Sayigh, this has been the main reason for failing to hold back the IDF:

> When an obviously superior force (in terms of numbers, armament, training, logistics and technology), such as the Israel Defense Forces (IDF), confronts a Palestinian enemy of extremely limited means and capabilities, such as the PLO, there is little question as to who will gain the upper hand on the battlefield.[30]

From the start, the scale of the attack caused severe logistical difficulties. For its assault on poorly trained fighters, the Israeli army had put together more than eight divisions, joined by massive artillery and air support, helicopter squads, marines, and naval units. While fooling the cabinet into believing that they were authorizing a restricted operation to clear the south of Lebanon of "terrorists," Sharon was preparing for a war of epic proportions. In many ways the hubris of his towering ambition caused his downfall. The need to hoodwink the cabinet presented grave difficulties as the IDF could not state its real aims:

> If the destination of the Northern Command was indeed Beirut, then the best and most efficient course would be to land contingents

> in the enemy's rear along the Beirut–Damascus highway, at key points along the north–south axes, and perhaps also at a number of spots along the coast.[31]

Such action would have given the game away, and the cabinet may have acted to stop the operation, so Sharon was limited to carrying out an upside-down pretend-plan, at variance with the real objectives.

Israeli intelligence reports prepared the IDF for a betrayal of the PLO by their allies. The Syrians and militias led by Walid Jumblatt had all moved out of the way, allowing the Israeli forces to concentrate on their target. However, this huge armed force was never really prepared for the fighting spirit of most PLO units. Though warned about the PLO readiness to fight and their ability to control narrow roads in the refugee camps, the IDF found clearing the resistance most difficult. It took longer to subdue the few defenders of Ain al-Hilwa, a large refugee camp on the outskirts of Sidon—eight days and nights—than it took to defeat the combined forces of Egypt and Syria in 1967. This stiff resistance stunned the front units and IDF command.

The War against Civilians

The IDF's barbaric treatment of civilians in Tyre and Sidon is worthy of note. The Zionist dovish writer, Reserve Colonel Dov Yermiya, wrote what he saw and it was published in Hebrew shortly after the war, shocking many Israelis. As Civilian Population Liaison Officer, he witnessed the bombing and destruction of purely civilian facilities:

> A PLO-run hospital suffered a direct hit, and a major part of it has collapsed. Expensive and sophisticated equipment is buried under the rubble, and the sickening smell of rotting bodies floats through the air ... We are beginning to deal with the water supply. All of the pipes in the city, together with those leading from the pumping station, are completely destroyed.[32]

Most of Yermiya's difficulties were caused by the cruelty and indifference of IDF officers and soldiers, who were incarcerating tens of thousands on the Tyre beach without food, water, or toilets, for many days. These soldiers

> are tensed for any act of revenge. Their mood, according to the cries and bits of conversation that I pick up, is volatile, extremist. Senior officers from the Military Administration, who arrive at the Regional Command Headquarters, the type who determine orders and attitudes, give vent in my presence to mean and poisonous remarks at the expense of the suffering population.[33]

Food and medicines sent to the port in Sidon by the Lebanese millionaire Rafic Hariri were blocked by the Regional Command, because allowing this

> will be used to prove the cruelty of the occupation ... Is there any necessity to prove our cruelty at a time when it is clear and demonstrable, as a result of the destruction, material, physical, and psychological, that we have brought upon this large population, and our inept and hostile lack of concern for their basic needs and their attempts at restoration?[34]

While the Regional Command frustrated his attempts to restore the water supply, they also refused to supply water to the parched crowd on the beach, and Yermiya seethed with helpless anger. His diary presented the inhuman behavior of the IDF to Israeli readers, as public anger was mounting after the Sabra and Shatila massacre.

By that time, few Western intellectuals were siding with the victim. An honorable example is the famous French author and playwright, Jean Genet, an early supporter of the Palestinian liberation as well as other anticolonial struggles. His descriptions of the Shatila camp after the massacres echo Yermiya's, with their lyrical power, and are shocking in their directness:

> For me, as for what remained of the population, walking through Shatila and Sabra resembled a game of hopscotch. Sometimes a dead child blocked the streets: they were so small, so narrow, and

> the dead so numerous. The smell is probably familiar to old people; it didn't bother me. But there were so many flies. If I lifted the handkerchief, or the Arab newspaper placed over a head, I disturbed them ... A photograph does not show the flies nor the thick white smell of death. Neither does it show how you must jump over bodies as you walk along from one corpse to the next.[35]

Genet's meetings with the survivors are, if anything, even more disturbing, without the finality and closure of death. Mentally destroyed victims were forced to relive the nightmare for the rest of their life:

> In the middle, near them, all these tortured victims, my mind can't get rid of this "invisible vision": what was the torturer like? Who was he? I see him and I don't see him. He's as large as life and the only shape he will ever have is the one formed by stances, positions, and the grotesque gestures of the dead fermenting in the sun under clouds of flies.[36]

Both texts, one by an Israeli reserve officer, the other by a French playwright standing in solidarity with the victims, converse with one another over the bodies of the dead and the suffering of the living. Unfortunately, Genet was unique among Western intellectuals, most of whom chose to remain silent. The horrors witnessed by Yermiya would be repeated later in Gaza, time and again.

The battles on the coastal route to Beirut were eclipsed by events on the second front—against Syrian forces in the Biqa Valley along the Beirut–Damascus highway. Sharon was in a bind; he had promised the cabinet there would be no attack on the Syrian force, and Begin made the same promise to Ronald Reagan. But striking the Syrians was one of Sharon's main objectives. As the Syrians well understood his game plan, their commanders had strict orders not to shoot, unless under fire, and then only with express orders. Sharon was courting the danger of starting a war with Syria without the knowledge or consent of the cabinet or the US administration. When the strike came, Reagan was on a tour of European capitals with Alexander Haig, who "was not the only senior American diplomat to come out of that week looking, at best, like an Israeli dupe."[37]

Indeed, while Israeli forces were attacking Syrian targets in the Biqa Valley on June 7, Haig was assuring journalists in Paris that there would be no such attack. Begin was repeating the same assurances in Jerusalem, only to hear a live radio report about it. It became clear not just to cabinet ministers, but to the Israeli population at large (and politicians elsewhere) that Begin was not in control of his army. Once the offensive had begun Sharon was in control, and no amount of political pressure could stop him. He had the army to remind Begin and his equally culpable ministers who was in charge.

Did Sharon Really Cheat the Israeli Cabinet?

The IDF senior command was fully aware of Sharon's plan to take Beirut and install Gemayel as president, while wiping out the PLO and the Syrian forces in Lebanon. Nevertheless the cabinet, which had not authorized either action, proved to be highly malleable and bought into Sharon's ruse, feigning ignorance and surprise. For them, as for Begin—presiding over a rightwing, militarist, and bellicose government—Sharon was a sorcerer's apprentice of unusual skills, the collective id incarnated into corpulent flesh.

Despite mounting IDF losses, not to mention thousands of civilians dying in Lebanon through indiscriminate bombing and shelling, the cabinet agreed fairly quickly to support the war. The later vilification of Sharon and pretense of cabinet ignorance was a shameful exercise in letting Israel's political elite off the hook. Indeed, after the Sabra and Shatila massacre, with Sharon's political stock at an all-time low, his cabinet colleagues and others within the IDF pinned everything on him; Sharon became the get-out-of-jail card for the leadership that had sent him to commit crimes on its behalf.

After the event, many participants were quick to argue that the IDF maneuvers were never cleared by the cabinet. Suffice it to say that at every single phase of the (allegedly unauthorized) war, no cabinet ministers found it necessary to resign, thus remaining active collaborators in Sharon's crimes. The same applies to the IDF leadership, who (with the exception of Colonel Eli Geva) followed Sharon like lovelorn lemmings, although they knew his actions were illegal.

The war progressed with a nod and a wink from the upper echelons, in a style inherited from Ben-Gurion in 1948.

Schiff and Ya'ari, the most well-informed and systematic analysts of this terrifying war, argue that Begin was cheated by Sharon throughout the campaign, though the evidence they present defies such a reading. Regarding the IDF's entering Beirut, they note:

> The prime minister of Israel, it seemed, was either ignorant of what his army was doing or was trying to conceal the truth—and doing a sloppy job of it at that. Menachem Begin emerged from the incident looking rather like a fool.[38]

A different reading of the same information leads to darker interpretations: Sharon was the willing hand of a grandiose, extreme politician who knew what he was unleashing but, driven by racialized hatred toward Arafat and the Palestinians, went along with this wish fulfilment. If Sharon was an odd apprentice, Begin had certainly made an incongruous sorcerer supreme commander of the IDF.

Other analysts were just as skeptical. Shimon Shiffer, political correspondent of Kol Israel, the official radio channel, noted:

> The war in Lebanon was, apparently, inevitable. It had its foundations in the deeply held world view of its architects, Menachem Begin, Ariel Sharon, and Rafael Eitan, and in the processes which shook the region since the signing of the Camp David agreements ... Begin lived his life as a confrontation with the world, with deep anger towards people and states which stood aside as millions of Jews were led to the crematoria. The only practical measure left for such a world-view is: force. Only Jewish might can assure the continued existence of Israel and the Jewish people.[39]

Such a stark reading of Begin's personality and its traumatic effect on his political thinking seems more astute than describing Begin as "a fool." This is also supported by Begin's style of leadership: "Begin has never been interested in detail. The decisions he reached were generalized. The detailed work he left for others, believing they will do their best."[40] This profile fits that of others responsible for genocide.

If the first, shorter part of the war had been bloody and difficult, the siege of Beirut promised to be much worse, and all intelligence reports supported this view. The reports have also highlighted the behavior of the Phalangists—massacres of Druze and Palestinian civilians, widespread rapes, robbery, and mass destruction of homes.[41] The IDF treated such reports with "understanding," because its own behavior in South Lebanon was no better. It became clear that while the Phalangists were unwilling to fight as soldiers, they excelled in postbattle atrocities. With Israel mobilizing more reservists for the battle of Beirut, the Israeli public, fed by foreign news snippets and revelatory details by reservists on leave, started to mobilize against the war; the public was more divided than during any past war.

Sentiments against the war intensified as preparations for "cleansing" West Beirut of "terrorists"—PLO fighters ensconced in the poorest part of the city—rose to a crescendo. This was not limited to civic society; some commanding officers at the front shared a deep unease about the war and its objectives, especially at a meeting of brigade commanders with Sharon and the IDF command: "The most outspoken opponent of an assault at this session was Col. Eli Geva, whose elite armored brigade had been picked to break into the city on one of the most difficult axes."[42] Geva's opposition was complex—he refused to endanger his men in a war that he could not justify and that, to his mind, would work to bring the case of the Palestinians into sharp relief in the international arena, which he opposed. Many officers shook his hand after the meeting, though remaining silent, then and later.

Geva returned to his brigade but found it impossible to act against his conscience; he could not bring himself to accept an unjust and unnecessary war and decided to resign his command:

> It was a radical step absolutely unprecedented in the history of the IDF, and a number of high-ranking officers tried to dissuade him. But Geva's mind was set, and finally he approached Drori [Commander of the Northern Command, and hence of the war] and asked to be relieved.[43]

Having resigned, Geva asked to be reassigned as a tank commander and fight as a soldier, but his request was denied. There followed a series of meetings with Sharon, Eitan, and Drori, in which they tried their best to dissuade him.

Destroying Beirut to Oust Arafat

When that failed, he was dispatched to see Begin, who failed to deter him. "Going into Beirut means killing whole families," Geva told the chief of staff, and he implied that it would destroy any hope of reaching an accommodation with the Palestinians in the West Bank. Eitan refused to discuss the political aspect of the issue and confined himself to the question of Geva's obligation to his men.[44] Geva was then summarily dismissed from the army and not allowed to return to Beirut to take leave of his soldiers. The former war hero became a nonperson.

As Israel tightened the siege of West Beirut, public opinion everywhere became hostile—the denial of water, electricity, and other essentials, including of medical supplies for hospitals, the food blockade, the frequent bombing are all acts that amount to war crimes.[45] Even Philip Habib, Reagan's special envoy to the Middle East and a staunch supporter of Israel, turned sour:

> The sights of women running with their children to find shelter from shelling and bombing and of the streets full of decomposing bodies—all these turned Habib into a biased observer in his relationship towards Israel and especially towards Sharon.[46]

Habib's disenchantment propelled Reagan into action. A plan was set to dispatch US Marines to Beirut to allow for the peaceful removal of the PLO from Beirut. This proposal was sent in great secrecy to Begin, for his eyes only, as Reagan needed congressional approval for a US force in Lebanon. Yet within hours it was made public on Israeli radio, leaked by Sharon's circle, which was concerned that it might end the siege. As a result, Reagan's move was torpedoed. Again, Sharon had managed to come out on top, keeping all the cards.

The last and most disturbing phase of the war started with the "success" of the United States and Israel, through the efforts of Philip

Habib, to force Arafat to surrender. By this time, the number of dead had reached some 25,000, mostly in Beirut, though the exact number may never be known. The starving of half of the capital, the rising death toll of babies who lacked not just milk or medicines but drinking water, the thousands of bodies littering the streets, while hundreds of thousands now lived underground, hiding from Israeli indiscriminate bombing and shelling, forced the PLO to sue for peace. The last straw was a series of brutal Israeli attacks during August on West Beirut that were allegedly opposed by the United States, though no action was taken against Israel, finally forced Arafat to leave Lebanon. The evacuation agreement included statements on the need to protect the "law-abiding non-combatant" Palestinian population of Beirut, but this commitment was never honored by either Israel or the Maronite forces,[47] which prepared to clear Lebanon of its Palestinian refugees.

Once Arafat agreed to leave, he was no longer able to protect Palestinians in Lebanon. Their lives and well-being now depended on their sworn enemies, the Phalangists and the IDF. Within a short while this precarious existence would be put to a severe test in West Beirut. But first, Israel and the United States had to deliver a victory for Bashir Gemayel, the only real candidate for the presidency of Lebanon. Various mechanisms were used to guarantee the prize—massive bribery, sheer violence, blackmail, and torture, including the flying of deputies to the Lebanese Parliament by IDF helicopters to guarantee their supportive vote. This was only achieved with little time to spare. Bashir Gemayel, the newly crowned president-elect, was installed by Israel as its imperial envoy. This achievement by Sharon impressed Israeli politicians, despite all their misgivings about him. He seemed to have pulled off a hat trick and appointed his own ally as president of an Arab country. Maybe the war was worth it after all.

Prelude to the Massacre

While it was clear that Gemayel's promises were worthless, events after June 6 established his refusal to fight on the side of Israel or to be a dependable stooge. After his election, assisted by his refusal to promise a peace deal with Israel unless "all Lebanese agreed" (code for the Muslim majority), it became clear that his actions in the Shouf and the Shia south region were focussed on reinforcing Phalangist rather than Israeli interests. In the Shouf mountains, a mainly Druze region, the Phalangists had moved in, instigating murder, rape, robbery, and torture. In the south, where Israel had for years maintained and trained the South Lebanon Army under Major Saad Haddad, the Phalangists moved in to displace Haddad, in violation of the agreement with Israel.

Such infringements quickly mounted and when the Maronite forces also attacked Palestinian civilians in the south, Israeli analysts belatedly admitted that Bashir was a loose cannon. Immediately after the evacuation of the PLO, Israel had worked toward evacuating all Palestinians from Lebanon, prohibiting the restoration of the destroyed camps, which Gemayel said would be turned into "'an enormous zoo.' As to the Palestinian inhabitants, he had a mind to 'load them onto air-conditioned buses' and dispatch them over the Syrian border."[48] In the same vein, he promised that by October 15, a month after his inauguration as president, "there won't be a single terrorist in Beirut," referring to Palestinians in the city.

For all the brash talk, the Phalangists were losing ground to the better organized Druze militia, which pushed them out of the Shouf, despite IDF support. Far from controlling the whole of Lebanon, the Phalangists were now confined to the same areas they had controlled before the war: the Jouniye enclave north of Beirut. In response, the IDF planners came up with increasingly hare-brained plans, including one to arm the Palestinians in the south as a counterforce against the Amal Shia militia, right after trying to use Amal as a force against the other militias had failed.

So, rather than working toward a united Lebanon, as Israel had constantly declared, the IDF tried to add two new militias that would work under its direct control: "The irony of it all was that, instead of creating new order in Lebanon, Israel seemed to be going out of its way to maintain the traditional balance of enmities."[49] *Pax Israeliana* in Lebanon worked out as more of the same, but worse.

By the time Bashir went to Israel for his last meeting, on September 12, 1982, what Sharon had to discuss with him would be realized a few days later, after his death. They spoke of the exact manner in which West Beirut would be cleared of Palestinians—a total negation of the US-brokered agreement for the evacuation of the PLO. This did not matter much to Sharon and Gemayel, for whom agreements were useful for getting what one wanted but were not binding. That the role of the Phalangists in clearing the camps was finalized at this point is important to note as Sharon later claimed that this was only agreed after Gemayel's death, in the "heat of the moment." This lie was interestingly countered by Sharon's own written record of the meeting, speaking of "destroying the terrorists' infrastructure so as to manifest the Habib agreement to the full."[50] However, on September 14, an explosion at the party offices in Ashrafiyya killed Bashir and twenty-six of his followers. Within forty-eight hours the agreement with Sharon became the trigger for both Bashir's death and the Sabra and Shatila massacre.

On the morning after Gemayel's assassination, Sharon gave the order for the IDF to enter West Beirut and surround the camps, letting the Phalangists "do their work." For two nights and a day they rampaged both camps, murdering, raping, and torturing thousands of people, protected by a heavy IDF cordon—thousands of soldiers with armor and tanks. The IDF forces were stationed yards away from the crimes, but somehow, despite reports spreading about a massacre within an hour of Eli Hobeika's (chief intelligence officer of the Phalangist Party, a known killer and torturer) force entering Shatila, they managed not to notice anything for thirty-eight hours of mayhem.

Those who had noticed, and tried to speak to their commanders, were rebuffed or told to go and "have a rest." Ron Ben-Yishai, a senior military correspondent, discussed this with General Amir Drori's aide very near the command post:

> [I] asked him whether he heard about the "nasty business" the Phalangists were up to in the camps. He patted my arm through the open window of his car but said nothing. I asked him again, and again I was rewarded with silence.[51]

The officers at the command post, some of the most senior IDF personnel, were fully aware of what was taking place inside the camps, but none of these officers did anything to stop the killing.

When, later on, Begin was forced by the largest demonstration in Israel's history to set up an inquiry into the affair, a solution was found to the systemic ailment of the political and military elites of Israel. The inquiry focused on Sharon and the Phalangists as responsible for the massacre, hence clearing the political and military leadership, despite the part they had played in authorizing the war and the massacre itself. Indeed, the Israeli public was an active partner in that war:

> After three months of a war that, despite grumblings from the press and the Parliamentary opposition, was accepted by a supportive public in Israel, the Cabinet chose to believe that the effects of the massacre would dissolve quickly as long as it was portrayed as a peculiarly Lebanese perversion—"goyim killing goyim," as Begin was reported to have dismissed the affair.[52]

While Begin could not care less about 25,000 Arabs killed by the IDF in Lebanon, he was very upset by the cost to the army:

> It drew Israel into a wasteful adventure that drained much of its inner strength, and cost the IDF the lives of over 500 of its finest men in a vain effort to fulfill a role it was never meant to play.[53]

While one may hold a somewhat different view about the roles meant for the IDF, especially in light of the 1956 war, it is clear that the high human cost to the IDF was weighing on the prime minister, ultimately leading to his early retirement.

The Aftermath: Sharon as the Fall Guy for the IDF

The Kahan Inquiry found that in addition to Sharon, Amir Drori, Yehoshua Saguy, and Amos Yaron, the senior commanders of the war "committed a breach of duty incumbent upon them"[54] and made some recommendations about their future roles in the IDF, though none had faced a court for their criminal behavior. Their senior commander and Sharon's main partner, the Chief of General Staff Rafael Eitan, was found to have acted inappropriately, though no punitive measure was enacted against him, either.

The inquiry was an exercise in imperial whitewashing; after analyzing the meaningless reports in a range of European and American press and media outlets, Eqbal Ahmad suggests:

> A dispassionate and sober look would have suggested to these instant historians that the Kahan Commission failed to fulfill its legal and human obligations in at least three fundamental respects: (i) it engaged in a politically motivated legal evasion unworthy of a commission which was in principle, judicial in character; (ii) although it was charged with examining "all the factors" connected with the massacre, it neither disclosed all the facts nor examined all the factors that led to the slaughter; (iii) most crucial, it did not assign legal and political responsibility in a way that could diminish the likelihood of a repetition of similar crimes.[55]

Sharon, the central figure of this sordid and terrifying episode, lost his position as minister of defense but faced no other sanction. Seventeen years later, he became Israel's prime minister, and initiated the Apartheid Wall, a landscape feature so massive it is visible from outer space. Before he took his exit in 2005 by entering a vegetative state, his fingerprints were writ large on the conflict, not just in Palestine, but everywhere in the Middle East. By the time of his delayed demise, he has become one of Israel's most loved and admired leaders.

That Sharon became so popular is not a measure of any meaningful change that Sharon himself underwent in his hospital bed, in a coma; he could not and would not change. What had dramatically changed was Israeli society; from the febrile polity rising against warmongering politicians and army commanders, Israel turned into a typical settler-colonial society—even more militaristic, aggressive, and inured, ready to support any and every move its leaders made, however unjustified, nonproportional, and brutal.

By the year 2000, most Israelis had only experienced life in the mini-empire of post 1967; they knew no other reality. IDF soldiers, without exception, had never fought against other soldiers, they had only confronted civilians and freedom fighters in Palestine and elsewhere. By that time, Israel's economy has turned into a combination of a military–industrial complex and a hothouse for high-tech security start-ups, initiated by the hundreds of companies set up by army officers. These innovations made use of their experience, selling destruction across the globe, undermining democratic rights through the use of surveillance technologies, recording and reporting the daily life of billions, under the deceptive trademark of antiterror measures.

The IDF learned some lessons from the Lebanon fiasco. Ehud Barak, Chief of the General Staff from 1987 to 1991, argued that fighting civilians required new methods. Adapting lessons from the first intifada, he developed his solution for the security of Israel: a "small, smart army" using new technologies to replace large formations, of the kind that fought so badly and so brutally in Lebanon. Under his guidance, investment went into developing spy satellites, drones, and software solutions dealing with the challenges of the twenty-first century.[56]

Like Dayan, Rabin, and Sharon, he had left the army to become a politician, hastening the decline of the Labor Party with scattergun politics, ironically paving the way for the election of Sharon in 2001. Barak's tenure also saw the war and occupation of Lebanon, started in 1982, coming to an abrupt end in 2000, as tens of thousands of unarmed Lebanese organized by Hezbollah managed to expel the IDF. Among the many witnesses on the Lebanese side of the border was the Palestinian leading intellectual, Edward Said, who was caught on camera symbolically throwing stones at the withdrawing soldiers.[57]

In many ways the 1982 war influenced Israel and the IDF more than any other. Never before did the IDF operate such huge formations in civilian areas or bomb and shell hundreds of thousands of civilians unperturbed, bringing about the death of many thousands, mainly noncombatants. The level of destruction in urban areas was without precedent, but with United States and European Union support, Israel could afford to wreak havoc with impunity. By branding this devastation and mass murder as "self-defense" and required to "guarantee Israel's security," a weapon more formidable than bombs, shells, and drones was created: a systemic argumentation of the exceptionalist positioning, framing Israel as an exceptional polity.

This was achieved by recycling Holocaust memes, presenting Israel as victim rather than perpetrator. Israel has in fact become immune to normative legal, social, political, or moral codes and systems of justice. The rightwing governments that controlled Israel for most of the last four decades—thirty-five out of forty-two years—invested not just in developing a militarized economy, but also in the massive propaganda machine that included an education system preparing Jews in Israel (and elsewhere) for their role of racialized masters in a *herrenvolk* apartheid democracy.

The role played by the IDF in shaping this collusive, servile, and collaborative society cannot be overstated. As most Jewish men and women serve in the IDF, Ben-Gurion's dream of a nation under arms was realized like never before; he could only hope and wait, but decades of acting as the sheriff of the Middle East have turned Israel, the recipient of more foreign aid than any other country for the longest period on record, into a strong and rich economy based on military exports and security-related products, as well as mass oppression of millions.

Israel used the IDF to turn itself into a world and regional power, well beyond what its size might suggest. By maintaining hundreds of nuclear warheads and more nuclear submarines than Britain or France, Israel is a power to reckon with, and many states depend on it for their military hardware, software, communications, and training of their armies and security services. After the 1982 war, the army had its state.

The Shortcomings of the IDF in Lebanon

The war proved beyond doubt that the IDF could successfully undermine all civic systems in Israel (such as they are) and bend them to its whims and objectives. More than any other leader in Israel's short history, Sharon managed to etch his style and methods onto the body of the Israeli polis. With his aggressive military policies, dynamic battle plans, settlement project, and Apartheid Wall, he influenced Israel's trajectory more than any politician before him, branding Israeli society with the hot iron of the IDF. Despite the demonstrations against him, he controlled the agenda and the future; without the brutality that he normalized, the incursions into Gaza in 2008, 2012, and 2014 and the Lebanon war of 2006 would have been unlikely. He inculcated a Spartan nation of soldiers in the racist arts of genocidal killing, enabled by a rare mechanism of inversion, a perpetrator perceiving and presenting itself as the "real" victim. Repeating lies often enough makes them "facts" for some, even for most, through saturation propaganda. The many supporters of Sharon crowned him Arik Melekh Israel (Arik King of Israel) and would be the foundation for his return to power in 2001.

Some resistance did, however, emerge including during the war in Lebanon from two different groups of women protestors: Women in Black and Mothers against Attrition. Both groups made important public interventions against the war in Lebanon and its continuation and contributed to its eventual ending. While both groups failed to offer unconditional solidarity with Palestinian women, they did create a radical space for opposing the war and other IDF atrocities.

The results of the 1982 war were momentous for both Palestine and the PLO. By expelling the leadership to Tunis and the fighters to eight different Arab countries, Sharon had achieved his objectives at a terrifying cost for Lebanon and its Palestinian refugees. The PLO was unable to keep in touch with Palestine in the way it did before, unable to continue the struggle in an effective manner, disconnected from the Palestine population. The first intifada, which started a few years after the 1982 war, resulted from this failure of the PLO and its armed struggle; Palestinians in the Occupied Palestinian Territories had risen because they understood they must act for themselves and could not rely either on their Arab brethren or on their own PLO, beaten, distant, and dispirited. Indeed, so disconnected was Arafat from the groundswell in Palestine that the news about the uprising caught him totally unawares and even doubting the reports. Helena Cobban notes that "writers judged that the PLO leaders were taken by surprise and have been trying to make up the lost ground ever since."[58] In an article by Tariq Kafala, he claims about the intifada: "It came as a complete surprise to both the Israelis and the PLO, at the time in exile in Tunisia."[59]

One of the ironic touches of the war was the emergence of Hezbollah as the strongest political and military force in Lebanon, out of the ashes of Amal—the smaller and weaker Shia organization. From then on, Hezbollah rose to become the effective voice of resistance in the country. This was a direct result of the war and the decades under Israeli occupation.

As with so many other junctures in the history of Palestine, before and after 1982, it fell to Edward Said to work out the lessons of this tragic episode. His first conclusion may sound obvious but had no effect on the PLO:

> At very least then, the Lebanese conflagration provides Palestinians with some urgent opportunities for reflecting on the future, and on those aspects of the past that directly affect the future. Lebanon was a disaster: there is no way of avoiding the facts, each of which, separately or as part of a whole, confirms a picture whose tragedy and loss exceed the events of 1948.[60]

Said considers the Lebanese period of the Palestine struggle as "the first truly independent period of Palestinian national history."[61] The loss of this achievement was tragic in the extreme and Said was not afraid to say so. He was also strong enough to ask:

> Was the end in Lebanon avoidable? Was the substitutive nature of Lebanon a necessary phase or a disaster in the long run around which we should have maneuvered? Were the fruitless but encumbering ties with various Arab states inevitable, or were they pursued as an end in and of itself? Was there enough understanding of the larger, the enormously complex global dynamic that involves the question of Palestine today? Were Palestinian politics and ideological struggle concentrated, directed, disciplined enough? Above all, has the new Arab environment of corruption, petrodollars and mediocrity—presided over by the United States—seriously, if not definitively—affected the Palestinian national struggle?[62]

Unfortunately, such questions did not seem to concern the PLO in Tunis or later when it returned to Palestine in the guise of the Palestinian Authority.

While the PLO did not learn the lesson, Sharon did. He dealt a decisive blow to the Palestine liberation struggle, and the leadership of the PLO could not even face the facts. And of all his would-be followers, Sharon prepared the ground for Benjamin Netanyahu, a master of acting the victim while perpetrating untold destruction. The 1982 war was the international stage on which the victimhood drama was played out with incredible success. Israel would never be the same again after this frightening episode. It has broken the PLO, but the price was the total dehumanization of Israeli society.

Jacobo Timerman, a renowned Argentinian journalist and editor of the Buenos Aires liberal daily, *La Opinion*, a courageous opponent of the military junta, was arrested and tortured before being deprived of his citizenship and deported. In 1979 he ended up in Israel, a place he cherished from afar. Timerman became a staunch opponent of Begin and Sharon after the start of the war in 1982 with his book translated into many languages.

Timerman may have been the most prophetic of all the Israelis who have commented upon the 1982 criminal adventure by Begin and Sharon; as opposed to the many who criticized them mainly for abusing the army and starting a war for dabbling in the affairs of Lebanon, Timerman understood that this war is also being used to irreversibly transform the nature of Israel itself.

Gazing well into the future, Timerman gauges the Begin government most accurately:

> It is evident to me that Israeli democracy is threatened by the Begin government, whose policy is not democratic and whose actions are establishing the basis for another kind of country: a totalitarian country which, like all totalitarianisms, cannot be likened to any other.[63]

That Begin and Sharon were laying the foundation for a very different society was not clear even to seasoned Israeli intellectuals of the period. Timerman did not share their predicament—he was a recent immigrant from a country that educated him in the convolutions of totalitarian democracy:

> I am thinking of the long period during which Juan Perón governed Argentina. Each of the three times he reached the presidency, he won in free election by secure majority—between 55 and 70 per cent of the vote. There was a parliament, political parties, and so forth; he never committed fraud in the elections. Yet the dynamic of his government smothered democratic life and undermined

> democratic institutions, until it became clear that he was using the democratic system for anti-democratic ends.[64]

Timerman was too sophisticated to accept the standard argument of Zionist apologists defining Israel as "the only democracy in the Middle East" by listing its apparent democratic indices. His prescient gaze pierces the smoke screens of the regime, discerning the long-term trends it would indeed acquire in decades to come:

> In the last few years, Israel has lost many of its democratic qualities, particularly since the Lebanon invasion. New concessions to religious groups, which in Israel are intolerant in contrast to their behaviour in the Diaspora, not only impede modernization of social life but reduce the scope of secular activities, especially in the crucial field of education. Economic policy is characterized by the irresponsibility of demagogic governments. Financial speculation takes the place of productive investment ... reduced, too, is investment for housing, highways and health care. Funds are used instead for the illegal establishment of businesses in the occupied territories.[65]

Timerman saw further than his Israeli contemporaries into a future where Netanyahu's totalitarian democracy of the extreme right, supported by racist religious zealots, would drive Israel toward becoming an apartheid state de jure.[66] Mordechai Kremnitzer, who comments on this move to legislate racist policies, himself gazes beyond the law into a murkier future: "Is this not tantamount to adopting the view that Arabs are trespassers in Israel? Does this not constitute their symbolic "transfer"—and the start of their actual transfer from the land?"[67]

Israeli apartheid, as opposed to the one introduced by South Africa in 1948, does not just legalize inequality and racism; it enables the continuation of the genocidal "cleansing" process started during 1948. The Begin–Sharon war in 1982 played a crucial role in enabling the new Israel, so accurately foreseen by Timerman in 1982.

7

Lebanon, 2006: Israel's Failed War against Hezbollah

In 1982 the IDF failed to make Lebanon into an Israeli latifundia run by a warlord in Beirut but had succeeded with some of its other objectives. The PLO was removed from Beirut and disbanded as a fighting force and focus for resistance. Not only were its physical emplacements destroyed; its archives, plundered by the IDF, as well as related organizations such as the Center for Palestine Studies, the Film and Photography Unit, and its vast educational and welfare system in Lebanon, lay in ruins. The PLO lost its apparatus of research, analysis, ideology, and propaganda and was never able to reconstitute it, not even when it returned to Palestine in 1993.

As a result, more than half a million Palestinians were left with no effective defense or political voice in Lebanon after 1982. This was a crucial defeat for the movement, but the shocked leaders had not come to terms with its magnitude. This had consequences for the Palestinian struggle that extended well into the next century.

The End of the Civil War

Israel continued to control a large part of South Lebanon until 2000, but it could no longer manipulate Lebanese politics in the way Sharon had originally planned. Instead, the state had retreated into the complex tribal politics of minority groups, none of which held the whip. And inevitably, outside influences such as Syria returned to assume a central role in the game of musical chairs played by political parties, religious communities, and maverick politicians.

Hezbollah, like Hamas, was a new development, a religiously based radical organization, opposed to historic alliances.[1] The old Islamic fundamentalists were anticommunists and Western-oriented, and employed by the United States in its wars against the Soviet Union during the 1980s.[2] But Hezbollah faced US and Western opposition and Israeli enmity from its inception, requiring a new type of organization. The "Islamic Revolution" was a new development, affecting the whole region, through an implosion of popular anti-Western Islamic fundamentalism in the whole of the Muslim world.[3]

Born out of the Israeli invasion, Hezbollah added a crucial element to the three-dimensional chessboard that Lebanon had become. It began with a confrontation with both Amal and the Communist Party, two organizations attracting many Shiites until 1982, and politically eclipsed both, partly through its links with the young Iranian Islamic Republic. Representing most Lebanese Shias, Hezbollah had the largest population base in the country, with Sunnis and Christians also joining. Much of its appeal came from its resistance to corruption and deal-making, in a country where corruption was normalized, where politicians lacked ideology and made do with racism and ethnic bigotry. The party was committed to the Shia community, but also to the cause of the Palestine refugees, and not only in Lebanon.

Its emergence as a major political force in Lebanon shuffled the political cards. Meanwhile, the civil war continued after the agreement with Israel in 1983, and the Maronites lost much of their power and became unable to block constitutional changes. In 1989 a complicated chain of events ended with President Michel Aoun leaving for exile in France after his anti-Syrian initiative misfired badly. This led to the signing of the Taif Agreement in October 1989, ending the civil war in 1990.[4]

Fifteen years of civil strife came to an end, even if peace did not descend overnight. The Lebanese constitution, supposedly crafted to offer continuity and stability, was used to bolster the Maronites by blocking any changes that might undermine their authority; now, this stranglehold was weakened by Hezbollah's emergence. Starting as a Shia electoral platform, the party quickly developed into a multifarious institution, including a school system, clinics, and social welfare foundations based on waqf (charitable bequest) and taxation, while its military wing was superior to the Lebanese Army. Hezbollah effectively became the strongest, most durable, and cogent Lebanese institution; by the late 1980s it dominated Lebanese politics. Due to Iranian tutelage, it also enjoyed the support of the Syrian regime.

This started an unprecedented power struggle: the rightwing Maronite-controlled bloc was supported by the United States, Saudi Arabia, and (most importantly) Israel. It now faced a substantial alliance: Hezbollah and Amal, the two Shia organizations, combined with the leftwing parties, led by the Communists and the Druze forces, and supported by Syria and Iran.

Hezbollah is hardly a leftwing party and is distinct from both its Iranian sponsor and its Syrian protectors.[5] Not adhering to the Islamic state model, and accepting most of the Taif Agreement as well as the amended Lebanese constitution, the party supports the multifaith, multicultural nature of Lebanese society. Achcar defines the Hezbollah ideology as "Khomeinism adapted to Lebanese reality":

> With its increasing insertion into the Lebanese social fabric, Hezbollah exchanged the fundamentalist programme of its foundational Khomeinist inspiration for adherence of a special kind to the Ottoman-inspired principle upon which Lebanese institutions are based: the "millet" system; in it, each religious community is autonomous in organising its religious affairs and—citizenship is mediated by the religious community.[6]

Most Lebanese life is still based on such foundations, with the secular state a remote, ineffectual, hostile, or indifferent entity, offering little in the way of identification or belonging. To belong in Lebanon is to belong to the whole through one's community, and Hezbollah is a prime example of this model. By not arguing for an Iran-inspired Islamic State, the party boosted its standing among other faith groups, strengthening its social base, originally exclusively Muslim and plebeian. In such ways, Hezbollah was "Lebanized," ridding itself of the Iranian tag.

Hezbollah's central political tenet is its opposition to Israeli occupation in Palestine, Syria, and Lebanon, which has touched a raw nerve, because the Maronite and some other Christian groups marked themselves as collaborators and co-perpetrators of the brutal IDF occupation. As Lebanon arguably suffered more than any other society apart from the Palestinians, such a policy stood to attract most Lebanese, even those who did not totally support Hezbollah's religious aims. After the privations of 1982 and the bloody record of the Maronites under Bashir Gemayel, no party in Lebanon could openly collaborate with Israel, so Hezbollah became a natural magnet for public identification, strongly supported by its freedom from corruption, nepotism, or cronyism.

Like Hamas in its early period, Hezbollah heralded a new type of politics, one that many in Lebanon yearned for. It boasted a military base that was stronger, more orderly, and more active than the Lebanese Army. As opposed to the PLO, which had no roots in Lebanon, Hezbollah was thoroughly Lebanese; in an interesting sense, and despite its sectarian identification, it was more Lebanese than any other political force in the country.

This was hardly understood in the West and Israel and, arguably, this is still the case. Israel has habitually looked at Hezbollah as it did upon the PLO before: as a terror organization, which needs a violent response. Due to the dominance of Israeli perspectives, this flawed understanding of Hezbollah has infected Israel's Western allies and to a degree is still dominant in the United States, the United Kingdom, and the European Union; by proscribing Hezbollah and Hamas as terrorist organizations, such governments have effectively removed themselves from any feasible political partnership in the Arab world, especially in Lebanon, Palestine, or Syria, not to mention a real role in resolving the Palestine question.

This played into the hands of Israeli politicians, from Ehud Barak to Ariel Sharon, Ehud Olmert, and Benjamin Netanyahu, but does little to support either peace or security in Lebanon. Nor does it satisfy the interests of the Western bloc, now again ensnared in a new cold war with Russia over control of the region. While Hezbollah grew in power and influence, the ever-twitchy Maronites were more divided than ever, prone to engage in unworkable schemes and partnerships. In this context, Christian politicians who could reach a rapprochement with the Shia and Druze stood a better chance than the sectarian Maronites, as direct Israeli involvement had waned, becoming toxic.

The status quo in Lebanon was shattered by the events leading to the 2003 Iraq war. Syria's president, Bashar al-Assad, retained his father's policies when he took power. He opposed the war as a neo-imperialist adventure. This led to a retrenchment of Syria's links with Iran and with Hezbollah, which also opposed the invasion. The result was a total break with the United States, its ally Saudi Arabia, and the other forces opposing Iran's power in the region. This abrupt change created a break between Syria and the Lebanese prime minister, Rafic Hariri, which later ended with the latter's assassination. Syria's stand cost it dearly, as the United States (together with the United Kingdom) was building a front against Iran, turning its attentions in that direction after destroying Iraq and becoming the occupying power. US forces were now the immediate neighbors of Iran and Syria. Between them, Iran and Iraq harbored untold resource wealth in the form of oil, which had brought the Western Alliance back to the Middle East, now that the Soviet Union had gone, leaving a political vacuum open to exploitation.

Israel's self-appointed role as the sheriff of the Middle East was hardly new. From its involvement in the failed 1956 tripartite aggression against Egypt to its destructive war in Lebanon in 1982, it presented itself as a willing proxy for Western interests, one ready to act in opposition to the Arab world. Such willingness to serve Western objectives is not prompted by charity. Israel is the largest receiver of American and European aid and has built a strong, aggressive war economy, based on exporting military and security hardware and training.

Urban Warfare and "Operational Art"

Even before the second Gulf War in 2003, Israel was at the forefront of developing methods, apparatus, and training for urban warfare, especially after its experience in Beirut and Gaza as theaters of conflict. Clearly, Israel had more recent experience than any of the leading nations in that area of warfare and had worked hard to develop the "art of war," as its military planners have dubbed it.[7] As both Weizman and Rapaport have noted, the IDF went through a phase of being influenced by the writings of Deleuze, Guattari, Debord, and other poststructuralist theorists in the development of its fighting methods. While it is unclear what such leftwing intellectuals may have thought of this use of their work by a brutal settler-colonial regime, it is certain that the use of poststructuralism has helped to turn an already brutal army into a lethal and merciless force. This is graphically demonstrated by the officer responsible for the attack on Jenin in 2002, in a revealing interview with Eyal Weizman:

> When he explained to me the principle that guided the battle in Nablus, what was interesting for me was not so much the description of the action itself as the way he conceived its articulation. He said: "This space that you look at, this room that you look at, is nothing but your interpretation of it. ... The question is how do you interpret the alley? ... We interpreted the alley as a place forbidden to walk through and the door as a place forbidden to pass through, and the window as a place forbidden to look through, because a weapon awaits us in the alley, and a booby trap awaits us behind the doors. This is because the enemy interprets space in a traditional, classical manner, and I do not want to obey this interpretation and fall into his traps ... I want to surprise him! This is the essence of war. I need to win ... This is why we opted for the methodology of moving through walls ... Like a worm that eats its way forward, emerging at points and then disappearing ... I said to my troops, 'Friends! ... If until now you were used to moving along roads and sidewalks, forget it! From now on we all walk through walls!'[8]

Weizman examines this bizarre adaptation of poststructuralist concepts and terminology in an attempt to uncover the functions of such an unlikely departure:

> Naveh, a retired Brigadier-General, directs the Operational Theory Research Institute, which trains staff officers from the IDF and other militaries in "operational theory"—defined in military jargon as somewhere between strategy and tactics. He summed up the

mission of his institute, which was founded in 1996: "We are like the Jesuit Order. We attempt to teach and train soldiers to think. ... We read Christopher Alexander, can you imagine? We read John Forester, and other architects. We are reading Gregory Bateson; we are reading Clifford Geertz. Not myself, but our soldiers, our generals are reflecting on these kinds of materials. We have established a school and developed a curriculum that trains "operational architects."

In a lecture Naveh showed a diagram resembling a "square of opposition" that plots a set of logical relationships between certain propositions referring to military and guerrilla operations. Labelled with phrases such as "Difference and Repetition—The Dialectics of Structuring and Structure," "Formless Rival Entities," "Fractal Manoeuvre," "Velocity vs. Rhythms," "The Wahabi War Machine," "Postmodern Anarchists" and "Nomadic Terrorists," they often reference the work of Deleuze and Guattari.[9]

Naveh, influential within the IDF before 2006, invented a new conflict and war theory based on philosophical writings in the second half of the twentieth century, defining his own art form, "operational art":

> Operational Art constitutes the operational principles of a military campaign. This concept reflects and incorporates fundamental tenets of the use of force and provides a common basis for commanders and soldiers for a better functioning of the complex army institution.[10]

How such elevated "arts" were practiced in the Nablus refugee camp can be gleaned from the testimony of a woman living in the camp:

> Imagine it—you're sitting in your living-room, which you know so well; this is the room where the family watches television together after the evening meal, and suddenly that wall disappears with a deafening roar, the room fills with dust and debris, and through the wall pours one soldier after the other, screaming orders. You have no idea if they're after you, if they've come to take over your

> home, or if your house just lies on their route to somewhere else. The children are screaming, panicking. Is it possible to even begin to imagine the horror experienced by a five-year-old child as four, six, eight, 12 soldiers, their faces painted black, sub-machine-guns pointed everywhere, antennas protruding from their backpacks, making them look like giant alien bugs, blast their way through that wall?[11]

For a number of years, this disturbing "philosophical" worm has burrowed its way into the heart of the IDF, directing not just analysis, but also praxis. "The military's seductive use of theoretical and technological discourse seeks to portray war as remote, quick and intellectual, exciting—and even economically viable. Violence can thus be projected as tolerable and the public encouraged to support it."[12]

This weird narrative has allowed the academic reification of murder and destruction in presentations by IDF officers at academic conferences and the conferment of university degrees in War and Conflict Studies. Such exercises further erase and normalize what are clearly war crimes, turning them into quotidian academic discourse. It reminds us, however, of the importance of "concepts" in IDF planning, and the role these played in earlier wars, such as in 1973. It is even more striking in 2006, as the defeat was caused by a small organization such as Hezbollah.

This process of modernizing the IDF included a whole new lexicon, also indebted to the 1991 and 2003 Iraq Wars with their notable linguistic innovations, such as "Shock and Awe" and "smart bombs." A spectrum of terms was introduced that spoke of the effects of the operation:

> Examples of the "effects" discussed were decapitation—attacking the enemy's leadership—since it alone sets the goals of the system and is authorized to alter them, and blinding—attacking the enemy's communications lines and senses—with the aim of denying him the knowledge of what is happening on the battlefield. In addition to "effects," including "paralysis," to be caused by massive carpet-bombing, new concepts developed in the IDF that were derived from systems analysis, such as "system idea," "levers," "consciousness," "breaking the enemy's logic," "designers," and so on.[13]

Interestingly, especially in light of close ties between Israel and the Islamic State of Iraq and the Levant (ISIL) in southern Syria, terms chosen were of bodily harm, disabling, and disfigurement. Arguably, there is a close relationship between "decapitation" and practices of beheading by ISIL, although Israel apologists would no doubt deny such obvious connections. The strange wedding of poststructuralism, systems analysis, business studies, and globalization with aggression could be described as one of the more violent, disturbing, and unique creations of Israeli culture, in turn shaped by the IDF. The work of pseudoacademics such as Naveh has been crucial for building the ethical foundation for mass murder, and as such, is a prime example of the integrative powers of Israeli society, the ability to recruit all layers of the polity, including intellectual workers—authors, poets, academics, researchers, media workers, the business community. War is the only activity in Israel that brings its Jewish citizens truly together. In this Israel is not unique; such are the achievements of all totalitarian regimes.

Another related "theoretical development" in Israeli military thinking, of a very different nature, has also taken place during the same period, changing the nature of planning and preparing battles. As armaments have become more and more technologically inclined and dependent, not to mention more expensive, the IDF, like some other dominant armies, found it needed to defend such spending as crucial to victory in the battlefield. Israel had a history of affecting battlefield results through massive use of air power, ever since the 1967 war. While the position of the chief of General Staff had never been filled by an airman, this changed in 2005, when General Dan Halutz, ex-commander of the air force, was promoted to that position. The appointment itself was premised on a paradigm shift in Israeli military analysis—the move toward a battle plan that was founded on air power as the main element of achieving a decisive victory:

> The first concept, which held an important place in the IDF before the war, was that the importance of the air force was growing dramatically while that of the ground forces was becoming negligible. Some even believed that the next war could be won using air power alone, despite the fact that there was no clear historical precedent for an "aerial victory."[14]

Meanwhile the aerial victory remained an IDF delusion; bombing was having a real effect in Lebanon. In his war diary, Rami Zurayk noted on July 15, a few days after the war started:

> Things were relatively calmer last night, and the lull is continuing. This is a typical Israeli tactic: bomb civilian areas causing as many casualties as possible, as if to say: "we mean business" and then to give some time for a few civilians to get out of the target areas with little more than what they can carry in a bag. Then destroy everything. This is what they did during the 1982 invasion.[15]

The Israeli Air Force could not "win" the war, but it could certainly destroy the life of millions of Lebanese civilians. This tactic doubled the budget of the air force, reducing funding for ground forces, and it became crucial to inflate the role and importance of the air force in order to justify the budgetary priorities.[16] Thus, a major general who has never commanded ground forces now become responsible for their operations.

Ironically, while Naveh's operational arts were gaining credit and support for Israeli outlawed operations, with Halutz promising easy victory without IDF losses, both seemed to have totally failed the IDF in the 2006 war on Hezbollah. Naveh himself admits that his contribution was somewhat problematic: "The war in Lebanon was a failure and I had a great part in it. What I have brought to the IDF has failed."[17] That the war was a failure did not mean that it was wrong to fight it, according to this hallowed Deleuzian strategist, but that his contribution did not help to win it. He and his team seem to be morality free in the most ethical tradition the IDF is capable of.

The War Starts

The trigger for the war, which was a long time in preparation (as it was in 1982) was almost too trivial to mention: an Israeli army patrol in Lebanese territory was ambushed by a Hezbollah force, resulting in a handful of fatalities. The force's mission, however, had been to abduct some soldiers alive for use as bargaining chips in the complex game of prisoner exchanges. Bearing in mind the great number of Hezbollah personnel regularly abducted by Israel, and the fact that they could hardly expect a proper judicial process once captured, it is hardly surprising that abducting Israeli soldiers became important for Hezbollah.

In preparation for this operation, the Hezbollah leadership did not remotely suspect that Israel would use this as their pretext to attack.[18] Hassan Nasrallah admitted, in an interview on August 27, 2006, that Hezbollah had miscalculated:

> We had not foreseen, not even to one-hundredth, that the hostage taking would lead to a war of that scope. Why? Because of several decades of experience, and because we know how the Israeli acts, it was not possible that a reaction to a hostage taking reaches such proportions, especially in the middle of the tourist season. ... If I had known that this abduction would lead to a war of such a dimension with one per cent probability, well, we would certainly not have done it.[19]

But Israel chooses its time and place, defying logical arguments or proportionality. In that sense, Hezbollah could do nothing to avoid the war, no more than Arafat could in 1982. Israel sought war and it now needed an excuse; and anything would suffice. That the attack was premeditated and long in preparation is clearly shown by comments from Ehud Olmert, then prime minister, and General Yossi Kuperwasser, then head of the research division of Military Intelligence, quoted in detail by, among others, Achcar and Warshawski.[20]

As the IDF now depended mainly on airpower, the operation was launched almost immediately with heavy bombing, shelling, and missile attacks. Indeed, the reserves were not even mobilized; the IDF command believed that the decisive effect of airpower made the use of ground forces unnecessary.[21] This proved to be an egregious error. Israel's declared war aims included "decapitation" of Hezbollah command structures, freeing the abducted soldiers, and changing political realities in South Lebanon, to impede growth of future opposition to Israel. In a sense, this was a replay of 1982—"correcting" its results.

The Gulf between Plans and Reality

Israel planned to turn "Hezbollah's mass base among Lebanese Shiites against the party,"[22] through massive bombing of the whole of Lebanon as well as inciting the Lebanese Army to fight Hezbollah. This involved dropping leaflets across Lebanon and destroying much of Lebanon's infrastructure. Instead, thousands were killed by bombing and shelling in the Shia villages of the South. In order to complete the misery of the Lebanese population, Israel implemented a total sea, air, and land blockade, making daily life almost impossible.

The gulf between what was planned and what was achieved was shocking. This was noted by the "number of commissions of inquiry set up in Israel, some of them under public pressure."[23] Indeed, the judgment of the army itself was most damning:

> The quiet prevailing on the northern border since August 2006 cannot change the fact that the Second Lebanon War was an abject failure. The sense of failure, stems from the fact that, Israel, with all its military and economic power, did not defeat Hizballah in the military confrontation nor did it score even partial achievement against the organization ... The sense of failure also comes from the fact that it did not achieve its publicly stated goals at the outset of the war, particularly the return of the kidnapped soldiers and the disarming of Hizballah in line with Security Council Resolution 1559.[24]

Indeed, the Winograd Commission was even harsher: "In most of the cases and the areas, the army displayed impotence vis-à-vis Hizballah in its ground operations."[25] Such words are poignant especially in view of the IDF terminology: "decapitating," "blinding," and "causing paralysis." However, the Israeli ground forces, used to the safety net of the Israeli Air Force and without rivals in the region, were quite useless in battle; indeed, they sometimes gave the impression of being blinded and decapitated themselves. Though arraigned against a tiny militia with less than 1 percent of the IDF firepower, they were seemingly powerless. They had the latest armor, but the basics were missing:

> Particularly vexing, however, was the fact that most of the sophisticated *Merkava IV* tanks were not equipped with basic protection against anti-tank missiles—from smoke canisters to camouflage. Because of cuts in the training budget in the years preceding the war, the tank corps did not train the soldiers in the use of canisters. Subsequently, the corps decided that, since the soldiers did not know how to operate the canisters, there was no reason to install them on the newly produced tanks. Of course, the root of the neglect lay in the concept that the ground manoeuvre belonged to wars of the past.[26]

Such IDF negligence was not unique to the 2006 war, but it had never been so widespread and pervasive.

The context is all-important. Ever since 1979, things seemed to be going Israel's way: the Begin–Sadat peace treaty, the destruction of the Iraqi reactor in 1981, the expulsion of the PLO in 1982, control of South Lebanon, the defeat of the first intifada, not to mention the signing of the Oslo accord without granting the Palestinians a single concession. The occupation running costs were now covered by the European Union and the United States. So by 2005, having pulled out from the Gaza Strip in order to lower the cost of the occupation and with Arafat a distant memory, having been (in all likelihood) poisoned in 2004, Israel had become the most stable, powerful, and wealthy nation in the Middle East, with a superior army in control of Palestine, with Syria and Lebanon unable to act. In this context, it was argued, Hezbollah was merely an irritating appendage of Syria and Iran—akin to mosquito bites that a lion ignores.

The first buzzing could be heard during the election in Palestine in 2006, after years of the corrupt Palestinian Authority doing Israel's bidding. The election was comparatively free, the first and last in Palestine. But conflict was stoked by Israel, which failed to see the coming disaster and fully trusted the Palestinian Authority under Abbas to win handsomely. When Hamas unexpectedly won, Israel now pushed Abbas and Fatah toward armed conflict with it, annulling the election results. Furthermore, Hamas lawmakers were imprisoned by Israel and the Palestinian Authority.

Ehud Olmert, who came to power after Sharon suffered an incapacitating stroke in January 2006, had no defense and security background, and his defense minister, Amir Peretz, was similarly unqualified. As a result both were prisoners of the IDF. Israel was in a febrile mood, and old plans to invade Lebanon were revived, a sure way of winning support for a lame government.

But the IDF was now a very different body from the victorious army in 1967; it had fought no wars against an Arab army since 1973. Since then, it has attacked civilian populations under siege or resistance movements like Hezbollah.[27] Many soldiers, especially reservists, understood that wars fought since then had been expansionist wars with crude political agendas. A new public perception had started to spread, dividing wars into "no choice" wars, ones with supposedly an existential raison d'être for the Zionist state, and "wars of choice" (*Milhemet Brera*), fought to support a political leader.[28] While cases of refusal to serve were on the rise, they never reached worrying proportions, and as the IDF was and remains the largest and strongest military force in the Middle East, such cases were dealt with through the IDF court system, without seriously weakening the army or its social narrative.

Under the surface, tensions were mounting and dissatisfaction was rife. Many young soldiers were unwilling to risk their life for a politician's sake. As the war against Hezbollah developed, with most Israelis believing it might be a short campaign (conveniently forgetting how long the 1982 war had lasted), there existed a lackadaisical attitude in the IDF, which produced the resulting chaos.

Unlike Hezbollah, which was fighting to protect Lebanese land and homes, the IDF's incursion was a colonial adventure, and soldiers understood this. What they had not anticipated was the failure of the various branches of the destructive IDF machine to break Hezbollah's fighting spirit. Israel was surprised by the agility and inventiveness of Hezbollah, its ability to fire large numbers of missiles, and its efficient defense system, as well as its fighting abilities; its highly motivated fighters managed to destroy the new Merkava IV tanks, supposedly indestructible.[29]

This was a most serious failing of Israeli military intelligence, as Achcar notes; this "is not the first time Israeli intelligence was caught with its pants down: it was surprised in 1973 by the large-scale offensive launched by the Egyptian and Syrian armies."[30] This may have happened due to IDF's inability to understand the motives, methods, and commitment of a religious resistance movement such as Hezbollah (or indeed, Hamas); it was thinking in terms of state armies—a failure of both operative and conceptual frameworks, difficult to overcome by a powerful military with enormous resources, with the unconditional support of the United States. Achcar locates the failure accurately ("it relates to colonial arrogance"[31]), of which the IDF is never short.

Of the many debacles, an especially challenging one was the failure of the Israeli Air Force to end the resistance in a few hours, as promised, or even in a whole month. It came as a deep shock to commanding officers and clueless politicians who sent them into battle. In consequence the IDF was forced to deploy large numbers of uncoordinated ground troops to try to liquidate the resistance, searching for ways out of the trap they had invented for themselves. Moshe Arens, a rightwing ex-defense minister and an acknowledged expert, wrote about Olmert, Peretz, and Livni a week into the war: "Here and there, they still let off some bellicose declarations, but they started looking for an exit—how to extricate themselves from the turn of events they were obviously incapable of managing."[32] This was especially hurtful and an acknowledgment that this was one war Israel could do without, even from a rightwing perspective.

But the bellicose trio was unable to extricate itself from the impasse, despite massive support from Washington, forcing the IDF to bolster the brutal, indiscriminate shelling and bombing,[33] further enraging international public opinion, and turning Lebanon against Israel ever more decisively. The Lebanese government actually started to convey Hezbollah's demands and positions to the UN, frustrating Israel's aim of dividing the country.[34] Pressure for an immediate cease-fire now threatened the continuation of the carpet bombing by the Israeli Air Force.

Nevertheless, at the UN, many resolutions were crafted but did not make the voting stage, because they failed to garner support from member states. As the ground offensive by the IDF badly failed and Hezbollah's countersheling of northern Israel, using long-range rockets, caused many deaths (ironically, mostly in Palestinian towns), the pressure by the West and especially the United States brought about Resolution 1701 on August 11. The cease-fire started the following day.[35]

Resolution 1701, proposed by the United States, was anything but balanced and had Israel's fingerprints all over it.

> It fails to condemn Israel's criminal aggression, mentioning only "Hezbollah's attack on Israel" and "the hostilities in Lebanon and Israel" [*sic*]. It demands that Israel cease its "offensive military operations" without even demanding the lifting of the blockade that it was imposing on Lebanon—as if a blockade were not a particularly offensive military operation in itself. And, worse still, the new UNIFIL—which, remarkably, was deployed only on the territory of the occupied country—was supposed to ensure that its zone of deployment was not used for "hostile activities of any kind": Resolution 1701 says not a single word about the protection of Lebanese territory against aggression by Israel, an occupying power in Lebanon for eighteen years.[36]

While few may be under the impression that the UN is fair and balanced, even at the best of times, this obvious injustice—turning victim into aggressor, with the aggressor deemed the "real" victim—was well beyond even the warped standards of the UN. In effect, the UN (in fact, the United States) had stepped in to achieve on behalf of Israel what the brutal attack on Lebanon failed to do. This, after more than a month of murder, destruction, and mayhem, there were over a million refugees, some abroad, and thousands whose homes had been destroyed who were unable to return. When a critical history of the UN is written at some future date, this resolution may appear as its moral nadir, together with the 1947 Resolution 181.

For Achcar, Resolution 1701 started a "continuation of war by other means."[37] As in 1982, what Israel could not achieve through war, it attained through diplomacy. Despite this, Hezbollah remained the strongest party in Lebanon, the only genuine mass movement. After its misapprehension about Israel's reaction to the snatching of soldiers, Hezbollah has been extremely careful not to hand Israel a pretext for another destructive invasion. Israel may choose to invade anyway, if it considers it to be practical, so Hezbollah cannot stop the next war from taking place; but it certainly is ready to inflict a high cost on Israel, higher than in 2006. Its military training and materiel are superior to that of the Lebanese Army, and since then, Iran and

Syria have supplied it with more missiles capable of reaching most parts of Israel.

The Efficacy of Hezbollah and the Damage in Israel

At the end of military activities in mid-August 2006, it certainly seemed that Hezbollah was the only effective organization in Lebanon, while the Lebanese government and the majority parties were mired in typical inaction, parroting Israeli propaganda against Hezbollah.[38] Nasrallah, though, was systematic in his approach:

> First, he promised all those who have lost their houses to start on their reconstruction at Hizbullah's expenses [*sic*]. The Resistance would also give the equivalent of one year's rent to all those people whose house had been totally destroyed. He also called for people, engineers and others, to volunteer their time and for traders of construction materials not to hoard materials.
>
> The second thing he said was in a very serious tone: It was shameful for people to start talking about laying down arms "when Israel still occupies our lands, and that there were people who have been trying to give Israel more than she has asked for." He finally added that we have first to build the nation in which we all feel secure, and then ask for weapons and not the other way around.[39]

Hezbollah was the only political force in Lebanon taking urgent social and political action when such was required, rather than staking meaningless, sectarian positions. Nasrallah was careful to act as a government, where no government has presented itself. Having damaged Israel and brought the war an end, Hezbollah was now attending to the rebuilding of Lebanon.

Israel also faced rebuilding; the cost in lives and property destroyed during the short conflict was enormous. It was the first war where civilians paid a higher price than the IDF. Northern Israel became prey to Hezbollah rockets, with normal life and work ceased until the end of the war, and people permanently trapped in shelters. Haifa, the third-largest city in Israel, was targeted the last day of the war. Hundreds of buildings were destroyed and almost a million Israelis sought refuge in the south. The poorest, unable to leave, were completely forsaken by the state.[40]

This was also pointed out by Uzi Rubin, the director of the Israeli program for missile defense within the Ministry of Defense, in a special report of the rocket attacks in 2006:

> When the outbreak of Second Lebanon War on July 12, 2006 elicited the anticipated reaction [Hezbollah missiles fired into Israeli territory] Israel's effort to block the attack, or even to lessen the severity of the damage incurred, proved, on the whole, almost as futile as in the 1980s and 1990s. The 33 day long rocket attack in the summer of 2006 was more extensive, more lethal and reached deeper into Israel than any of its predecessors.[41]

Rubin also pointed out that Hezbollah had "managed to sustain a lethal rocket campaign through the entire war, its launchers falling silent only when Israel ceased its military operations."[42] He further notes the "remarkable ingenuity on the part of the Hizbullah in terms of their preparation and operation."[43]

The research described concealed missile launch systems enabling Hezbollah's uninterrupted operations throughout the war, despite sophisticated Israeli locating technologies and spy satellites. It is clear that Hezbollah had internalized the lessons of the Viet Cong campaign, fighting a superior enemy with enormous resources, and had developed its own solutions in the Lebanese context. In the event, not only did Hezbollah survive the war relatively unscathed, but also exhibited better fighting strategies than any of the armies involved in the conflict.

What is emerging from recently published research is the depth of the Israeli debacle. Hezbollah's use of short- and medium-range missiles and rockets had harmed Israel in ways it has never before faced. Around 200,000 fled to other parts of Israel—some never to return. "About 2,000 dwellings have been destroyed or severely damaged"[44] as well as many civilian infrastructure facilities of all kinds.

> An estimated one million Israelis were compelled to stay in or near shelters during the entire campaign. The towns that suffered the most hits, Kiryat Shmona, Nahariya, and Safad, turned into ghost

> towns, with many of the remaining inhabitants living in communal shelters, public services barely functioning, no traffic lights, and a complete cessation of commercial activity.[45]

Many of the IDF's failings, as well as successes of Hezbollah, were clear early in the war, but gave rise to little Israeli protest in comparison to 1982, when the anger that erupted after the Sabra and Shatila massacre led to the sacking of Sharon. Achcar and Warshawski note that due to deep processes transforming Israeli society since 1982, widespread protest against unjust wars was no longer likely or even possible.[46] The whole social register had moved to the right, and in the wake of September 11, 2001, Israel had reconceived itself as the sharp end of the struggle of "Western civilization" against "Islamist terrorism."

Indeed, this position is recognized by the unlimited support the nation enjoys not just from the United States but from Western and European nations as well. The role of Israel as the Western regional guard dog has been imbibed and internalized by most Israelis, with the terminal decline of the liberal left in Israel, and expansion of extreme rightwing parties, which have ruled Israel for decades. This meant that during the fighting, there was no large-scale public action against the war. Olmert knew he was safe with most of the Jewish public, including the traditional "left," which offered him unwavering support. This situation still pertains and if anything, public support for wars and virtual war crimes has increased in ways never seen in Israel.

Nonetheless, public anger over the miserable failure of the offensive fuelled a political storm once the IDF had left Lebanon. Journalists were baying for blood, as were Olmert's political opponents, in his party and others. Criticism was not slow to emerge:

> Olmert failed in setting the initial goals of the war, in comprehending the implications of the military moves, in the freedom of action he continued to grant the General Staff despite seeing before him how its expectations disintegrated, and in authorizing a ground offensive on the eve of a cease-fire agreement. This is enough for any decent person to conclude that the position of prime minister is simply too much for him.[47]

It also seems that the public, waiting quietly while the army fought and failed, was persuaded that the war was not only a failure, but totally unnecessary—they were angry with the politician who had led them into this quagmire.[48]

This was true even during the fighting. Leading journalists and editors started calling for the IDF to leave Lebanon immediately. Nachum Barnea, a senior columnist of the daily *Yedi'ot Ahronot*, demanded that the IDF retreat without delay, in an article headed by the provocative title "Run, Ehud, Run." "Barnea recommended that Olmert cut his losses and 'hightail it' out of Lebanon. The IDF, he wrote, was not winning the war and would not win it in the coming days."[49] This article, coming almost four weeks into the bungled war, influenced public opinion against Ehud Olmert.

Nonetheless, Olmert continued hedging his bets, vacillating between supporting a ground offensive into Lebanon (suggested by most of the generals from the first day of the war) and delaying it, due to the conflicting view held by the Chief of General Staff. Major General Halutz had placed total trust in the air force completing the task, despite all evidence to the contrary.

Only a day later, another influential journalist, Ari Shavit, "published a response in *Haaretz* to Barnea's column of the previous day: "Olmert Has to Go." According to Shavit, the prime minister

> is allowed to decide on unconditional surrender to Hezbollah. This is his right. Olmert is a prime minister that the press invented, that the press defends, and whose regime the press protects. Now the press says to him "get the hell out." This too is legitimate. Not smart—but legitimate. But one thing must be clear—if Olmert leaves during the war that he initiated, he cannot continue as prime minster for one day more. There is a limit to his chicanery. You can't lead the whole nation into war with the promise of victory, and then suffer a humiliating defeat and remain in power.[50]

As things turned out, Shavit's challenge was optimistic—Olmert would remain in power until 2009 and leave office not because of war crimes he had committed, but for corruption charges.

The ground offensive was launched at the worst possible moment, after thirty-one days of bombing and shelling that had not broken the fighting spirit of Hezbollah. In the north of Israel, the population was forced to stay in their shelters for four weeks; destruction was widespread, and scores of civilians and soldiers were killed. On August 6, 2006, a Hezbollah rocket landed on kibbutz Kfar Giladi killing twelve soldiers, and later the same evening a deadly rocket attack on Haifa, the largest town in the north, killed three civilians. A disproportionate number of victims in northern Israel were Palestinian, due to the lack of adequate bomb shelters. Olmert's war has led to a critical deterioration of daily life in Israel, rather than an improvement: "The IDF spokesman's surveys also showed that public confidence in the army, which had peaked after the first week of the war, had plummeted from grade C to grade F."[51]

Indeed, this information, fed to the cabinet, may have weighed heavily in the decision to reject the efforts of Secretary of State Condoleezza Rice to bring about an immediate cease-fire on terms dictated by Israel. The cabinet was fighting for its political future, and in that game the lives of Lebanese, as well as Israelis, were small change. They could not afford to leave Lebanon without some kind of "victory," even an illusory one.

Olmert decided on a ground offensive, which stood no chance of success, rather than face accusations of chickening out. An attack that needed some weeks to fulfil its objectives started hours before the UN was due to vote on the cease-fire agreement. By the time it was debated, the ground forces were in more trouble than before, getting nowhere fast. In one spot, the Saluki River crossing, eleven officers and soldiers died in less than an hour, including two company commanders, and more than fifty were wounded by accurate Hezbollah fire. In the last two days of the war, thirty-three officers and soldiers were killed and many more wounded.[52] Not only were the Israeli ground forces nowhere near achieving their objectives, they were proving that Hezbollah fighters were better motivated and trained and that they could not be defeated. The IDF needed out fast.

The Aftermath

The result of this brutal and useless war was an object lesson. More than 1,200 Lebanese civilians and foreign nationals were killed by the IDF bombings, and 4,400 were injured, while the whole infrastructure lay in ruins, just over two decades after the 1982 destruction. More than a million people became refugees; many found their way abroad through Cyprus, some never to return. Hezbollah lost between 260 and 400 fighters, depending on whose numbers one wishes to trust, and an unknown number of civilians in the south disappeared. In Israel, forty-five civilians died and 1,500 were wounded. Thousands of properties were hit, many totally destroyed. The IDF, which initiated this conflagration, lost 120 officers (some very senior) and soldiers, and 1,244 wounded. The Lebanese army itself had fifty soldiers killed and 100 wounded.

The figures certainly tell the story of an IDF debacle, but the failure was felt on many levels, well beyond the military establishment. The war had clearly been gratuitous, an ineptly planned and executed exercise of self-delusion by Israel's military and political leadership. Lest it be thought that this judgment is biased, one need only examine the verdict of some of the more than fifty official commissions of inquiry that looked into this war in Israel. (It is indeed astonishing that there were more than fifty such investigations.[53]) It is beyond the scope of this chapter to delve into such inquiries. It will suffice to quote sparingly from the main investigation, the official Winograd Commission of Inquiry, reporting in late April 2007. While one cannot suspect the commission of being unbiased or hostile to the IDF, it is fascinating to read its conclusions, especially when one considers how hard Olmert had pressed to make the report "acceptable": "Decisions surrounding going to war, said the commission, 'entailed the worst kind of mistakes,' the responsibilities for most of which lies firmly with Olmert, Peretz, and Halutz."[54] Among the conclusions of the Winograd Commission was the following:

> The decision to respond with an immediate, intensive military strike was not based on a detailed, comprehensive and authorized military plan, based on careful study of the complex characteristics of the Lebanon arena. A meticulous examination of the characteristics would have revealed the following: the limited ability to achieve military gains having significant political-international weight; a military offensive would inevitably have led to missiles being fired at Israel's civilian north; there was no other effective

> military response to such missile attacks than an extensive and prolonged ground operation to capture the areas from which the missiles were fired ... [The high price of this] did not enjoy broad support. Cabinet support for this move was gained in part by the use of ambiguity in presenting goals and ways of achieving them, which made it easier for ministers with different or even contradictory attitudes to support them. The ministers voted for a vague decision, not understanding or knowing its nature and implications. They authorized ... a military campaign without considering how to get out of it ... Some of the war's declared goals were vague and unachievable.[55]

Most of the above could have also been taken straight from the 1982 inquiry report. It seems that neither the military nor the civilian elite in Israel were capable of thinking outside the box of military violence. Bearing in mind this was a cover-up, it is hardly surprising that it left Olmert intact as prime minister; the IDF, too, survived unchallenged, despite all the detailed criticisms directed at it. One can assume that if and when Israel is again faced with the same or similar choices, it is programmed to make analogous mistakes.

When one examines the war in hindsight, it seems clear that for Olmert and the IDF, the need for the war also stemmed from Israel's part in US strategic aims in the Middle East, which assigned Israel certain roles, in return for generous and unquestioning support. With Iran supposedly becoming the main adversary of the United States in the region after the demise of the Soviet Union, the United States was looking to Israel to "sort out" Hezbollah as Iran's proxy, as noted by Krauthammer:

> The defeat of Hezbollah would be a huge loss for Iran, both psychologically and strategically. Iran would lose its foothold in Lebanon. It would lose its major means to destabilize and inject itself into the heart of the Middle East. It would be shown to have vastly overreached in trying to establish itself as the regional superpower. The United States has gone far out on a limb to allow Israel to win and for all this to happen. It has counted on Israel's ability to do the job. It has been disappointed.[56]

While Iran is not in any way comparable to the Soviet Union of pre-1989, it is a substantial foe, well-armed, motivated, and organized, well-resourced and with combat experience. That Israel had failed so badly, not in a confrontation with Iran itself, but with its small proxy in Lebanon, fighting a guerrilla war, was indeed a bitter disappointment to Washington, and some in the State Department must have questioned the wisdom of the investment.

For the Lebanese, the shock was even sharper than in 1982. The country's infrastructure now lay in ruins again; its main cities, having struggled to rebuild after the earlier devastation, were back at the starting point—whole neighborhoods flattened, a million and a half internally displaced citizens, many who had lost their homes. The war-weary Lebanese felt the Sisyphian effort of rebuilding had been dealt a mortal blow. Beirut, at one time the nerve center and media hub of the Arab world, may never return to this unrivaled position. With much of the Arab world recently devastated by war, Beirut joins other Arab capitals in a tale of woe and ruin. The United States has had center stage in this landscape of failed states, internecine conflict, and the destructive agenda of global control, with Israel the local agent of this project.

Ironically, Israel failed in Lebanon for the same reasons that the United States and its servile allies failed in Iraq in 2003 and ever since. Through an analysis built of little knowledge and even less understanding of the region despite massive resources expended, with US objectives of controlling the Middle East militarily overriding any other approach, using Israel seemed the only way of achieving a strategic defeat of Iran. Countries have the choice of either agreeing to US/Israel demands or facing subjugation, destruction, and occupation.

In this monochromic landscape, there is no space for other modalities of action and no role for dialogue. The willingness to dialogue is conceived as a weakness. In many ways there is a direct link between the aims of the failed Lebanon war in 2006 and the impasse brought about by Donald Trump, with his refusal to extend the Iran Nuclear Agreement negotiated by the Obama administration and the re-imposition of sanctions on Iran. The methods have not changed, the target is the same, and the depravity just as prevalent. One may assume that this method will produce the disappointing results it did before.

As in 2006, the United States is reluctant to employ US troops and prefers its proxies, Saudi Arabia and Israel, to do the actual fighting against Iran, and face the repercussions and obvious risks thereof. Thus, it seems that the United States has also learned very little from its incursions and wars in the Middle East and keeps repeating old mistakes in new guises—the blind are leading the blind. To have this strong military and political partner supporting Israel in anything it may wish to enact is likely to lead only to more bloodshed, more unconsidered military action, and widespread destruction of the kind we see already in Iraq, Syria, Yemen, Afghanistan, and may also witness in Iran, as the two accident-prone partners prepare their next military adventure.

PART II

THE ARMY AND ITS STATE

8

The Armed Settlements Project

Today, more than ever, settlement in the territories endangers Israel's ability to develop as a free and open society.

—Zeev Sternhell

Most Israelis still believe that the 1967 war was forced upon them and that Israel had faced an existential threat, despite clear evidence to the contrary. Recent publications on the fiftieth anniversary of the war have clarified that not only did Israel never face such a threat, but it was indeed Israel who initiated the war. After describing the main points of the Israeli narrative about the war, historian James North notes:

> This Mainstream Narrative remains unchallenged in the popular imagination, 50 years later. Just the other day, a *New York Times* reporter stated as fact that in 1967, "Israel defied annihilation by its Arab neighbors."[1]

This was pure invention, but this false formula keeps popping up. Patrick Tyler has argued that the decision in 1967 to launch a preemptive war was hardly a reaction to the moves of Israel's enemies.

For Tyler, it represents the continuation of the provocative policy of escalating tension toward a war that would correct the 1948 "errors." For Tyler, it also represents the decisive win of the generals against the politicians—what he calls a move from the "rise of the generals" toward "military revolt"[2]—a quiet takeover of the political agenda away from politicians like Levy Eshkol and Moshe Sharett

> who believed in a broader devotion to statesmanship as a means to avoid war and resolve conflicts through diplomacy, negotiations and compromise. Instead, with its pre-emptive war on June 5, 1967, Israel set a precedent that is still followed a half century later.[3]

This began many decades before 1967. Nur Masalha reminds us that during a discussion in 1938 (when Zionism controlled a mere 7 percent of the country), Ben-Gurion noted:

> This is only a stage in the realization of Zionism and it should prepare the ground for our expansion throughout the whole country through Jewish–Arab agreement ... The state, however, must enforce order and security and it will do this not by moralizing and preaching "sermons on the mount" but by machine-guns, which we will need.[4]

Arguing that the Arabs already have many states, Ben-Gurion repeated at a confidential meeting on December 17 of the same year that the whole of Palestine belongs to the Zionists.[5]

Benny Morris agrees with this evaluation:

> Zionist mainstream thought had always regarded a Jewish state from the Mediterranean to the Jordan as its ultimate goal. The vision of "Greater Israel" as Zionism['s] ultimate objective did not end with the 1948 war. The politicians of the Right, primarily from the Revisionist Herut Party, led by Menachem Begin, continued throughout 1949 and the early and mid-1950s to clamour publicly for a conquest of the West Bank.[6]

Such deep-seated convictions led to the 1956 war and the conquest of Sinai—simply because this was an expansion that Britain and France could agree to. The real target was the West Bank, and this depended on an enabling historical juncture. In 1967, Nasser unwittingly handed Israel an opportunity to realize its territorial imperative. The generals, now firmly in control after Dayan's appointment as defense minister, saw an opportunity and were not about to waste it.

~

The victory in 1967 was a surprise for the Israeli public, frightened by the waiting period before the war and by political talk of extinction and existential danger. This public, shocked and surprised by the victory its army had achieved over the Arab armies, had taken its eye off realities in the general jubilation that ruled the day.

Not so the Israeli government. Less than two weeks after the war ended, the cabinet made a number of clearly irreversible decisions: the whole Haret al-Maghariba (Moroccan Quarter), the neighborhood bordering the Wailing Wall, was razed to the ground with thousands of Palestinian inhabitants thrown out to make way for the massive open piazza now abutting the Wall. This was clearly an illegal action, not to mention immoral and inhuman. The inhabitants were given a few hours to leave their homes and never received compensation or alternative housing. But this was a sign of things to come: in the next five decades it provided a model for Israel's occupation regime in the Occupied Palestinian Territory (OPT): a regime in defiance of the Geneva Conventions, the relevant UN resolutions, and basic human rights of the Palestinian population.

Immediately following the destruction of the Moroccan Quarter, Israel annexed East Jerusalem, a move that was universally condemned. Another pattern was established: after such violations by Israel, international players, such as states and international organizations including the UN, the Red Cross, UNICEF, and UNESCO voiced verbal or written condemnations but did not follow up with action when their protest was ignored. Not even the mildest sanctions were considered to compel legal behavior. This normalization of the occupation and its iniquities enabled Israel to not only continue the occupation for five decades but also to expel a quarter of a million Palestinians, to incarcerate the rest of the population in less than 10 percent of the territory of historic Palestine, to deny human rights, political rights, the right to employment, education, property, and water, and the right to free movement, as well as all other basic human rights.

Since then, Israel has built an enormous network of illegal settlements, covering the OPT with more than 750,000 Jewish settlers and a network of roads that exclude Palestinians. It took control of all resources, periodically denying Palestinians water and electricity, and limited the supply of food and medicines for long periods.[7] It slaughtered many thousands of Palestinians for no apparent reason.[8] It installed hundreds of checkpoints disabling daily life and destroyed Palestinian infrastructure—water, electricity, telephone lines, and roads. To cap it all, it has built the longest and largest separation wall in modern times, one clearly visible from outer space. The separation wall is the most visible trace of the policy of the occupation—the separation of the Palestinians from their land.[9]

None of the many crimes committed by Israel in Palestine are incidental or arbitrary; such behavior is calibrated to make life in the OPT insufferable, to force a gradual emptying of the land (in other words, the continuation of the Nakba by other means). This long-term plan was worked out in 1963, according to documentation found by Ilan Pappe, in order to work out the legal codex for the occupation of the rest of Palestine.[10] This work did not go unused—it has been fully implemented since 1967.

A short while after the Israeli cabinet moved to allow settlements, some members—Menachem Begin, Yigal Allon, and later Shimon Peres—initiated the illegal settlement movement. At the time, ethnic cleansing options were also considered but could not be agreed upon, so other "solutions" were adopted. Israeli leaders believed that the international community would be unwilling to confront it over land expansion, provided that Israel did not declare de jure annexation of the OPT. The West Bank and the Gaza Strip were considered occupied territories, while the 1948 conquests were all recognized as an integral part of Israel. As the OPT population could not be ethnically cleansed, there was also no way of integrating it as Israeli citizenry—at least not if the Jewish character of Israel was to be retained.[11]

With mass expulsion ruled out, the control of the territories was decided upon, not for the short term, but as a permanent solution, which meant dispossession of the land resources of Palestine. This could not take place without populating the areas with Jewish settlers, thus continuing the age-old Zionist credo—Israel stretches as far as Jewish settlement does. While full annexation was impossible unless Israel was prepared to give full civil rights to the population under occupation, annexation of the land while leaving the inhabitants in a juridical limbo was the chosen solution. Israel would then populate the territories with Jews while enforcing its rule over the Palestinians through a brutal military regime. Thus, the long-term goal of ridding Palestine of its indigenous population and settling it with an Israeli colon would be enacted.

Most Palestinians were to be forced out by the worsening conditions—so believed the Israeli government; presumably they would go to the surrounding Arab countries, as happened in 1948. And in time the world would come to accept Israel's control over the whole of Palestine. In the words of Neve Gordon, the process undertaken was moving through stages, from colonization to separation:

> As opposed to the colonization principle, which was rarely discussed, the separation principle has been talked about incessantly ... while the first is interested in both the people and their resources, even though it treats them as separate entities, the second is only interested [in] the resources and does not in any way assume responsibility for the people.[12]

This is a rather gentle, though accurate, description of the situation in the OPT. As opposed to Gordon's view of a shift, I believe that much of this was planned and represented a gradual development rather than a sharp transformation.

As mentioned earlier, this preplanned and carefully implemented project was also paired with what was euphemistically called a "peace-making process"; while this process never envisaged Israeli withdrawal from the 22 percent of Palestine that Israel conquered in 1967, it managed to assuage its Western allies. A resolution of the conflict, Israel claimed, depended on the Arabs, who could be relied upon to reject it, thus removing the danger of real peace and the need to relinquish the OPT. This worked after 1948—why would it not work after 1967? But what if the plan would not work? What if the Palestinians would not leave as planned?

Ilan Pappe points out the Israeli dilemma: the wish to combine holding the territories with the expulsion of the Palestinians. As there was no way that a forced expulsion would be successful immediately after June 1967, Israel was faced with an impossible conundrum: how to achieve both aims?

> The minuets [*sic*] of the meetings are now open to the historians. They expose the impossibility and incompatibility of these two

impulses: the appetite for possessing the new land on the one hand and the reluctance to either drive out or fully incorporate the people living on them, on the other. But the documents also reveal a self-congratulatory satisfaction from the early discovery of a way out of the ostensible logical deadlock and theoretical impasse. The ministers were convinced, as all the ministers after them would be, that they have found the formula that would enable Israel to keep the territories it coveted, without annexing the people it negated and while safeguarding the international immunity and reputation.[13]

Some five decades later, it is clear that the settlements plan has been a great success. There are now almost 800,000 Jewish settlers living in the West Bank and the Golan Heights, with huge sums invested; numerous army camps, outposts, and checkpoints set up; Jewish only roads in place; and the massive Apartheid Wall (also called the Separation Wall) dividing Palestinians from their land and separating towns from one another. Israel has smashed the West Bank into hundreds of little pockets, separated and surrounded by settlements, army camps, and apartheid roads.

The Palestinian dream of building a state(let) on 22 percent of Palestine lies in tatters, further from realization than it ever was. The broken bits of Palestine are ruled (if that is the appropriate term) by the unelected Mahmud Abbas, with seventeen different Palestinian security forces guaranteeing the security of Israelis by methodically suppressing the Palestinian population and any vestiges of resistance. Hope for change is all but gone; Israel is supported by a Western consensus of a unique kind, supplied with arms and the largest unilateral capital transfers from the United States. Ever since the start of the so-called Arab Spring, the Israeli-Palestinian conflict is off the agenda; the Palestinian conundrum has been replaced by concern about "Islamic terrorism" and the huge numbers of refugees streaming from conflict zones ignited by the West in Africa, Asia, and the Middle East, as well as by the large numbers of economic migrants that brutal globalism has catapulted from their bottled economies. Israel seems secure in its illegal occupation and the denial of rights to over four million Palestinians, operating with impunity in Gaza and elsewhere, with periodic large-scale massacres of civilians going unpunished.

By employing legal travesties and lacunae to secure the occupation and justify the persecution of the indigenous population, Israel has broadly followed traditions of control and oppression established by totalitarian regimes during the 1930s.[14] The General Security Service has operated in the OPT as a rogue organization—until 2002, it "had no authority to conduct searches, to carry out arrests, or to launch an independent investigation. The secret organization's existence, actions, and power were, consequently, the result of unwritten agreements between it and other state authorities."[15]

This was part of a regime of deception and deliberate confusion and a system of arbitrary and unpredictable control mechanisms that were all designed to subjugate the Palestinians and bewilder the wider world.[16] A crucial part of the war was the economic assault enacted against the OPT population. Israel has strictly constrained and later disabled any process of industrialization or economic autonomy, not to mention independence.[17] The West Bank has been turned into an economic cripple, when before 1967 it was the most vibrant element of the Jordanian economy: "In 1968 the deficit was $11.4 million, and by 1987 it had grown to $237.3 million."[18] The OPT were forcibly turned into dependencies by Israel's actions. Everything was controlled, including

> the types of fruit and vegetables that could be planted and distributed, and ... an array of planning regulations ... determined where crops could and mostly could not be planted ... The objectives ... were to create dependency, to undermine development and competition, and to facilitate the confiscation of land.[19]

Israel used an antiquated Ottoman regulation that allowed the confiscation of unplowed and untilled land, so limiting cultivation was crucial for confiscating agricultural land. This and the draconian restrictions on movement have also greatly increased poverty: "The poverty rate rose from 36 percent at the end of 1995 to 41 percent at the end of 1997."[20] The process was very fast; by 1980 Israel had confiscated most of the arable land in the West Bank—more than 20 percent of all land area, to build fifty-seven settlements on it.[21] Since then, Israel had seized more than half of the area, and the number of settlements is in the hundreds, leading to a great decrease in land tilled by Palestinians.[22] Since then, this has dramatically deteriorated. Israel now directly controls 59 percent of the West Bank (labelled Area C in the Oslo Accords),[23] all contiguous and well-connected by Jews-only roads, while Areas A and B are divided into 131 clusters cut off from one another by IDF checkpoints. When East Jerusalem, illegally annexed in 1967, is added, Israel controls more than two thirds of the West Bank.

From the beginning, the settlements were considered an extension of the IDF, a human tripwire and a vanguard of Israel's control of the OPT.[24] Gordon notes the main aims of the settlements:

> First, they were part and parcel of the mechanism of dispossession and helped transform the legal confiscation of land into a concrete reality. Second, ... the settlements and settlers within them served as a civilian apparatus to monitor and police the Palestinian population. Finally, the settlements in the West Bank were part of Israel's defense line against external enemies, deployed in order to help the military guard the border, secure roads, and ensure internal communications.[25]

Put simply, the settlements were the crucial component of securing the territory.

The settlements, once a series of isolated outposts, are now home to more than a seventh of the Israeli-Jewish population, or, put differently, the same number of Palestinians who lost their homes in 1948. If, in the past, the settlers did not have a distinct political voice, now they are represented by the strongest and most decisive group of racist politicians in the Israeli Knesset. Settlers now control key ministries, the government and cabinet, and the IDF.

In the April 2019 election campaign, the number of parties that directly represent the settlers shot up from three to six, with one of the new parties, Otzma Yehudit (Jewish Power) being none other than the Kach banned party, now invited by Netanyahu into the coalition he planned after April 9, 2019.[26] The representation of the settlements in the next Knesset is much larger than their electoral size, as the recent elections in September 2019 have proven. Before the elections, Netanyahu announced that he plans to annex huge parts of the West Bank.[27] While it currently seems Netanyahu may not be there to carry his promise out, other Israeli leaders may do so.

The settlers are heavily armed; protected by the IDF and paramilitary Border Guard, they roam all of the West Bank at will, stealing, burning, pillaging, killing, and maiming. As a result, the West Bank has been turned into the Wild West of Israel, where the law only exists to confiscate land, punish the indigenous, and protect criminal acts of various kinds. Every new wave of legislation makes the life of Palestinians even less viable. One need not even mention Gaza—blockaded, starved, destroyed, disconnected, dehydrated Gaza. It lacks drinkable water, medicines, and electricity; its soil has been poisoned, and its population, that once exported vegetables and flowers to Europe, now lives on international handouts and is controlled by Israeli guns, helicopters, gunboats, bombers, and drones. Palestine seems to have been defeated beyond any hope of restoration. With the election of Donald Trump in the United States, even such handouts are now being axed, leading to starvation.

However, bearing in mind the difficult dilemma described accurately by Pappe, one is reminded of the main failure of Israeli policy: After seven decades of Zionist atrocities, the Palestinians are still there. More than six million of them live in their own country, some as second-class Israeli citizens, but mostly as an occupied and oppressed population lacking rights altogether. Most have never left their country and have nowhere to go, assuming they are willing and able to leave. Though their life has become impossible, their spirit of resistance has not been defeated—one has only to look at images of the Great March of Return in Gaza during 2018.

The difficulty facing Israel in 1967 is still there, only more present. After five decades of brutal occupation, the Palestinians seem to have defeated Israeli politicians. By learning the lesson of the Nakba, Palestinians are on no account willing to leave their homes. Through their *sumud,* or steadfastness, despite the inhuman conditions enforced by Israel, Palestinians are defeating the most important aim of Zionism—emptying Palestine of its indigenous population.[28]

What options are left for Zionism in the face of Palestinian endurance? It seems there are precious few to choose from.

First, the unlikely option of Israeli withdrawal: Israel may come to accept that the West Bank is indeed Palestinian and vacate the territories. This option, allowing for a Palestinian mini-state to be proclaimed through some kind of agreement with Israel, would be a limited practical choice, if it included evacuating the settlers, including in East Jerusalem—which is what Palestinians have demanded for many decades. One may assume this is totally unlikely because no government in Israel has ever examined such an option. According to Benjamin Netanyahu, this will never be discussed:

> "There will be no more uprooting of settlements in the land of Israel. It has been proven that it does not help peace," he said. "We've uprooted settlements. What did we get? We received missiles. It will not happen anymore. And there's another reasons that we will look after this place, because it looks after us. In light of everything that is occurring around us, we can just imagine the result," he said, citing threats to Israel's Ben-Gurion International Airport and a main highway that runs along the border with the West Bank.
>
> "So we will not fold. We are guarding Samaria against those who want to uproot us. We will deepen our roots, build, strengthen and settle," he said, using the Jewish name for part of the West Bank.[29]

In short, it is hard to see how, in the foreseeable future, Israel can be forced to withdraw; the willing parties for such an act of international justice do not exist, hence this option can be sadly discarded.

The second option is to make a Palestinian "state" out of those areas in the West Bank where Palestinians remain in civil control, as well as the Gaza Strip, which in total comprise less than 9 percent of Mandate Palestine. This is obviously a pitiful, even ridiculous outcome: the areas are dissected by territory where Israel maintains full control over security, planning, and construction, which is a larger area, hence making this "mini-state" a series of isolated tiny enclaves, without geographic contiguity and totally unviable. While Israel has not agreed to such an option, no Palestinian leader in his right mind could accept it. Thus, this option can be also discarded as unlikely.

The third option involves trying to continue the current impasse—Israel controlling the whole territory, building more and more settlements, ejecting more Palestinians from their homes, biding its time—until it judges the conditions are right for a second Nakba, expelling all or most of the Palestinian population.

This is the option chosen by Netanyahu over the last decade. This option will leave four million Palestinians as stateless and lacking any rights, so it is likely to trigger a violent or even desperate reaction, including ousting of the current unrepresentative Palestinian Authority (PA) administration and its replacement by a radical popular front. Israel is clearly aware of such dangers but until now has seemed ready to risk this, trusting in its ability to suppress any uprising with military violence or to use it as a pretext for ethnic cleansing. A version of this option was pointed out by Jeff Halper in 2006, writing on "strategizing Palestine." He notes that the function of Sharon's Apartheid Wall is to declare permanent borders unilaterally:

> The final stage in strategizing Israel—the phase that would end the conflict in Israel's favor—was conceived by Sharon and has been publicly adopted as the agenda of the next Israeli government by his successor, Ehud Olmert. Hamas domination of the PA eliminates the PA as a "partner for peace," thus justifying a unilateral coup de grace: declaring the route of the wall the permanent border of Israel, thereby annexing Israel's major settlement blocs and creating a truncated Palestinian state.[30]

It is clear that this strategy devised by Sharon is one that Netanyahu sees as his starting position, rather than its culmination; Netanyahu and Lieberman, Naftali Bennett, and the other rightwing extremists in government see the area within the wall as safely secured. Safeguarding the changes over the last decade and a half, since Sharon started the project, they are using PA ineffectiveness to gain the rest of the West Bank. Again, the problem is not military control of the land, it is the removal of Palestinians from it. That is also why the Wall has many unfilled breaches, to allow further growth for settlements that will take more land from Palestine.

Just before the March 2019 Israeli elections "Prime Minister Benjamin Netanyahu said Saturday that he would start to extend Israeli sovereignty over the West Bank if given a fourth consecutive term."[31] This ploy meant that he would keep the size of the Likud in the following elections on September 17, 2019. One can just imagine President Trump signing on the bottom line, as he had before, on various issues promoted by Netanyahu.

The fourth option may be termed Nakba 2.0—using or manufacturing an event of Palestinian resistance in order to trigger the wholesale expulsion of the OPT. One of the main supporters of this option is Avigdor Lieberman, the defense minister from May 2016 to November 2018, when he resigned from the post. He had been speaking of ethnic cleansing and "population transfer" for two decades and may well force such a situation were he to gain control again, but the Knesset is full of even more extreme politicians. Because the destruction of four million Palestinians is near impossible (one certainly hopes so), they may replicate actions taken in 1948 and 1967. A number of factors have to be in place for such an atrocity to succeed:

- The Palestinians have to "play their part" and leave, willingly or otherwise. This is much less likely than it was in the earlier expulsions.
- The IDF has to be relied upon to do this without serious dissention in its ranks. There is no reason to foresee humane behavior in the IDF.
- The 1.8 million Palestinian citizens of Israel will have to be very tightly controlled so they will not oppose the ethnic cleansing, together with the microscopic leftwing Jewish minority, which may join them in protesting such illegal measures. There may well be problems in controlling this large group of Palestinians.
- The Arab countries around Israel have to accept the expelled Palestinians. It is unclear if Israel can force the Arab states to play their part and take the refugees, as they did in 1948; it is most unlikely that they will do so, even though they probably would not take action against Israeli atrocities or try to stop them militarily.
- The international community, such as it is, will have to be relied upon not to intervene, and the UN will have to collaborate by staying inactive.

While some of these conditions can clearly be achieved, others are much more difficult. It will be irresponsible of the international community not to realize that such actions are being planned in Israel. The evidence from the Knesset and media is clear enough: the context that might enable and excuse such actions is being planned.

There is a fifth option: after a period of conflict and pain on both sides, both leaderships may find that the only solution to life in lasting peace and prosperity is the most obvious one—a single state in the whole of Palestine. This must be a secular, democratic, nonconfessional civic polity, enabling the return of such Palestinians (and their families) who have been expelled, or their compensation, if they so choose. Such a society will have wide backing from most members of the UN and will depend on both sides giving up on nationalist exclusivist agendas while treating all people in Palestine as equals. Zionism must be abandoned, as well as its myriad civic and military colonialist and racist institutions.

At the moment, support for this option from Palestinians is growing, but very few Israelis view this as either positive, or possible, so this is not a realistic option for the near future; ironically, this was the original PLO solution, abandoned in 1988 under enormous US pressure. However, once the rest of the options are exhausted, and most are nearly dead right now, what else remains on the negotiating table?

Jewish Israel will have to give up on its apartheid policies, on every single shred of racialized inequality. Readers may view this as improbable but should remember that most people believed South African apartheid would never be defeated and that the Soviet Union was a permanent feature of modern history. Political change is a complex result of social will and political pressure, and if both exist, this option may indeed find expression. Such a move will have to be initiated by external forces—the international solidarity movement; the Boycott, Divestment, and Sanctions campaign; a large number of UN members; and, finally, the main powers. This is what happened when South African apartheid was brought to an end, but it was a long and messy process, delivering a partial and imperfect solution, due to pressure by the United States and its financial interests. One hopes that such a solution in Palestine may be more equitable and bring about real change in Palestine—something not yet achieved in the new South Africa.

Israel is facing a set of tough choices. Unless the Palestinians and other states play into Israel's hands, the task of emptying Palestine of its inhabitants remains very difficult. After five decades of occupation, with almost a million Israelis living in settlements, most Palestinians stay put, despite the atrocities; for them, *sumud* means that Palestinians are not going to repeat the errors of 1948. But in spite of such difficulties, the third option as a conduit to the fourth one remains the Israeli preferred line of action, and one cannot see this changing in the foreseeable future.

This means that everything must be done to frustrate such nefarious and illegal plans. This will remain a tall order indeed. Consider the context: a fractured, conflicted, and disheartened Arab East; Europe in financial difficulties; rising racist nationalism; growing Islamophobia; the United States moving inexorably to an even more militaristic, racist, and conflictual modus operandi; China facing serious financial difficulties; the rest of the international community showing clear signs of Palestine fatigue; and global movement toward the extreme right. Israel may yet have the upper hand through international criminal neglect, a repeat of the pattern of 1948 and 1967.

Indeed, the events that followed the so-called Arab Spring of 2011 have almost guaranteed that Palestine will be removed from the international agenda for a long time, seemingly displaced by the destruction of leading Arab polities and the termination of their democratic struggles, as well as the neocolonial moves by the main players, the United States and Russia, who again vie for influence in the terrifying political landscape of the traumatized Middle East. While one would be foolish to predict the results of such developments, it is doubtful that much good may result from such a chaotic and conflictual set of political factors. One certainly hopes to be proven wrong.

9

The First Intifada and the Oslo Accords

The conditions described in the preceding chapters created a powder keg in the OPT after the 1967 war. Palestinians could no longer realistically hope that a resolution of their plight might be brought about by Arab armies that have been so decisively defeated or by other Arab polities. Other routes were needed.

The first line of resistance was the increasing activity of the PLO, set up only a couple of years before the 1967 war, now under the command of Yasser Arafat, a younger, more energetic leader. Bases were established in Damascus, Jordan, and Cairo, and the new wave of the 1967 refugees, imploding the large refugee camps around Israel, also swelled the ranks of the main resistance groups: Fatah, the Popular Front for the Liberation of Palestine, and the Democratic Front for the Liberation of Palestine. A series of spectacular operations was set in motion, including mainly passenger plane hijacking; while not militarily successful, they had placed Palestine on the international agenda. The local population in the West Bank and Gaza offered shelter and support to the guerrilla fighters, and the years following the war were spent by IDF developing "antiterror" methods, with enormous expense and substantial manpower investment.

On the whole, those troubled years shaped not only Israel, but the modern world beyond its borders. The methods, strategies, equipment, and procedures developed have been exported and marketed by Israel globally, becoming standard methods of securitizing against a growing wave of terror and guerrilla activities. Indeed, Israel has managed to turn its adversity into a powerful, lucrative world-famous brand:

> As a country embroiled in a potentially debilitating, endemic, resource-draining and alienating conflict, Israel has managed, as Klieman[1] shows, to parlay its advantages into potent assets. Not only does Israel possess broad military and securocratic experience, but its global reach—if measured in depth of involvement with regimes and nonstate actors throughout the world—might even rival that of the US.[2]

Looking at the growth in guerrilla and terror activities since the 1960s, it is clear that Israel had something real to offer. Such growth was not limited to the PLO, Palestine, or the Arab world; many other organizations across Europe, the Americas, and Asia adopted PLO tactics of resistance, building urban guerrilla armies in some European countries and elsewhere. Some anticolonial organizations like the Irish Republican Army also used such methods and supported the PLO in various ways. Even before the current spread of urban terror in the West and the Arab world, by the activities of al-Qaeda, al-Nusra, ISIL, and others, terror acts were a delimiting, damping factor of global capitalism. Such organizations were facing a large, powerful group of Western states, all intent on eliminating threats to their continued domination. Israeli methods were seemingly successful in defeating the small groups of guerrillas that emerged in the 1960s and 1970s. Every airport, every large public event, every government office had adopted and installed Israeli-developed "solutions," leading to an enormous growth in Israeli industrial production and export of counterinsurgency, antiterror, and surveillance methods and equipment. Israel has become the antiterror global expert. While the PLO had not been eliminated, it was neutralized as an efficient arm of fighting the occupation or liberating Palestine. The Palestinian population had found ways of surviving the occupation despite suffering enormously, transforming beyond recognition in ways the PLO did not fully appreciate. Palestinian communities built new self-confidence, no longer relying exclusively on the PLO for changing the situation.

Israel, which found itself in control of almost two million additional Palestinians in 1967, moved to exploit the situation. The Palestinians, citizens of Jordan (in the West Bank) and Egypt (in Gaza), were a captive market for Israeli products; no other options were available, and the Israeli shekel was the legal tender. Not only was Israel enjoying an end to the economic downturn that afflicted it before the war, but it had now acquired a reserve army of labor (and consumption). Palestinian peasants, who have traditionally exported foodstuffs to Jordan and Europe, found employment as day laborers in Israeli factories and farms; later, they also worked in the many settlements built across the Green Line for much lower salaries than Israeli workers. The economic miracle of Israel's mini-empire was built on the backs of Palestinian farmers, turned into a subproletariat without rights.

But this double exploitation had other consequences. By moving freely within Israel, Palestinians came to know Israeli society—its strengths, weaknesses, fears and expectations, its army, language, and culture. They also came to join forces with their Palestinian brothers and sisters—the "1948 Palestinians," the second-class citizens of Israel. For the first time since 1948, Palestine was under a single state control. Palestinians could visit most of Palestine, with some exceptions.

This was a unique and exciting situation. It meant that Palestinians lost their long-held terror of Israeli society, better understanding its workings. They also connected with radical groups acting against the occupation, making common cause with them. A civic resistance arena was growing throughout Palestine, across the forcibly erased Green Line, and new partnerships flourished. Civic, nonviolent direct action was revived and intensified during the first two decades of the occupation, which many Israelis liked to describe as "enlightened occupation," not a phrase any Palestinian would consider using. Many Israelis on both right and left felt that the occupation was functional, providing an interim solution to the Palestine question—the right, because it planned to control the territories, the left because it was unwilling to oppose Zionism and thus was unable to develop a long-term just solution. Some organizations with Israeli and Palestinian membership began operating, and some business partnerships took place across the Green Line, giving the impression that a solution that maintained Israeli occupation was possible.

Throughout the period, the IDF had continued its repressive practices, hunting for pockets of Palestinian resistance on either side of the Jordan River and instigating a reign of terror in the Gaza Strip, under the guidance of Ariel Sharon, commanding officer, Southern Command. Despite collecting huge sums in taxes and customs from Palestinians under occupation, Israel did nothing to alleviate their difficulties. No schools, clinics, hospitals, labor exchanges, libraries, or other social facilities were built with tax income; funds were used to bolster the occupation and to build military camps, massive prisons, checkpoints, and special Jews-only roads in the West Bank and Gaza, barring Palestinian use. Banks and industrial plants could not be established, and new homes could not be built in many parts—a difficulty still afflicting even Palestinian citizens of Israel. Under the boot of the "enlightened occupation," Palestine was seething with anger and resistance, a powder keg ready to ignite.

The daily contacts with Israelis had a complex influence on both societies. Palestinian workers became the mainstay of many Israeli industries, working in areas where Israeli Jews would not deign to participate and accepting very low pay. This both increased the number of workers in modern industries and proletarianized a rural economy further, by taking most young peasants off the land. Trade unions were set up as the workers found that the Histadrut, which was happy to take their money, would not represent them. The Histadrut had a racialized pay scale, Arab pay being officially much lower for the same work: "Sadly this is a discrimination that the *Histadruth* has not only failed to challenge but has in many cases encouraged. A 1989 report found that *Histadruth* companies had the worst record of systematically excluding Arab workers."[3]

This led to thorough radicalization of Palestinian society, as well as rising expectations, having witnessed the Israeli higher standard of living, financed by their toil and exploitation. All attempts to resolve the iniquities of the occupation had been unsuccessful, leading to further social strife and deep unrest. For much of the Palestinian population, daily existence was becoming increasingly difficult:

> The uprising occurred in an economic setting in which many middle- and lower-class Palestinians found themselves suffering from several years of severe financial hardship. Dramatic price drops, particularly in agriculture and amputated international markets, caused enormous strain on the local economy.[4]

In this tense atmosphere, a little spark was all that was needed to set off the explosion; this happened in December 1987, when a Palestinian girl was killed by an Israeli driver in Gaza, in what otherwise would have been considered a routine traffic accident. The bottled anger has been released, igniting an uprising across Palestine. Multifaceted national protest, mainly through nonviolent direct action, engulfed the OPT, lasting, in different forms until the Oslo agreements of 1993. Over a thousand Palestinians were killed by the IDF, mostly children and young adults, over twelve thousand were injured, with many losing mobility, vision, and mental functioning—hit by live ammunition as well as rubber and plastic bullets, as lethal as live ammunition. Despite the presumed order to shoot at the lower body, or to shoot in the air above the protesters, most injuries were to the head, demonstrating the deadly intentions of the IDF.

The new form of resistance was a far cry from the old attacks by the PLO. In the front line were not a few trained, armed guerrillas, but the whole population of Palestine, for the first time; in their ubiquitous and unremitting ad hoc protests against the IDF, they did not use firearms but rather stones against armed mobile Israeli squads. The intifada, as it became quickly known, was led by the youth of Palestine and women, organized in local and regional committees; support offered to the youth resisting the soldiers involved paramedics evacuating the wounded, supplying them with food and water, printing and distributing leaflets and publicizing the *shuhada* (martyrs) through large posters with their name, picture, and date of their death.

When considering the intifada as a phase in redefining Palestinian identity and resistance, some have pointed out the surprising similarities to the struggle of 1936–39, as well as obvious differences:

> The most striking conclusion is the large number of general similarities between these two manifestations of Palestinian national consciousness. The two most significant differences between the two uprisings, however, are that the Intifada generates a deeper and prolonged Palestinian national coherence across all classes than did its predecessor. And second, it clarified and crystallized Palestinian opinion, which, in conjunction with other events, helped to create a historic compromise in Palestinian public policy.[5]

The unarmed struggle, typifying the intifada, is oddly missing from these main differences listed by Stein, a significant omission. Arguably, that was the difference between the rebellion of 1936–39 and the PLO struggle. After all, the 1936 action against the Mandate started as a civic campaign of strikes and demonstrations, which, when brutally suppressed, developed into wide-ranging armed resistance, albeit an unequal struggle against an immeasurably stronger force, assisted by Zionist auxiliaries. In 1987 and later, the protest was a nonviolent direct action. This profound method emerged from a crucial realization by the Palestine population—not shared by the PLO despite the strong links—that a change in Palestine cannot rely only on armed struggle, crucial as that is. Israeli military strength and substantial Western support meant that without direct political action combined with an international civic campaign, change would be extremely unlikely.

Despite the unarmed nature of the resistance, the IDF could hardly control the whole of Palestine, unprepared as it was for this protest. Tens of thousands were arrested, either for throwing stones, being members of community organizations, or being suspects. No family in Palestine remained unaffected. New prisons were hurriedly constructed and existing ones enlarged to deal with the influx of detainees. The defense minister, Yizhak Rabin, called upon the IDF to "Break their bones!" in a famous outburst on television, and this is exactly what the IDF went on to do:[6]

> An Israeli colonel accused of ordering soldiers to break the limbs of Palestinians testified today that beatings were "part of the accepted norm in that period" of the Palestinian uprising. Testifying in his own defense, Col. Yehuda Meir told three military judges that his superiors did not question the beatings because "there was nothing special in it ... There was nothing out of the ordinary. Meir testified Thursday that former Defense Minister Yitzhak Rabin gave orders in January, 1988, to break the bones of Palestinian inciters as punishment.[7]

Many such instances were caught on video, with cameras sent by international organizations.[8] The material, shot by news cameramen and local activists, demonstrated the brutality of the IDF soldiers;

many documentaries on the topic, reaching international audiences, caused Israel political damage but did not stop IDF brutalities.[9]

The order to break the bones of Palestinians was issued a few days after the intifada started. The IDF was seriously encumbered, with extra units mobilized to try and stem the widespread protest, without success. Israeli society was rocked by the protests. Most Israelis believed their own propaganda that they were conducting an enlightened occupation, so were unprepared either for protest or for its nonviolent nature. The IDF turned into a glorified riot police, with images of its brutality on world screens. Solidarity protests were held around the globe, and Israeli policymakers gradually realized that they did not know how to address the rage in the streets of Palestine. At every new funeral of a martyr, others were shot by the IDF. In order to avoid such funerals, the IDF, already wallowing in war crimes, started snatching dead bodies from the streets or bodies of the wounded who died in hospital, and burying them in a "cemetery of Numbers," a secret cemetery where Palestinians were buried in unmarked graves.[10] This severe crime against basic human rights is ongoing; Palestinian bodies continue to be secreted and buried in unmarked graves, with the army refusing to release them to grieving families:

> Amid the panicked public atmosphere that resulted when terrorist attacks began taking place within the 1967 Green Line Israel about three weeks ago and following political pressure on cabinet members, the security cabinet decided to vary its standard practice and begin delaying the transfer of the bodies of terrorists killed while committing attacks to their families.[11]

Other measures were used by Israel to try and end "disturbances"; the measures were unsuccessful but had an enormous effect on Palestinian life. In the first year alone over a thousand Palestinian homes were illegally demolished by the IDF, properties were confiscated, tens of thousands of trees were burned (by IDF and settlers), heavy fines were imposed on families of youths active in protests, and heavy taxes were imposed on the whole population.[12] The IDF also had some "scientists" at the Haifa Technion design a specialized vehicle, produced in some numbers, which catapulted large amounts of gravel toward protesters, causing many injuries. The IDF started using water cannons against protesters, flooding whole neighborhoods considered hostile with putrid "skunk" water. None of these measures ended the protests.

The extraordinary reaction by the IDF led to wide-ranging support for the protest by all political parties, organizations, and groups in the OPT. Despite constant communication with the PLO in Tunis, the protests were led by a local coordination committee, the Unified National Leadership of the Uprising (UNLU), which included representatives from the main secular political organizations at the time: Fatah, the Popular Front for the Liberation of Palestine, the Democratic Front for the Liberation of Palestine, and the Palestine Communist Party. The UNLU had a somewhat strained relationship with the two Islamic organizations: the Muslim Brotherhood (which morphed into Hamas in 1988) and the Islamic Jihad, mostly supported in Gaza.[13] Indeed, Hamas did not immediately join in the struggle: "By contrast with Jihad, the Brotherhood [as it then was] spent the first few months sitting on the sidelines, continuing its campaigns for 'proper Moslem social behavior' while the Intifada raged on around them."[14] This may well have been because the Israeli secret service had a role in the setting up and financing of the Islamic movement as a counterweight to Fatah, which they wished to undermine and discredit. Despite such differences, actions on the ground were synchronized at many points, as local activists worked together irrespective of what the leadership decreed, united by the prevailing conditions of struggle.

The most outstanding difference between earlier actions and the first intifada was probably that the bulk of the resistance fell on the young and the women of Palestine, for the first time since 1948. Most men active in the resistance were already in prison; people organized themselves in local committees, confronting IDF efficiently with the weapons David used against Goliath—the sling, stones, and rocks—facing tanks and modern armaments, dying and being injured in large numbers. Over 1,200 were killed by the IDF in the six years during which the intifada raged;[15] in the first two years alone, some 30,000 children were injured.[16] Although they confronted a modern, highly organized and battle-hardened IDF, Palestinian children managed to injure some 1,400 Israeli soldiers—an impressive achievement by any measure, bearing in mind they were unarmed.[17] The active partaking by women and young girls was a new and unsettling feature for the IDF; army actions against women and girls looked even more damning on television screens. For the women of Palestine, this was the high point of the struggle; their contribution was crucial. The coming together of the whole community was a new feature of the conflict, one that the IDF did not foresee and could not prepare for. The intifada persuaded Israel that it could no longer control Palestine through established means. The lower morale of IDF units and the unsettled state of the economy, hit by the withdrawal of Palestinian labor, were signs of the intifada's effectiveness; a new set of policies and means was clearly needed, if the occupation was to continue and more settlements constructed.

Repercussions of the First Intifada

The most striking result of the intifada did not occur in Israel, but rather at the PLO command level. Realizing the importance of the civic struggle and its implications, the PLO had changed its focus, and in late 1988, a year into the intifada, it declared support for the partition of Palestine into two states, a Jewish State and an Arab State, in effect recognizing Israel. This support of the PLO for the "two-state solution," proposed in 1947 at the UN General Council, was not shared by Hamas, however. When asked about his reaction to the PLO declaration, Hamas Sheikh Ahmed Yassin said: "I support and I oppose. I approve of the establishment of a state, but I refuse to relinquish the remaining territory of my homeland, Palestine. Hamas ... will not negotiate as a substitute for the PLO,"[18] thus reinforcing the PLO's leadership position.

The PLO's radical departure was a humbling admission that the world had indeed changed since 1948 and that its ineffective position in Tunis did not allow it to espouse solutions without support on the ground in Palestine, where people were struggling daily against the occupation, and paying a high price. The move, directed at the Israeli leadership, indicated that a peace agreement, one on Israel's terms, was possible. Typically, Israel rejected the move as meaningless; it would do nothing to weaken its control of Palestine.

The people leading the intifada may not have shared the PLO's belief in early negotiations; they had more direct experience of Israeli intransigence. They realized early on that this popular struggle would be long and demanding, needing a new mode of organization. This was stated overtly in one of the serial UNLU leaflets, printed and distributed across the OPT:

> Let all suitable organizations such as committees and units be formed in every area, on every street, and in every city, village and camp in order to pave the road towards general civil disobedience. Disobedience means boycotting all enemy organs. It means boycotting the enemy economically and not paying taxes ... The disobedience will be a strong blow to the enemy, its economy, and its plunder of our people's wealth and resources.[19]

The struggle no longer depended on a few, far away fighters of the PLO; the forces of the intifada were the Palestinian people, the whole country its theater of action, the international media acting as a focusing lens, amplifying the effect. The enemy, no longer limited to the IDF, with its customary brutality and superior weapons, was recast as racialized Zionism, its history, its ideology, its injustices, its exploitative strategies, and its uses of capitalism and colonialism as powerful means of subjugating and breaking down the resistance of Palestinians. The people of Palestine, under illegal and brutal occupation, were now in direct conflict with Zionist occupiers—those directing, managing, financing, justifying, and benefitting from the dispossession of Palestine. The leadership understood that one does not fight symptoms, but rather root causes. This deep change in the contours of the conflict took two decades to mature, a period during which Israel bolstered its oppressive machinery; ironically, it also prepared the Palestinian people for the task before them: defeating the project of colonizing Palestine. Now the Palestinian people were uniting in their struggle against this repressive apparatus—the racist, exclusivist society of Israel, with its myriad means of persecution and exploitation—from the military and financial to the judicial and cultural, all used in the service of maintaining control of Palestine.

This was the real sea change in the conflict, eventually understood by the PLO, which, by issuing its declaration, attempted to seize control of the struggle from the UNLU and its local committees (or, put differently, from the people of Palestine). The PLO was deeply wary of being written out of history by the very people it had for so long been fighting for and on behalf of, to the best of its limited ability. As a centrist organization (it could not be otherwise), it was opposed to losing control to the population it distrusted. Hence started a complex and at times very stressed relationship between the PA (which has in many ways replaced the PLO, though it still exists, mostly symbolically) and the Palestinians—one which in many ways continues to this day.

Ever since the mid-1980s, it became clear to the Israeli authorities that some new form of control was needed in the OPT, one which ideally did not involve the IDF in daily contact with Palestinian populations. It was also evident that Palestinian workers, which acted as the Israeli reserve army of labor in cities and towns, would have to be replaced. In a move to undermine their livelihoods, Israel no longer allowed Palestinians to work as laborers in Israeli population centers; they would be replaced by other workers.

The result was a massive importation of manual workers from all over the globe. Hundreds of thousands of workers from Eastern Europe to Sri Lanka and from Africa to China were lured to Israel after the first intifada, only to find themselves employed as indentured labor, little more than coolies. They were housed in decrepit shacks hastily constructed in fields and near factories, locked in for the night on pain of immediate deportation, their passports held by employers, making it impossible for them to change employment. Most of these workers came from centers of mass unemployment, their pay even lower than that of the Palestinians they replaced. The problem of employing the Palestinians was thus resolved, but this solution created further problems.

There remained the difficulty of controlling over two million Palestinians, who learned their real power through the intifada. This was exacerbated by mild but insistent international pressure, mainly by the United States and the European Union, against the continued building of settlements, illegal under international law and the Fourth Geneva Convention. This was when the so-called two-state solution (a very attenuated version of the Partition Resolution 181) gained traction as the favorite resolution of the conflict. The resolution had called for a Jewish State comprising 55 percent of the country, but it was now thought that returning Israel to the 1967 boundaries, which gave it control of 78 percent of the country, would be a fitting solution; the other 22 percent could be given over to a mini-state of Palestine, with international security guarantees. The six years of the intifada with its damaging results strengthened this consensus. Observers who feverishly wished for such a solution, more inclined to ignore facts on the ground, had hope that Palestine could become a small but independent state.

The PLO's change of direction, as signalled in the 1988 Declaration of Independence by the PNC in Algiers, was a sign of the times. The beleaguered organization, strongly swayed by the intifada, isolated in Tunis without potential for effective resistance, had also suffered from murderous Israeli raids, with many of its top officers assassinated and with no capacity for retaliation. It was gradually coming around to the view that in order not to disappear altogether into irrelevance, it needed to buy into the new realities, abandoning armed resistance for political and diplomatic action so as to remain in some control of events in Palestine. This new willingness to end the conflict through negotiations was shared by their Arab and Western interlocutors; there seemed to be an opening for a negotiated solution of the conflict. Ironically, the success of the unarmed resistance by the whole Palestinian people proved the efficacy of the nonarmed approach, even to the guerrilla forces in Tunis and Syria. With a green light for talks, Western powers were only too ready to apply mild pressure on Israel to move toward a negotiated solution. The change was greatly boosted by the dramatic transformation of world politics in 1989, when the Soviet Union and its appendages evaporated overnight; the victory of Western capitalism was seemingly all but complete, and Fukuyama's end of history was celebrated. Soviet opposition to an unjust solution was no longer an issue, which created a real opening for forcing Pax Israeliana; this was desirable for the Western powers as they prepared for a new Middle East consistent with the new world order.

The pressure on Israel started on May 22, 1989, when US Secretary of State James Baker told American Israel Public Affairs Committee —the Zionist lobby organization in North America—that Israel must abandon its expansionist policies and that the new president,

George H. W. Bush, intended to have a different approach from that of Ronald Reagan.[20] Bush was, after all, ex-director of the CIA, and hence, somewhat more "enlightened" than the previous administration, as the CIA had consistently advised that a more balanced approach to the conflict would be preferable for US regional interests.

This eventually led to the Madrid peace conference, a short while after the Western "victory" in the first Gulf War, in March 1991. Israel was at the time asking the United States to support its international loans through US guarantees, a procedure that had played out many times before. President Bush surprised many, not least the Israeli government led by the rightwing extremist Yitzhak Shamir, by refusing to provide such guarantees unless and until Israel agreed to participate in the Madrid Conference. In September 1991 Bush had asked Congress to freeze the guarantees for three months, intensifying the pressure.[21] Within Israel, political pressure was also being exerted on Shamir to force him to join the talks. The Labor Party realized that Israel might with US support impose its solution of choice, being more practical than the doctrinaire Shamir toward the planned talks. The party's rising star, Yitzhak Rabin, was an early convert to the virtues of an agreement, realizing that the PLO was an empty husk, unable to offer resistance to Israeli control of Palestine. What started as US pressure on the unwilling Shamir turned into a full Israeli project of collaboration with the United States in putting pressure on the PLO and its leader, Yasser Arafat, forcing him into an agreement he could not refuse but also did not fully comprehend.

For the PLO, staying in Tunis when the Palestinians were in the streets was not an option. While it was not able to direct the action, it kept in close touch through its component organizations. Either the PLO would take part in what transpired in Palestine, or it would disappear as an arm of resistance. The intifada was led by young activists born in Palestine, shunning armed struggle as a method of resistance. The PLO was under siege, and Rabin and his US partners were perfectly positioned to exploit this.

Israel had other reasons for wishing to change the circumstances in the OPT and readjust the control mechanism. The intifada had exhausted the IDF and Israeli society, and the huge financial cost of keeping tens of thousands of reserve soldiers as well as the large permanent army in the West Bank and the Gaza Strip on what was in essence police duty, with little success and international anger growing, was draining. If what the United States and the European countries were suggesting could be used to resolve both these difficulties, then Israel would be more than ready to return the PLO to Palestine. Israel would, in the person of Shimon Peres and his minions in Oslo, force an agreement leaving it holding all the cards and drastically reducing financial, military, and political costs.

After the rigmarole and fanfare of the Oslo Accords and the US ceremonies were over, with excitement around the world rising as the "most intractable conflict" was about to be resolved, the veteran Palestinian academic and activist, Edward Said, world famous for his work on orientalism and cultural imperialism, had been the first to properly examine and analyze the accords from a Palestinian perspective. In a *London Review of Books* article he proved to be the most far-sighted of all commentators on the peace accords, exposing them for what they were:

> First of all let us call the agreement by its real name: an instrument of Palestinian surrender, a Palestinian Versailles. What makes it worse is that for at least the past fifteen years the PLO could have negotiated a better arrangement than this modified Allon Plan, one not requiring so many unilateral concessions to Israel. For reasons best known to the leadership it refused all previous overtures.[22]

Reading Said today is a stirring experience. His insight and clear thinking are astounding, in the face of the prevailing nonsensical commentary at the time and since. Said refused to join the choir of those who shunned reality and praised Oslo without confronting it as the disaster it was:

> In order to advance towards Palestinian self-determination—which has a meaning only if freedom, sovereignty and equality, rather than perpetual subservience to Israel, are its goal—we need an honest acknowledgment of where we are, now that the interim agreement is about to be negotiated. What is particularly mystifying is how so many Palestinian leaders and their intellectuals can persist in speaking of the agreement as a "victory." Nabil Shaath has called it one of "complete parity" between Israelis and Palestinians. The

> fact is that Israel has conceded nothing, as former Secretary of State James Baker said in a TV interview, except, blandly, the existence of "the PLO as the representative of the Palestinian people." Or as the Israeli "dove" Amos Oz reportedly put it in the course of a BBC interview, "this is the second biggest victory in the history of Zionism."[23]

Such stark truths are denied by some even today; that Said was clear about them at the time and was considered a traitor to the Palestinian cause by many for pointing it out, is indeed tragic. When speaking of the unequal agreement, he also refers to the Palestinian willingness to end the intifada:

> It would therefore seem that the PLO has ended the intifada, which embodied not terrorism or violence but the Palestinian right to resist, even though Israel remains in occupation of the West Bank and Gaza. The primary consideration in the document is for Israel's security, with none for the Palestinians' security from Israel's incursions. In his 13 September press conference Rabin was straightforward about Israel's continuing control over sovereignty; in addition, he said, Israel would hold the River Jordan, the boundaries with Egypt and Jordan, the sea, the land between Gaza and Jericho, Jerusalem, the settlements and the roads. There is little in the document to suggest that Israel will give up its violence against Palestinians or, as Iraq was required to do after it withdrew from Kuwait, compensate those who have been the victims of its policies over the past forty-five years.[24]

Said was also clear about the alternative to Oslo; at the beginning of his book *End of the Peace Process: Oslo and After*, he tells of a meeting at Columbia University Department of Journalism, where he was invited to give his views on the potential for peace. Being badgered continuously by three supporters of Israel, he was then asked what his own proposal for resolution of the conflict would be. He did not hesitate:

> When I was asked for an alternative I said that the alternative has been there from the very beginning: end of occupation, removal of

> settlements, return of East Jerusalem, real self-determination and equality for Palestinians. I had no problem at all with the prospects of real peace and real coexistence and had been speaking about those for twenty years; what I, and most Palestinians opposed was a phony peace and our continued inequality in regard to the Israelis, who are allowed sovereignty, territorial integrity, and self-determination, whereas we are not.[25]

Such a simple formula, based on international law, was never under discussion at Oslo or later; the power represented by Israel and its allies would not entertain it. We know today that every single prediction by Said has indeed come true, and Arafat, bamboozled by the publicity and spectacle of the Camp David agreement, the Nobel Peace Prize, and his return to Palestine, had closed his eyes to the abomination that was the Oslo Accords until he could no longer hide from reality and initiated the September 2000 intifada, sparked by Sharon's provocative visit to al-Aqsa. Until the day he died, Arafat and his cronies refused to admit that their real function in Palestine was policing on behalf of Israel. Thanks to the PA, Israel no longer needed to police the OPT or budget the operation; it was now off the international agenda, free to exponentially increase settlement and continue confiscating Palestinian land. The expense of policing would now be covered mainly by the European Union, supporting the inflated PA security apparatus. Arafat would be free to control "his own mosquitoes," as he revealingly remarked. The PA was not a state, controlled nothing real, and existed mainly as an agency of Israeli security and a machinery to enrich the cabal around Arafat, through exclusive service contracts in the OPT. This was a small price for Israel to pay, far outweighed by all the benefits it received.

The Oslo Accords and the agreements that followed were never intended to set up a Palestinian state, nor did they claim to do so. They made the PA the arm guaranteeing settlements' security. Israel retained the right to intervene wherever and whenever it saw fit and did so numerous times ever since. Palestine was practically sold out by the international community, which supported the Zionist project. The Palestinians were now in a worse situation than ever before, with Israel controlling the whole country and others meeting the great cost of policing. The Palestinian economy more or less evaporated, with unemployment in Gaza reaching 60 percent.[26] Even freedom of movement within the West Bank did not exist, as hundreds of checkpoints and Jews-only roads crisscrossed the territory and IDF incursions, extrajudicial killings, and house demolitions were daily events. The four million Palestinians living in Gaza and the West Bank lack basic human rights, many are unable to travel to any part of the world, and cannot meet their families or hold a position of employment outside their locality, while students are unable to reach their universities abroad or in Palestine and people needing medical treatments are similarly refused.[27] Israel periodically destroys the infrastructure and lives in Gaza, where two million civilians are trapped in subhuman conditions, and this illegal and terrifying situation continues despite most countries claiming that the occupation is illegal but often actively assisting it. This occupation has now become the longest in the twentieth and the twenty-first centuries and shows no signs of ending; indeed, as more and more land is illegally confiscated and more settlements are built, Palestinians are squeezed out of land and home, adding to the millions of refugees already living in camps abroad.

In essence, the 1948 Nakba has never ended, its reverberations traveling like constant tremors through the region, torn apart by other developments. As Israel has never been held to account since 1948, there is little incentive for Israelis to even consider a just peace in Palestine, when all the cards are in their grasp.

Edward Said did not hesitate to blame the Palestinian leadership for its betrayal of millions of their own people in the OPT and beyond. Despite the great disparity of power between Israel's rulers and Yasser Arafat, he is courageous in openly and clearly presenting Arafat ("Mr. Chairman-President" as he calls him) as the collaborator he was:

> What I find unforgivable is that in all this he [Arafat] has appealed not to his people's best instincts, but to their worst ... The various beatings, tortures, closures of newspapers, and summary arrests have induced an atmosphere of fear and indifference: everyone now looks out for himself. At times I find it hard to believe that this is happening to a people who fought stubbornly against the British and the Zionists for so long, but who seem to have given up all hope and all will to resist the extraordinary disasters visited

> on them by their leadership, which cares not a whit for anything except its own survival.[28]

Said, the passionate fighter for Palestinian democracy, was deeply offended by Arafat and his sycophantic coterie—as far from the people of Palestine as one could get—and this despite (or maybe because of) the public respect and devotion toward Arafat on his return to the West Bank. Said correctly delineates the antidemocratic fault lines in the Palestinian leadership, explaining why such a leadership cannot properly represent the population it claims to lead:

> What is symptomatic about the Palestinian Authority's mentality is its total inability to answer criticism, or seriously engage with its critics, whose number is growing as the situation deteriorates ... Arafat and his advisers have closed themselves to their own people. They have no conception at all either of accountability or democratic and free debate. The worst of all is that in his disastrous policy of capitulating to the Israelis and then signing all sorts of crippling limits on his people into agreements with his occupiers, Arafat has mortgaged the future of his people to their oppressors.[29]

These lines were written during October 1995 and were presented as treason by the leadership Said described as unable to face criticism. Said was isolated in the Arab world where intellectuals were taken in by the propaganda and hoped for a genuine solution to the conflict resulting from the deeply flawed Oslo process. Nowhere was this truer than in Egypt, where the population had tired of conflict and war and the great social cost this had entailed. Said was not an ivory-tower academic, after all; in the late 1970s he presented a plan for a political solution to the conflict authored by the US government to Arafat who rejected it, though fifteen years later he would agree to a much worse "solution" in the Oslo Accords:

> To take one example of which I have personal knowledge: in the late Seventies, Secretary of State Cyrus Vance asked me to persuade Arafat to accept Resolution 242 with a reservation (accepted by the US) to be added by the PLO which would insist on the national rights of the Palestinian people as well as Palestinian self-determination.

Vance said that the US would immediately recognise the PLO and inaugurate negotiations between it and Israel. Arafat categorically turned the offer down, as he did similar offers.[30]

That the first intifada (a grass-roots uprising which had developed totally innovative and effective modes of civic resistance to brutal occupation) had led to the Arafat regime in Palestine, riding roughshod over all the achievements of the uprising, replacing the popular struggle with undemocratic betrayal—that was Said's most painful assessment of the situation following the Oslo Accords. In pointing out the betrayal and the dictatorial traits of the Arafat regime, Said was well ahead of the curve, almost alone in so doing. In the years since it has become clear that his prognoses were, if anything, not pessimistic enough. The result of Oslo was a great victory for militarized Zionism and can be clearly seen today:

- No end to the occupation and control of Palestine by the IDF. Even when it evacuated Gaza in 2005, Israel had not relinquished control; if anything, it was now in total control of the lives of two million Palestinians who were living in a massive ghetto, without recourse to any form of international justice, law and order, aid, medicines, or food. Over five thousand Palestinian civilians have died in IDF attacks on civilians in the last decade alone.
- Israel stopped paying for controlling the Palestinian population—the European Union pays for the PA security services securing the settlements, rather than their own people. The periodic attacks, mainly on Gaza, are financed by special legislation in the US Congress and special presidential bills passed on such occasions. Thus, Israel has shed the main cost of the occupation.
- Israel controls the Palestinian economy, environment, resources, communications, borders (though they do not exist), traffic, and movement in the OPT, as well as financial transfers, such as the value-added tax that Palestinians are paying to the Israeli exchequer. Through the imposed Israeli shekel, Israel has a stranglehold on Palestinian existence; it controls water resources, the airways and the radio spectrum, the telephone and cell phone spectrum and apparatus, all international exit

and entrance points, and the hundreds of checkpoints on all roads. By such means, Israel is able to stop Palestinian social and economic development, and to force the population into an underdeveloped, disenfranchised state, which has now lasted for five decades.

- Israel controls the PA, through its close relationship with the United States and the European Union. The Palestinian movement for independence is intercepted and annulled through the political and diplomatic power of the Western bloc at the UN. This has frozen even the timid attempts to revive the process of conflict resolution.
- Israel controls "negotiations" for the "final status" of Palestine through the same means. By refusing to meet, or by subverting any planned negotiations through subterfuge and delay, it has managed more than twenty-five years of paralysis—no retreat to the Green Line, while increasing the settlements by some 200 percent and disallowing any building by Palestinians in the areas it controls. This has limited Palestinian "control" to the Gaza Strip and less than 40 percent of the West Bank, or less than 9 percent of Mandate Palestine. Such "control" is further compromised by the free movement of IDF extrajudicial killing squads in areas under the nominal control of the PA, not to mention the frequent major incursions into Gaza.
- Zionist parliamentarians from all parties are now able to pass racist, antidemocratic legislation removing even the few rights Palestinian citizens of Israel held. More than sixty laws limit this population, annulling their cultural, political, and educational rights and placing them within the apartheid structure that incarcerates Palestinians in the OPT. This extends racist state control, already total in the OPT, into Israel's pre-1967 boundaries. In essence, it completes Zionist control over Palestine.

Most of these successes could not have been imagined at the time of the first intifada, as Israel was reeling from the shock of independent, civic direct action, searching for a way of returning to full control. Arafat, pressured by Israel and the Western bloc, supplied that crucial mechanism; as in certain martial arts types, his energy was used to overpower him.

It is important to point out another result of the Oslo Accords, no less important than the ones listed above. The PA managed to undermine the political grassroots movement that brought about the first intifada, a popular movement with great achievements in local democracy and civic management. By instigating a government of corrupt officials with no recourse or possibility of recall by popular vote, the PA has stripped its captive population of any real hope for democratic change or a just peace in Palestine. By abandonment of the cause of *awda* (Palestinian return) and the refusal to run elections either for the PA and parliament or for the Palestine National Congress, it has managed to decapitate the local leadership that emerged during the intifada and disable all democratic political forces, perpetuating its illicit rule. It is difficult to foresee a positive change on the horizon or to see from where such change might emerge.

The crashing defeat of democratic forces following the uprisings in the Arab world, as well as the smaller tremors in countries in both the Mashriq and Maghrib, have also put an end, at least for now, to any cross-fertilization or collaboration between democratic Arab forces. What started as a new wave of pan-Arab, popular, democratic political energy has, on the whole, strengthened totalitarian leaders, with Tunisia being the only partial exception. Such a dramatic failure of the popular will has further damaged Palestine, where the likelihood of an uprising against the PA was already very low; nations under brutal occupation can hardly rise against their own dictatorial leaders. Until these contextual elements of Middle Eastern politics radically change, it is difficult to see the potential for transformation in the Palestine situation or a serious move to a long-term, just resolution.

With the arrival of President Donald Trump to the White House, it has become even more difficult to anticipate positive changes. While President Vladimir Putin, an ally of Trump during the election campaign, has found it necessary to state his support for a Palestinian state while speaking to a summit of the Arab League in Cairo in March 2015, Trump nailed his colors to the Zionist mast; his own racism and hostility toward Muslims matched those of his ally, if indeed there are any such sentiments left at the high echelons of Russian diplomacy, as there was no real follow-up to Putin's announcement or any other signs of policy change in that direction.[31]

Indeed, Trump claimed to want to end conflict and wars, but during the first two years of his presidency he introduced great dangers into world politics. That his envoy to the Middle East, his son-in-law Jared Kushner, refused to put any real pressure on Israel to return to the peace track or end settlement building is evidence enough of the direction the United States may take in the next few years.[32]

The move of the US embassy in Israel to Jerusalem in May 2018 and the recent ratification of Israel's annexation of the Golan Heights were the clearest signs that Trump does not even pretend to be balanced or evenhanded.[33] He moved the embassy because his "friend," Prime Minister Benjamin Netanyahu, asked him to and because he could; that such moves flew in the face of US policy and international law seemed to spur him into action. While Obama was visibly troubled as he supported Israeli infractions, Trump is visibly elated to support a regime that operates in a way similar to his own, one that may have inspired him.

10

Israel's Military-Industrial Complex

In Israel, old weapons never die. They don't even fade away; they just get recycled to some remote corner of the Third World.

—Benjamin Beit-Hallahmi, *The Israeli Connection: Whom Israel Arms and Why*

It is almost unimaginable that one of the smallest states on earth with no obvious mineral wealth could become a leading armaments exporter within a few decades. For Israel to achieve this status, it had to put arms production at the very heart of its society, industry, and identity. The beginnings of the Israeli military–industrial complex (MIC) are the results of the violence innate in the settler-colonial nature of the Zionist project. For a tiny minority to take over the country and expel the majority, the use of violence and the force of arms is nonnegotiable.

In Chapters 1 and 2, I argued that the British Mandate enabled the move from Hashomer and the Jewish Battalions toward the large and powerful Hagana. The riots in 1921 and 1929 ensured that the development of a national Zionist militia in Palestine was accelerated, in preparation for the battles over Palestine. Because such an army was illegal under Mandate regulations, so was its arming and training. Although the Mandate authorities were quite relaxed about this development, it was still difficult to import arms illicitly. From the early 1920s, the need for local production was therefore evident and urgent.

With the arrival of Jews from Germany and Austria after 1933, the situation dramatically changed again. Within a few years, the

Zionist sector industrialized the cities. Most workers, until then employed in agriculture, found work in the thousands of new factories built by Jewish immigrants fleeing the Third Reich. The nature of the Ha'avara (Transfer) Agreement, signed by the Zionist organizations with the Nazi authorities in Berlin, meant that the only way for German Jews to bring most of their capital with them was by importing it as German industrial goods. This caused a flooding by industrial machinery, including some of the most advanced machinery of the strongest industrial state in Europe.

The new industrial plant paid for itself by supplying the more than 100,000 British troops stationed in Palestine with everything from food to live ammunition. While the British knew that the industrial plant that supplied their forces was also producing arms on the side, they ignored it, as long as the arms were not used against Mandate authorities.

To begin with, Hagana made do with stealing whatever they could from British bases in Palestine and Egypt, from light arms and vehicles to explosives and munitions. In the chaos of Arab insurrection followed by World War I, this task was easier than it should have been, and there were always British soldiers prepared to assist the process, through ideological or financial motives.[1] During the war years, both illicit production of munitions and covert purchases of arms enormously increased, especially after the Hagana set up Rekhesh, an arms acquisition organization.[2] While Rekhesh dealt with buying or stealing arms, the Taas secret industrial plants produced everything from light weapons to medium-range mortars, with stockpiles being kept both in Europe and Palestine, readied for the coming conflict.[3] This effort paralleled the great wave of Palestine Zionists joining the British forces after 1939, meant to provide military training to over 40,000 Jewish males from Palestine, as well as accessibility to arms and munitions.[4] Many of those soldiers became experts in various branches of armament servicing and repair, which they would put to good use after the war, in the service of Hagana (and later, the IDF).

Among the refugees and volunteers arriving after the war were other experts and seasoned soldiers from a variety of countries in eastern and western Europe, as well as the Americas, with unique skills acquired during the war years. They would all contribute toward building the arms industries, supporting the fighting, first against the

British and then against the Arabs of Palestine and the rump Arab armies, which joined the war in mid-May 1948. While both Britain and the United States joined the UN-promoted armament-blockade on both sides, much materiel was smuggled from a variety of countries, mainly to the Zionist forces. For example, by the end of the war in late 1948, the IDF had more than a hundred illicitly acquired planes, of varying capacity and vintage.[5] Ben-Gurion was not then invested in building a powerful air force—later he would be persuaded to do so by Shimon Peres.[6]

Arms and munitions production had been a crucial factor in securing the victory for Zionism in 1948. Israel had the upper hand both numerically (in combatant numbers) and in quality of arms, as well as an edge in modern command and supply systems. But the initiation of a policy of modern arms production did not arrive until 1953, when the IDF was reimagined as a fast-moving, offensive military force. At that point, Ben-Gurion decided to invest heavily in building arms production industries. He was able to do so because Israel had what is described as a double-budget system:

> The ordinary state budget was financed by revenue internally generated through taxes, while the separate development budget was financed through foreign sources, such as donations from world Jewry, the sale of independence and development bonds abroad, and loans and grants from foreign nations.[7]

Such grants included the enormous 1952 Reparations Agreements with West Germany, the great majority of which went into financing the defense and security projects, including the most costly: the nuclear weapons project. This gave the small state enormous resources that it could never gather from its citizens. Furthermore, the development of such pet projects was kept secret, out of the state budget and without public scrutiny of any kind. It was also supported by generous French grants and loans between 1955 and 1967, and thereafter, through the unilateral transfers of US military and civilian aid contributions. In its seven decades of existence, Israel has received more foreign aid in various forms than any other state in history. Indeed, we may not even have enough information about the true size of such support.

The main areas to which this funding was devoted in the mid-1950s were the nuclear weapons program; the metallurgical infrastructure enabling Israel's arms production, based at the French-designed and operated plutonium nuclear reactor at Dimona; the large Bedek plant (later Israel Aircraft Industries) for servicing, modifying, and producing aircraft; the Tadiran electronics industry plant, producing communications and other electronic equipment for the IDF and other bodies (later to include Elbit); Taas, the general arms production arm of the ministry, with a wide range of products; and Soltam, a Histadrut steel production plant, making mortars, heavy guns for artillery and tanks, and munitions for the IDF. An additional beneficiary was Emet, the research and development branch of the IDF that was moved to the Defense Ministry, which developed new types of weaponry, and later became Rafael (the Weapons Development Authority).

There were many other, smaller bodies that were financed to produce, modify, and export arms of many kinds, all contributing to a network already worthy of the moniker Israeli military–industrial complex. Indeed, by the mid-1950s this was becoming Israel's largest industrial sector and its main exporting sector. This could never have happened without the enormous financial support Israel received from the West, with the funds used almost entirely to support military and weapons production.

This success was possible due to some strategic decisions taken as early as 1953. The first was to use a public–private investment model for these bodies so that the financing came both from the funds controlled by the government as well as foreign investment, mainly by wealthy Jewish businessmen in the West, who saw the potential of such industry. This meant that the enterprises worked not just for the IDF but also for the international market, where the need for arms was greater than ever.

It is interesting to examine some of the early international markets Israel sold arms to. By 1954, Israel was able to sell tens of older aircraft after it had received its first jet fighters from Britain. By agreement with Taas, Burma purchased thirty overhauled and serviceable Spitfire 9s from Bedek for use against its Communist insurgency.[8] (Recent news reports tell of Israeli arms supplied to Myanmar in its war on the Rohingya Muslims.) The sales of the older planes helped to finance the new fighters.[9] At about the same time, Israel was selling Uzi submachine guns to many countries in Europe, the Americas, and elsewhere. From such humble beginnings rose the mighty Israeli MIC. By 1981, the industry was responsible for employing more than 300,000 workers, according to the *Financial Times*—one quarter of Israel's workforce.[10] The figure is much higher now, but exact numbers are hard to come by.

Like other secondary industrial players, Israel concentrated on improving and upgrading products that other states with substantial industrial systems had invested untold sums in developing, adding value rather than creating it. Israel had one important factor working to its advantage: it fought more wars than most countries, possibly more than any other country in modern times, so testing weapons was an ongoing exercise.

In addition, there was no shortage of Jews working in related industries elsewhere who were prepared, under certain conditions, to provide crucial details to the state of Israel. This sharing of information (illegally gained or not) allowed Israel to become specialized in improving weapon systems. As state-run enterprises arms producers had no competitors, and each controlled a sector of the technological market. The enormous sums invested in weapons manufacture soon brought results and, as it were, the economy abandoned oranges for hand grenades.

Another important factor was Israel's isolation in the Middle East, surrounded by Arab states that were more or less hostile. This meant that even the smallest items had to be imported from distant producers, and there was an imperative to produce locally. The inability to produce simple items like batteries, which are time dependent and deteriorate quickly, has been a major factor pushing Israel to set up Tadiran, for example. Munitions could not be left to long and precarious supply chains, so Soltam was set up to produce most IDF needs. Aircraft had to be serviced in Israel, as did tanks and other vehicles; all these demands and circumstances converged toward an independent weapons-production capability, the Israeli MIC.

Because of the enormous overhead involved in such an undertaking, export potential was crucial if the system was to be sustainable. While Israel may not have planned to become a major weapons developer, obvious commercial and technological vectors propelled it onto the world stage from the start. Israel became a specialized war economy, depending on and benefiting from armed conflict. War became its raison d'être and its organizing principle.

Israel's main weapons were still bought from its supporters in the West, sometimes employing a roundabout route in order to hide the transactions. For example, during the 1960s, West Germany supplied most of the heavy armament for the IDF; it was the conduit through which the United States channeled arms to Israel, so as not to get in trouble with its Arab commercial-oil suppliers in the Middle East. Nonetheless, during the mid-1960s, even West Germany had reason to stop supplying Israel directly because it worried that this might prompt the Arab countries to recognize East Germany. As a result, West Germany changed the supply route, agreeing to finance Italian and French arms for Israel so as not to be a direct supplier.

These were all internal NATO arrangements. The armaments in most cases were US-made NATO supplies, and the agreements were kept secret for many years.[11] During that period, and specifically between 1956 and 1967, Israel's main supplier was France, which modernized all IDF materiel, not to mention its building of Israel's nuclear facility. This lavish support was only abandoned when the United States replaced France as Israel's protector and purveyor after the 1967 war. This was made possible as the Middle East became the sandbox of the Cold War—a theater of conflict where both blocs were playing proxy wars, with the United States supplying Israel and the Soviet Union supplying Egypt and Syria. The *Wall Street Journal* noted in 1981 that "Israelis complain that in criticizing Israel's hawkish military policies, the US overlooks the fact that Israel has served as a kind of 'combat laboratory' for US weapons development."[12] Israel could not lose under such conditions and by 1981 it ranked fifth in global arms sales, after the United States, the Soviet Union, France, and Britain.[13]

Such incredible success also created problems. As its industries reshaped and upgraded systems purchased in Germany, France, the United Kingdom, and the United States, much friction was created when these were then resold as the IDF moved to newer models. In this way, Israel was competing with states and companies, selling an improved version of the product it purchased, arguably undermining the original producer.

For example, certain US items were sold to Israel with the understanding that they should not be passed on to a third party. It would behoove Israel to honor this arrangement, but its repeated failure to do so has led some US presidents to delay the provision of a new or secret item.[14] However, more often than not, America has relented. In this way, one may consider Israel an "honorary" member of NATO rather than an outlier.

No case exemplifies the power and influence garnered by Israel in the international arena, and especially in the United States, as the Iran-contra affair of the mid-1980s. On the face of it, the United States and Israel lost their investment when the Shah left Iran. The complex machinery of repression headed by Savak was dismantled, only to be replaced by a parallel Islamic system. The Iranian armed forces, which until 1979 had had close links to the IDF, were purged of their Western and Israeli influences, with Iran beyond the pale of its former allies. But Israeli planners did not relent; they understood Iranian hostility toward Arab countries and planned on using it to their advantage. The links they had built with army officers may not be easily reactivated, but they might become the foundation for a future investment. The Mossad also built on cracks and fissures within the Islamic Republic itself—Western-oriented, market-friendly officials who might be helped to topple Ayatollah Khomeini and his henchmen.

Thus begins a fascinating chapter of intrigue, corruption, bluff and double-bluff, greed, byzantine machinations, spying, and wishful thinking of stupefying magnitude. A trio of major Israeli arms dealers —the Iranian–Israeli Ya'akov Nimrodi, the American–Israeli Al Schwimmer (developer of a major chunk of Israel's MIC), and the British–Israeli spymaster David Kimche—together with Iranian arms dealer Manucher Ghorbanifar and Saudi billionaire arms merchant Adnan Khashoggi worked out a grandiose plan.

The plan, in outline, was ingenious; Iran, at war with Iraq, had run out of spare parts for its US-made weapon systems, and hunted everywhere for parts for their Phantom fighters, Western-made tanks, and missiles. Thus, the cabal of arms dealers agreed in mid-1981 to present Western-oriented, highly positioned Iranian officials with a secret deal. They offered to supply crucial items to Iran to enable it to win the war with Iraq; these officials would then be perfectly placed for the coming struggle over the leadership in Tehran, as Khomeini's life neared its end.

Ghorbanifar persuaded the dealers that these Iranians trust him implicitly and their group would greatly benefit from indirect access to US military supplies in the coming power struggle in Tehran. Neither the United States nor Iran would consider direct contact for this purpose, because Iran was strongly anti-American, and the United States was forcing the Europeans not to supply arms to Iran. At the same time, Nimrodi, Kimche, and Schwimmer won over the Israeli government (Prime Minister Shimon Peres, Defense Minister Yitzhak Rabin, and Foreign Minister Yitzhak Shamir). National Security Advisor Robert McFarlane (and later his replacement, John Poindexter) sent Ghorbanifar to Iran to sew up the deal with his contacts, mainly through Ayatollah Hassan Karoubi, Khomeini's adviser for many years.[15] The possibility of finding a source for the missing parts excited the ayatollah, who undertook to meet with the greedy plotters in Europe.

There followed the most complex negotiations over fifteen months. The arms were supplied by Israel, mostly Hawk surface-to-air missiles (SAM), only to be discovered by the Iranians to be outdated. Iran refused to pay for the missiles, and they ended up back in Israel, to be replaced by others channeled by the United States through Israel. The negotiations were further confused by the US demand that Iran force Hezbollah in Lebanon to free some hijacked Americans, including William Buckley (the CIA chief in Beirut), who were held for a long time at secret locations.

As the secret negotiations failed to deliver any of the goals of the parties, the White House moved to change horses. They picked Oliver North, a US colonel, working at the White House, who connected this deal with arming the contras in Nicaragua. If this sounds bizarre, much worse was to come. North was persuaded by the young and inexperienced antiterrorism advisor of Peres, Amiram Nir, to partner with him and drop the large group of co-conspirators and arms dealers. North invented the perfect solution: the United States would sell arms to Iran at inflated prices and use the money to arm the contras, already supplied by Israel with various items, mainly light weapons.

The labyrinthine details of this whacky saga are beyond the scope of this chapter and have been covered in a number of publications already. But it demonstrates that Israel performed complex roles, acting as a prosthetic limb of the US government. No favor is too extreme or unthinkable when the US masters are asking.

A group of ayatollahs in Tehran blew the operation's cover in order to harm the political faction conniving with Israel and the United States, removing them from the struggle over Khomeini's replacement. All hell broke loose in Washington, during the congressional elections in November 1986. The hearing by Congress made North the fall-guy for the American side, and he landed in jail.

Why Is Israel the Largest Recipient of US Aid Per Capita?

The Israeli armament industry would never prosper without the massive support of the United States. But one should never assume US support for Israel is charitable. The United States depends on Israel as its main power base in the eastern Mediterranean: it is safer than Egypt, more secure than Saudi Arabia, and closer to Europe and Russia. It is difficult to separate US and Israeli foreign and military policies—one is the continuation of the other. Though often Israel does not deliver (the United States felt it did not get an appropriate return on its investment during the 2006 Lebanon War, for example), on the whole, US presidents found it impossible to deny Israel the support it craves. A president who refuses to back Israel requires much courage, as Barack Obama found out in 2015, struggling to ratify the Iran Nuclear Agreement with Benjamin Netanyahu arguing against him in the Senate and Congress. The Israel lobby is the strongest and most capable on Capitol Hill,[16] hardly ever failing to deliver, supported as it is by the powerful security industry as well as the Bible-belt preachers and assorted alt-right billionaires. Israel found the levers to push within the US political machine and it excels at using them, exacting a high price for tasks it executes on behalf of the administration.

Thus, the sword was turned into a powerful financial and political device, well beyond its military value. This also means that Israel has a great investment in conflict making: crisis, war, and adversity strengthen its industrial core, making it more lucrative. The raison d'être of Israel has turned full circle, from defending itself and its ill-gotten territorial gains since 1948—to becoming a purveyor of conflict and means of destruction and "security," including advanced methods of surveillance, spying, and controlling, not to mention subjecting, subjugating, and destroying—the whole range of activities that undemocratic regimes (and even some who claim democratic credentials) are invested in and depend on.

Much of Israel's MIC sales are in the Third World, democratic or otherwise: "Mention any trouble spot in the Third World over the past ten years, and, inevitably, you will find smiling Israeli officers and shiny Israeli weapons on the news pages."[17] That leaders of impoverished and underdeveloped nations spend the meager resources of their societies on arms and conflict is bad enough; that Israel has become rich and powerful in part by purveying armaments to such politicians is doubly shameful. It became the frame of Israeli politics—war and conflict are the preferred methods of dealing with reality and its challenges. We may well wonder, with Jeff Halper:

> So why, then, does not Israel end its conflict with the Arabs and Muslim worlds, in particular by ending the occupation? Why did it not agree to a two-state solution back in 1988? Most pointedly, why does it *choose* to be the most militarized state in the world?[18]

Halper identifies a type of existential lethargy or stasis: "Israel lives and has always lived in a state of deep cultural militarism."[19] Israel has used such methods even before it was born to resolve its relationship with the Arabs of Palestine, as well as those around it. Most of the creative energy of political Zionism, as well as its cultural dynamism, went toward the production of an efficient, deadly, and ruthless sociopolitical-military machine; its wherewithal—physical, cultural, and ideological—remade the society in its own image; "doing" and living conflict created a society that feeds on aggression, thriving by it. This, more than the utilitarian conversion of adversity into advantage, is its raison d'être. When society is deliberately transformed into an efficient fighting machine, conflict is its organizing principle.

But why did Israel turn itself into a difficult-to-maintain militaristic enterprise? The answer must be sought in Zionist history of the first half of the last century. By 1929, more than three decades after the initiation of the Zionist project, the movement managed to purchase less than 5 percent of arable Palestine. The Yishuv remained a small minority with no chance for realizing the dream of a Jewish State without non-Jews marring the landscape. It was clear that it would not be possible to acquire the whole country, or even a substantial part of it, through land purchases, an almost impossible project by then; the only means for controlling Palestine and ejecting its population was military force, as noted by Wolfe.[20] Zionism and then Israel had to engage in militaristic enterprises or give up the ghost—it was the only way to win. When such an approach is adopted, changing course is not an option. The injustice and brutality involved in achieving the forced expulsion of the Palestinians set in motion further acts of violence, leading Israel deeper down a "reduced option" roadway. More pacific routes were abandoned as unfeasible and could not be retrieved. Militarism is a one-way road toward even deeper militarism. For this pattern to change, everything needs to change, and that has not been an option.

By becoming dependent on the United States for technologies it could not manufacture, Israel accepted a high level of dependency, carrying certain risks. This has been historically mitigated by a number of factors, such as by keeping its own technological capabilities up to date and producing most of its own materiel. By modifying and improving American weaponry, Israel is fulfilling a crucial role for (and has become an integral part of) the MIC of the United States, as many of its design improvements have been incorporated in product upgrades.

The United States has paid for Israel's MIC to develop innovative systems of its own, later to be adopted by the US military. One case in point is the SAM Kipat Barzel, developed in Israel against rockets dispatched from Gaza and Lebanon, for which extant US systems did not offer satisfactory solutions. The development had lasted almost a decade, and, although it offered only a partial solution, it was accepted by the United States for further development. Such partnerships have reinforced Israel's position with the United States and its NATO partners and have further secured the generous support it enjoys. They have also underwritten the political and military carte blanche enjoyed by Israel.

It is impossible to cover Israeli nuclear weapons development in this chapter, especially because it is obviously not part of the market economy of the MIC. It will be useful, though, to relate a historic shibboleth that divided the Israeli elite around this topic. During the early development of the nuclear program, two camps formed for and against evolving nuclear devices as costs climbed higher than even externally boosted invisible budgets could sustain. Once Ben-Gurion, the originator and main proponent of the nuclear option, left the government in 1963, he was replaced by Levy Eshkol, and the cabinet decided to slow down the Ben-Gurion plan but retain a nuclear option, promising the United States to demote its nuclear development in return for US guarantees of conventional arms supplies and political and ongoing support for the territorial status quo.[21]

Levy Eshkol, Yigael Allon, Golda Meir, and Yitzhak Rabin believed Israel could not support the dual-track development of both a conventional armed force and a fully fledged nuclear option. In their view, continued nuclear development might seriously harm the IDF, hampering its supplies through shortage of funds, as the Dimona project costs soared. A nuclear option, especially during the 1960s, was out of the question for most nations; only four superpowers had successfully implemented it. There is no reason to think that Israel could develop this option more cheaply or quickly, and it was considered a white elephant by the new government. The fact that slowing it down was also benefiting the modernizing of materiel through US agreement was proof that the cabinet was right.

This also meant, ironically, that Israel could invest much more in the development of the MIC and hence become more independent, while the cost of the nuclear option was shackling it more decisively to its Western providers. The supporters of the nuclear option were Ben-Gurion and his political allies, Shimon Peres and Moshe Dayan, then out of favor. Interestingly, they still won in the long term, and by 1966, Israel was ready to produce its first nuclear devices.

The capitalization of the industry, which is what the cabinet was worried about in connection with the nuclear option, was enabled mainly by the reparations, as previously mentioned. In the years 1953–64, the level of industrial investment in the MIC went up some 900 percent (in 1967 prices)—a very steep, persistent rise.[22] This also helped to stem Jewish unemployment. This high level of capitalization did not stop Israel from dramatically increasing its arms imports, which rose from $31.2 million in 1955 to $249.2 million in 1968. Unilateral transfers enabled simultaneous growth of both tracks while allowing for no drop in the standard of living—a unique achievement in such a small economy.

Thus was built the economic model for a war economy, perfected during the first decade and a half of Israel's existence and applied ever since. With demand for advanced arms increasing further after the end of the Cold War, and certainly since 9/11, Israel was able to weather the financial storms of the new millennium better than larger and richer economies, which were more exposed to market forces. As pointed out by Shir Hever, Israel's MIC, as opposed to the American or British ones, remained centralized and state-controlled even after partial privatization, so that it was more immune to market rises and drops, as well as remaining a nationalized asset.[23]

With new military campaigns initiated (mainly by the western alliance) in the Middle East and Africa, and to a lesser degree by de-communized Russia, selling arms and "security" was guaranteed to bring substantial returns on investment. With Israel using the Middle East as its war lab, its business spread further than ever before. Apart from selling the weaponry, it also used its "tested in action" label to sell security hardware and software for spying on city populations, industrial security, internet surveillance and training of armies, police, and security forces—both public and privately owned. Stockholm-based Peace Research Institute (SIPRI) produced a table showing the top 100 arms-producing companies table for 2017; three Israeli companies made the list:

33. Elbit with $3.395 billion
47. IAI with $3.538 billion
51. Rafael with $2.258 billion

The figures are from the most up-to-date SIPRI database at the time of writing.[24]

Numerous other Israeli companies sell arms and training bringing enormous wealth to the Israeli economy. The three companies above have earned almost $10 billion, and the total for 2017 was probably in the neighborhood of $13 billion—a colossal income for a country the size of Israel.

Another SIPRI statistic is the apparent decline of Israeli military spending since the early part of the century. From 8.5 percent of GDP in 2003, it fell to 4.7 percent in 2017—a steep drop.[25] The explanation is simple—since the early part of the century, other countries have been bearing the brunt of the cost of the occupation; the United States, European Union, Japan, and some countries in the Gulf have subsidized Israeli control. The rest of the story is substantial growth in unilateral transfers since that time, mainly from the United States and European Union, as well as the continued rise in Israeli GDP as a result of arms exports. Thus, despite the actual defense spending of Israel over the same period rising slightly from $15.591 billion in 2003 to $16.489 billion in 2017, the percentage of GDP has actually fallen—because the unilateral transfers are not included in the budgetary accounting, and Israeli GDP keeps climbing as a direct result of arms exports.[26] Indeed, so successful has Israeli arms sales been since the 1990s that six nuclear submarines have been ordered from Germany since 2012. Israel is now the equal of the United Kingdom in its ability to deliver multiple nuclear warheads.[27] Indeed, one of the main corruption charges against Netanyahu, the prime minister responsible for the submarine purchase, is related to this deal.[28]

More importantly, the IDF and its appendages, now civilian, privatized companies, have been the major industrial force in Israel's financial life. Initially this activity was also channeled through the many combines and conglomerates of the Histadrut. Later, with Israel becoming a market economy, the MIC shifted toward heavily subsidized private ownership, as in the United States. Such conglomerates are massive income generators, buffeted by IDF orders and foreign sales, and by capital transfers from the US federal budget. Israel, the smallest society to enjoy one-way disbursements from Washington, also benefits the most from US aid and defense budgets, and such payments have steadily grown, to a staggering $8.5 million each day, or a total of $3.1 billion in 2014. According to a Congressional Research Service report, the 2015 budget gave Israel 55 percent of US global military aid.[29] This enormous direct boost of Israel's military and related industries gives Israel an inflated, artificial profit and finances much of its production for export, which also includes huge sales to the US defense and security industries and military. The stake of the IDF and MIC in Israel's financial and industrial life is constantly expanding, with income and exports growing, having now reached more than $12 billion—how much more is unknown, the figures being secret. High-ranking officers fill most senior posts, not only in the MIC but at most other industrial concerns. The managers have come exclusively through army ranks, but now, with academia totally integrated into the MIC, scientists in key positions hold parallel university posts. Kimmerling supplies evidence for the trend of the MIC becoming the lead industrial and financial institution in Israel in a comparative table giving figures for 1976–77. Since his work was published in 1985, this trend has increased exponentially. Israel leads the world in all three measures: percentage of the workforce, of expenditure per capita, and as part of GDP.

The ten countries with the highest per capita spending on defense in 2009 are presented below:[30]

Country	*Per capita spending on defense (in US$)*
United States	2,140
Israel	1,882
Singapore	1,593
Saudi Arabia	1,524
Kuwait	1,289
Norway	1,245
Greece	1,230
France	977
United Kingdom	940
Bahrain	912

Per capita military expenditure on defence in 2009, in US$
Source: Stockholm Peace Research International Institute

Such figures do not include many other, hidden spending categories, such as research and development, where Israel is at the top of spending per capita, as well as security industries. If these were included, Israel would lead the world in its military expenditure per capita. This is unlikely to change.

Political Objectives of Israeli Arms Trading

The sale of arms has been used by powerful nations to control and influence weaker and dependent ones, especially in the Global South. The purveyors of deadly armaments provide a security net for such tyrants (protection from civil unrest and popular uprisings) so that they may continue plundering their nation unhindered. Such cases proliferate. Israel has provided armaments to dictators and strongman leaders on four continents—Europe, Asia, Africa, and South America—bolstering not just Israeli financial interests, but also its political power, as well as the interests of the United States, its paymaster.

After many decades of this activity, Israel has become a purveyor of death and destruction, a beneficiary of suffering and violence. Beit-Hallahmi, paraphrasing Marx, defines it accurately:

> Existence determines consciousness, and the existence, the immediate experience of most Israelis does not include the Holocaust. What it does include is the life of settler-colonialists and the constant war against the natives. The Jewish history of persecution and oppression is remote. The war against the Palestinians is close and permanent.[31]

In Africa, Israel supplied and supported Idi Amin of Uganda and "assumed full responsibility for the development of the Ugandan armed forces in 1966."[32] Zaire's Mobutu Sese Seko, who plundered his country for more than three decades, was one of its best clients. Mobutu himself was, like Idi Amin before him, trained in Israel, as were large numbers of his soldiers, and he later used Israelis to train his forces. In Ethiopia, Israel supplied and trained Emperor Haile Selassie's armed forces for decades and provided the same service for the Marxist regime that replaced him: "Whatever Israel was doing for, and with, the Marxist government of Ethiopia has been coordinated with the United States. According to the *Economist*, the United States has used Israel to 'keep a channel to Ethiopia's Marxist leaders.'"[33] Beit-Hallahmi also notes that the CIA financed Israeli operations in Africa:

> The total paid by the CIA was estimated to be in the millions—perhaps as much as $80 million ... A top-secret White House document, dated May 23, 1967, reveals that the United States increased its allocation to Israel for operations in Africa by $5 million for 1967.[34]

The same friendly services were extended to other repressive governments in Africa—Sudan, Morocco, and a clutch of colonial regimes: Algeria under the French, Rhodesia, and of course, the main client of Israel on the African continent, and one closely connected politically to Israel: apartheid South Africa.[35] The relationship with the apartheid state included not just the normal items required by a repressive regime but encompassed also the nuclear arena. Israel has extended its support to sharing its secret nuclear development with South Africa, presumably also sharing the testing of at least one nuclear device, in 1979; the testing in the ocean off the South African coast was discovered by a US spy satellite.[36] The Israeli nuclear program was indebted to South Africa. Ten of the twenty-four tons of uranium used for manufacturing weapons-grade plutonium in Dimona were imported in 1957 and possibly 1962 from the apartheid pariah republic.[37]

The close collaboration between both countries covered all areas—military, financial, cultural, political, academic, and scientific. The closeness between both countries was not a result of their strict anti-Communism and rightwing politics but was founded on their geopolitical realities: two settler-colonial states surrounded by a large number of hostile political entities, with both facing a wide-ranging political and commercial boycott. Only the fall of the apartheid regime brought about the end of this love affair. The words of the South African prime minister in 1960 have a strange resonance today, describing the similarities between the two states:

> Prime Minister Vorster goes as far as to say Israel is now faced with an apartheid problem—how to handle its Arab inhabitants. Neither nation wants to place its future in the hands of a surrounding majority and would prefer to fight.[38]

One aspect of the relationship with South Africa is of intense interest for us: the close collaboration with a regime that is not only very racist but led at the time by avowed Nazis and antisemites like Vorster himself and many of his ministers. Hating Jews did not affect their admiration for Israel. In a sense, says Beit-Hallahmi, "One can detest Jews and love Israelis, because Israelis somehow are not Jews. Israelis are colonial fighters and settlers, just like the Afrikaners. They are tough and resilient."[39] Breyten Breytenbach, quoted by Beit-Hallahmi, points out that this closeness is not obvious: "They have the greatest admiration for Israel, which has become ... White South Africa's political and military partner in 'the alliance of pariah states.'"[40]

This certainly rings some bells; in the last few decades, but especially since the return of Netanyahu as prime minister in 2009, Israel was closest to the governments and regimes in Europe where antisemitism is rife—Poland, Hungary, Slovakia—and in some cases also supported their leaders in antisemitic campaigns, such as the one led by Victor Orban against George Soros in Hungary during 2018. Just recently, words resembling Breytenbach's were published in many articles in *Haaretz*, when Netanyahu invited such leaders to a special meeting in Israel: "For Israel, those parties' friendly attitudes toward the Jewish state, and their hostility toward Islam, appear to be a seductive proposition. Even when, in those same right-wing parties, there are deeply entrenched anti-Semitic views."[41] These "infractions" are nothing of the sort, nor are these attitudes new—they are as old as Israel. Antisemitism has been a rather interesting ally of the state—the most important link between such racists and Zionism is a shared interest in getting Jews to leave their countries and emigrate to Israel. Beyond that, Islamophobia provides a platform they agree on.

Other nondemocratic regimes were also important partners. Typical were the Hutu rebels in Rwanda, responsible for the genocide in 1994. Israel supplied the murderers with the rifles and ammunition. Israel's support for genocide has become normalized. In 2017 it supplied arms to the Myanmar killers of the Rohingya.

In Latin and Central America, the same pattern was followed. General Pinochet in Chile was supported in every possible way by Israel.[42] The military regimes in Haiti, Argentina, Paraguay, Nicaragua (where, after the fall of Somoza, Israel continued to supply the contras), Honduras, El Salvador, and Guatemala were all bolstered by Israeli arms and training.[43]

The word "training" may be somewhat misleading, however. Beit-Hallahmi points out that in most cases, we meet Israeli mercenaries fighting in the service of repressive rightwing regimes against the population:

> Israeli mercenaries have become notorious; according to some estimates there are over a thousand such freelancers trading in blood and money, usually working for regimes that do not have popular support ... Most soldiers of fortune today are from Britain, the United States and Israel. In terms of population, Israel is the nation most overrepresented in the marketplace of mercenaries.[44]

Moreover, Beit-Hallahmi also points out that such mercenaries, which in most cultures are the dregs of society, in the Israeli case come from the elite: "Israeli mercenaries are unique; they come from among the finest of Israel's military."[45]

Thus, the export of death is not limited to materiel or training—the very elite is exported as Israeli ambassadors of destruction. A case in point is the famous Amiram Nir, who, after being exposed as the linchpin of the Iran-contra affair, has become a roving arms dealer and mercenary; he was allegedly killed in a plane crash in Mexico in 1988.[46] Evidence points to his having been murdered after the plane crash; his son has claimed that the CIA and George H. W. Bush were responsible,[47] while others blame Mossad.

In Asia, Israel was barred from supplying India with arms, because of Gandhi's reaction to the Palestine Nakba. This changed once the Bharatiya Janata Party joined the government in 1996; Israel started supplying India with arms. With the party's victory in 2014, both governments shared a rabid Islamophobic agenda, and India has become one of the main clients of Israeli arms, with multibillion-dollar purchases since then, especially in 2017, when it became the largest client of Israeli arms.[48] India follows a rich history of Israel's supplying arms to dictators and military rulers, in Burma, Taiwan, Singapore, South Korea, the Philippines, and Indonesia, all anti-Communist bastions fighting leftwing uprisings or neighbors. Similarly, the Sri Lankan ethnic cleansers, the Afghani mojahideen, and now even the

People's Republic of China has been purchasing armaments from Israel, in service of their "security" policy, such as repressing the Muslim Uyghurs. China's purchase of advanced weapons such as spying systems has been worrying the United States, because these weapons include technology that originated in the United States.[49] Chinese purchases included the Phalcon spy aircraft—an Israeli development of the US AWACS system—which certainly concerned the United States.

Israel's relationships with and sales to more than 160 states across the globe are clear indicators of Israel's global reach. It is clear that without its massive and sophisticated MIC, there would be no way for Israel to build and maintain such a network of financial, military, and political links. Its clients are certainly not inclined to censure its behavior against the Palestinians or vote against it at the UN.

The system described above, through which Israel supports US interests around the world, does elucidate its great value for US policy and political power. Nothing short of such global reach and advanced capabilities could command the huge sums that the US has invested in Israel since 1970—more than in any other country, big or small. Beit-Hallahmi describes the Israeli arms policy as a war with the Third World.[50] As a militarized settler-colonial entity, Israel assists the United States in arresting and reversing the process of decolonization in the Third World, through supplying the forces that oppose this process or that use newly gained "independence" as a vehicle for control and plunder. Israel is the enabling mechanism of many if not most of such regimes:

> Israel is always ready as a military supplier of last resort to desperate regimes, and its competitive advantage, so to speak, in the field of military exports extends to battle-tested and proven military software and counterinsurgency techniques, which enjoy a worldwide reputation for excellence in covert operations, and they have been exporting their secret-police experience, which is much in demand all over the world.[51]

Thus does Israel assist the global processes of disabling citizenship, workers' rights, civic action, or the struggle for political rights across the globe, systematically assisting the powers that be in controlling, oppressing, and eliminating resistance. For Israel (but not only for Israel), selling arms promotes a rightwing, pervasive antirights ideology and practice, shaping the global political landscape and serving and perpetuating injustice and brutality.

Past patterns of such collaboration between extreme and undemocratic regimes were studied by Meyer, Klieman, and Beit-Hallahmi, and given the apt label Pariah Club.[52] In the 1970s, this group of nations included Israel, Iran, South Africa, South Korea, and Taiwan, with additional members being added all the time as they were taken over by military regimes: Argentina, Chile, Bolivia, Brazil, Brunei, Philippines, Saudi Arabia, Singapore, Thailand, Zaire, and Uganda. All such countries are client states of the United States, but some were too extreme for the US Congress to continue arming them, and so they needed other suppliers.

All such regimes were shunned, at one point or another, by the international community (such as it is), and many have faced boycotts and sanctions; their survival strategy depended on collaborating with one another (hence the Pariah Club moniker). Most such countries controlled enormous mineral resources: gold, copper, diamonds, phosphates, uranium, oil, and rare earth elements required by advanced technologies, with minerals unevenly distributed. Others control unique industrial skills and capabilities that are necessary for other pariah countries. It was logical that this group, dependent on the United States but also wary of its fickle support, would come together to protect themselves from the shifting dispositions of US politics. If Iran has much oil, but other members have none, then they might exchange armaments (Israel) or uranium (South Africa) for it; South Africa and Zaire have large deposits of diamonds, and they need the Israel diamond industry to make them marketable. Such exchanges of commodities and services on the periphery of world politics have been the mainstay of these regimes for decades; they have been able to successfully face boycotts and sanctions, mainly because of the continued support of the United States and other pariah countries.

The preceding explanation is, to my mind, somewhat tenuous and even "organic" in its inclination, and so I would like to offer another interpretation of the same phenomenon. The club is described as having emerged from the political quagmire of rightwing regimes around the globe, but this emergence cannot have been mere happenstance. This was less an accident and more the result of the careful planning of the United States, through the State Department and the CIA, of a model of interdependence, rather than one of total dependence on Washington—politically, financially, and militarily. Such a modus operandi is much more sophisticated, achieving the same results with less expense and much less exposure for the United States.

If the members can support one another, systemic life is created by the network, which does not need continuous US life support at so many locations on earth. One may consider Henry Kissinger one of the architects of this web of interdependence, who participated in creating the conditions for it. We can see how this system functions in such cases as the US support of links between South Africa and Israel, or Israel and Iran, and later on, Israel and Egypt, or Israel and Saudi Arabia. In a way, this is one of the main reasons for the defeat of the Soviet Union, which was unable to build a similar system of global cooperation of leftwing states to support a global model of socialism.

Historically, the British empire has pioneered such interdependence, especially in the production of foodstuffs.[53] Once set up, such a complex network tends to look after itself through self-interest, and if it is well-designed, it will have longevity and be able to withstand occasional crises. In the case of the Pariah Club, it certainly did; it was designed to assist the demise of the Soviet bloc, and it played its part well.

The current iteration of this organizational principle can be observed in a new network woven by the Trump regime—Israel, Brazil, Poland, Hungary, India, Myanmar, Sri Lanka, Argentina, Saudi Arabia, and Egypt. New applicants for membership include the post-Brexit United Kingdom, Austria, Slovakia, and Serbia, to name but a few. As in the past, the United States is well ahead of Russia in organizing its pawns on the international chessboard. A crucial part of the new club, as was the case with the older one, is the armament industry, apart from the bloodcurdling crisis of basic commodities, which is awaiting the world due to climate change, such as the shortage of clean water or the scarcity of arable land to feed the world. The club is arming itself to the teeth, preparing for the new conflicts, so the MICs of Israel and the United States are stronger than ever.

Are Armaments Really Being Phased Out?

The global arms market suffered a downward blip after the 2008 financial crisis, as nations were temporarily unable to continue inflating their security spending. Some futurologists had declared an end to conflict, like the earlier end of history hailed by Francis Fukuyama. Sadly, such sightings of a peaceful future were somewhat hasty and not well-founded, as it turned out. Together with other rosy-spectacled prognostications of the new millennium, the internet, and the digital economy—the virtual reality of an end to conflict turned out to be just that. Virtual.

Israel suffered decline like other major suppliers. Israeli arms exports have been shrinking since the all-time high of 2012. A number of commentators have argued that arms exports have peaked and will now continue to decline, offering a range of explanations, including tidings of peace and prosperity built on digital engines of regeneration, which totally ignores the sobering realities of a planet being run down by its inhabitants. As it happens, the decline was quickly reversed, and by 2017, the last year for which data exists, a sharp hike in arms exports to a new high of $9.2 billion was recorded. This is probably the clearest evidence that universal peace and harmony are some way off, and that passengers on the planetary Titanic are being diverted by amusement on the doomed voyage toward water shortages and peak conflict.

In fact, claims of peak arms sales for Israel were erroneous, as sales have climbed even higher in 2018, though final figures are not yet available. One must also remember that such figures are only approximate, and that the real figures are substantially higher, because they do not include a number of categories, such as arms and security training and related spy and security technologies. The range of clients—most members of the UN—and that of products exported guarantee Israel a growth factor. It is interesting to note where most of the clients are based:

> Fifty-eight percent of these exports went to Asia and the Pacific, stemming primarily from the $2 billion defense contract Israel signed with India. Under the agreement, Israel Aerospace Industries will supply India with advanced Barak 8 air defense systems worth

> \$1.6 billion including missiles, launchers, communications devices and command, control and radar systems. Next in line is Europe, which took 21 percent of Israel's defense exports, followed by North America, Africa and Latin America.[54]

India is not only a huge client, it is also a relatively new one. Israel is now moving to become India's largest supplier of new weapons technology, together with Russia. In 2012, for example, Russia supplied 80 percent of the arms bought by India and 78 percent of the arms bought by China.[55] In recent years, Israel has overtaken Russia through very large arms deals.[56] There was a large sales spike in 2017, as Israel sold even more arms in Asia:

> The largest distribution of sales, some 58 per cent was in Asia-Pacific, with Israel's top three customers all originating from the region; India soaring ahead at \$715 million, with Vietnam at \$142 million and Azerbaijan at \$137 million in second and third place respectively.[57]

This is especially interesting because for decades it was the Middle East that led the world in arm purchases. It is now being replaced by Southeast Asia, Africa, and South America, despite the great turmoil in the Arab east following September 2001. This should certainly sound the alarm for future conflicts in these regions.

The Functions of the Arms Trade for Israel

Some very basic needs of the Israeli society and economy are served and delivered by specializing in arms and security, all working together to make it the heart of the social order and one that cannot be easily replaced. The first, historical rationale was fear of unemployment, which caused emigration of Israelis to other countries with larger economies. By employing the skills developed in the budding arms industries, Israel could support and develop an industrial base offering full employment, supplying the IDF and lowering overhead on weapon production. This helped to cover the high investment on research and development, which normally only very large and rich states were able to afford. At the time, Israel's balance of payment was disastrous, making it hard to sustain essential imports, such as arms, chemicals, metals, oil, foodstuffs, and medicines. Exporting arms was an opportunity to earn foreign currency to finance such imports. It would take some decades for this particular difficulty to disappear, mainly through the US unilateral transfer system after the 1967 war.

Second, arms supply can be used as a tool of diplomacy and political influence. This was one of the main uses of the arms exports of Israel in its early period, directed at garnering political support in the UN and the international arena, in an environment hostile to Israel because of popular support of Palestinians' rights. Another political end served by arms exports was the ability to join the United States as an agent against the Soviet Union and achieve influence in the Middle East and elsewhere. States that could not be supplied openly by the United States could count on Israel, and so could US policy makers.

All this made Israel a crucial appendage of the US system of influence and control. An example was the support Israel gave since the 1950s to the Kurdish resistance forces in northern Iraq, supplying and training them with modern arms in order to undermine the Iraqi regime. This long-term relationship influenced the US position toward supporting the Kurdish claims for a statelet of their own in the region—even as Kurdish claims managed to derail US aims in Iraq and Syria in the last decade.

Part of the political mission of the Israeli MIC was the maintenance of the Pariah Club. This type of collaboration with the United States has helped in building Israel up as the main partner of the United States in the Middle East and beyond and in guaranteeing the large sums it continues to invest in the IDF and its appendages. Israel has become the main technological partner of the MIC in the United States during the last few decades, not a typical position for a small country to attain and hold for decades.

The arms exports also supported exchanging technologies and resources, leading to further development of weapon systems. This was the case with South Africa and its supply of uranium and other elements required for the Israeli nuclear program, and both countries benefited by collaborating on the development of nuclear weapons and testing them. In the case of Iran, a major oil supplier to Israel in the late sixties and seventies, advanced nuclear and other weapons technologies were similarly exchanged.

The Israeli MIC boosts all high-tech industries in Israel; it supplies the components, systems, software, hardware, and (most importantly) skilled technical personnel. The MIC has become the major partner of Israeli academia, and universities could scarcely survive at the level they are used to without deep collaboration with the IDF. The deep synergy between the Israeli polity and the arms industry is part of the wider social synergy around the IDF and its violent and conflictual energies.

For this reason, one can hardly expect Israel to move away from its functional policies of support for armed conflict in the region and beyond. Such policies are not merely part of the Israeli system—they *are* the system, and its deepest rationale. They are totally unlikely to be replaced by peaceful and nonconflictual policies and industries in the foreseeable future.

11

The IDF Today

The IDF has been transformed over the last two decades, especially since the signing of the Oslo Accords. These changes in the nature, size, and character of the IDF were driven by the shifting challenges since 1967, when the IDF started functioning as an army of occupation. The last time the IDF faced a foreign army in major combat was in 1973. The ambiguous result of that war, with the clear achievements of the armies of Egypt and Syria early on, helped to move Egypt toward the Camp David agreement, initiated by President Anwar Sadat in his visit to Jerusalem in 1978. The peace agreement with Egypt, underwritten by the United States, removed the largest and best equipped and motivated Arab state out of the conflict.

From then on, Israel did not need to invest resources in combat with Egypt. The peace agreement with King Hussein, which followed, removed Jordan from the conflict. Syria and Lebanon were also neutralized, all of which led to the IDF concentrating on its main task: fighting a Palestinian population of (then) over two million civilians, without an organized or functional leadership, army, or any other civic or national government. The IDF became the Israel Occupation Forces.

Controlling a civilian population under occupation differs from waging battles against state armies. Enormous investment was required to build up the new system of control—scores of settlements and army camps were built over the decades all across the OPT, as well as in Sinai and the Golan Heights, which was illegally annexed in 1981, without any noticeable reaction from the international community. The new settlers were, like the old ones in the 1920s, 1930s, and 1940s, mobilized armed civilians securing the occupied territories, marking with their bodies and their luxury homes the new, expanded zone of the Zionist project.

IDF camps in the OPT, therefore, had multiple aims. It made the terrain known to army recruits and controlled the captive population through brutal measures, such as the Military Government and Emergency Regulations, inherited from the British Mandate of 1935 and the 1945 Emergency Regulations. Such illicit legislation was used for defending illegal settlements, redrawing the map by confiscation of private and public land.

The army soon built a system combining features adapted from similar trouble spots, such as Vietnam and Algeria. A wide network of informers, agents, and agent-provocateurs was set up to collect information. New vast prisons were built to hold more than 10,000 prisoners, and at the height of this conflict, more than double that number. Many prisoners were (and are) held without charge or due process as administrative detainees, whose detention may be extended indefinitely, without a right of appeal or even knowing the charges. A sophisticated network of interconnected, computerized checkpoints, roadblocks, electronic surveillance mechanisms, fences, and other means of limiting freedom/s of the population was built, making sure every single Palestinian was continuously tracked.

This system totally failed to warn the IDF of the most original of resistance efforts. This failure came to light during the first intifada that erupted in December 1987 in Gaza and quickly engulfed the whole of the OPT. Tanks, nuclear submarines, long-range missiles, nuclear devices, and F-16 fighters were of limited value in this conflict with civilians—mainly boys and young men and women, demonstrating and stone-throwing against the strongest army in the Middle East. Israel had no inkling that an uprising would emerge; it was genuinely spontaneous.

The PLO in Tunis was as surprised as the IDF when the intifada started and quickly spread. It was the result of genuine popular anger and frustration after two decades of occupation, with no hope for a peaceful resolution of any kind. It involved the whole population in a newly found sense of pride and purpose, offering the ability to resist peacefully and effectively. The shocked IDF, never expecting such resistance from an unarmed civilian population, acted with great brutality. More than 500 children were shot dead, with many thousands wounded badly; images of the uprising flooded screens everywhere, building support for Palestine and a deep revulsion toward Israel's transgressions to quell it.

This coincided with Israel's becoming part of the neoliberal globalized economy, with the IDF becoming a target of modernization and "efficiency" drives. It fell to Ehud Barak, chief of staff from 1991 to 1995, to suggest the solution: a small, smart army, less dependent on huge formations and more reliant on new technologies. Ben-Eliezer described the changes:

> The military reform was evident in a series of changes that Barak introduced. On taking over as chief of staff, he declared, "Anything that doesn't shoot will be cut," and immediately launched a series of measures that were intended, partly symbolically, partly concretely, to demonstrate an effort at economizing and the IDF's metamorphosis into a "small, smart army," in Barak's words.[1]

To achieve his aims, Barak built up the Staff and Command College, advanced the encouragement of outstanding company commanders and an ethical code for the IDF, and acted to raise the quality of the officer corps.[2] Some have claimed that such changes in the army were enabled and maybe even forced by the Oslo Accords. In truth, by the time that the changes were being enacted, Israel has already signed peace treaties with its main enemies, or had so hurt the rest of them that a new interstate war had become unlikely. Barak himself claimed that the transformation was part of the changes in the nature of modern warfare itself—the wars of the twenty-first century, he claimed, would be "pushbutton wars," and hence required that the army be revolutionized.[3]

Ben-Eliezer notes:

> Under Barak, this "new war" technological approach was reinforced by concrete steps, such as the development of spy satellites, an antimissile project, the use of helicopter gunships for infantry tasks, the development of RPV (remotely piloted vehicle, or drone) and the enhancement of the ability of headquarters to control the forces in the field through electronic devices and plasma screens.[4]

The changes initiated in the early 1990s proved the mainstay of the IDF in the coming decades and are forming the dominant outlook in military and political chains of command.

By transforming the army into a technological force, using advanced science to achieve its mission, Barak and his followers have also changed the makeup of the IDF, making the role of scientists, technologists, and academics central for the routine operations of the IDF. What has also become normalized is the growing privatization of many parts of the army and the transformation of sections into commercial companies. Private conglomerates were brought in to run parts of the operations, such as checkpoints, military prisons, camp construction, and the Apartheid Wall (separation fence in Israeli parlance), thus freeing more personnel for security duties, such as armed raids on Palestinian homes or the periodic attacks on Gaza.

The economics of this transformation is of interest. Since 1994 and the setting up of the PA to oversee securitizing the West Bank and the Gaza Strip, the IDF was able to leave most of the Palestinian areas A and B and rely on the PA's seventeen security forces, which were, and still are, mainly financed by EU subventions. Thus, the great cost of securing the OPT, which Israel has had to meet since 1967, was removed by the European Union and the United States. Israel continued its frequent raids, based on the IDF's plethora of spying and surveillance methods, covering every single Palestinian home. But the cost of managing the OPT has been substantially reduced, as well as the danger to the IDF. This, combined with the captive market of the OPT, has transformed the Israeli economy, making it relatively immune from the global crisis of capital at the start of the new millennium.

A simultaneous development, which has transformed the IDF beyond recognition, was the gradual replacement of the kibbutz-based military elite with one based on the settlers and Habayit Hayehudi (Jewish Home). This party, which has a lead in the settlements of the West Bank, is made up of religious, extreme rightwing, neoliberal nationalists who have become the mainstay of career officers of all ranks within the IDF. In the past, the officer corps had been left-leaning nationalists who belonged to the Mapai, Mapam, and Ahdut Ha'Avoda parties or strong and committed kibbutz movements; now they emerged from the neoliberal elite of the right and religious parties.

The combination of religious rightwing nationalism, racist and extreme ideology, laissez-faire economics, and a strong base in the OPT has had a long-term transformational effect on the IDF, pushing it even further toward the professional, technological, ideological, and extreme–nationalist force it has become. This new army mainly represents the interests of the elite in the OPT—the highly subsidized, racist nouveau riche of Israel, living beyond its "borders," in the mini-empire under its control. This process was unavoidable; the "left" lost control over the political agenda and its economic muscle after the spread of privatization and market economy. As a result, the Kibbutz movement lost its Zionist social function, becoming just another (quite affluent) small part of the new social structure of Israel.

The growth of this new elite, emerging from the IDF but branching out into science, technology, and finance, has meant that the old elite felt increasingly threatened. What has not changed much is the fact that the underclass structure of Arabs (Palestinian citizens of Israel), Mizrahim, Africans (Jews from Ethiopia), and to a lesser degree "Russians" have been largely barred from membership in the new elite, which, like the old one, is mainly Ashkenazi-based. In a sense, this social upheaval (whose poster boy is Benjamin Netanyahu) has created great tensions in Israeli society, taking it to the brink of chaos during the tents' protest of the summer of 2011.

Here, again, Netanyahu used the IDF to quell the protest, not by attacking the protesters, as was done in Egypt, but by drafting them into reserve service for two massive attacks on Gaza in 2012 and 2014. The attacks were supported by more than 90 percent of the (Jewish) population, but by the time the soldiers were demobilized, their difficulties returned, as well as their anger (although it was not sufficient to unseat Netanyahu, who won the March 2015 election by a comfortable margin).

This process of transforming the IDF into a front and base of national-religious ideology/praxis is continuing apace, with Netanyahu—a secular Jew pretending to be religious when it suits him—pushing the IDF further into the religious camp in order to satisfy his partners on the religious right. Indeed, this issue raises its head daily with the most obscure demands made upon soldiers by their commanders. Recently, such an incursion by officers caused an angry editorial in *Haaretz*:

> The report in Sunday's *Haaretz* that Home Front Command commanders had confined to base soldiers who refused to wear a kippa or a skirt when teaching in state-religious schools should be a warning light for the public. The kippa is not part of the Israel Defense Forces uniform and the army has no authority to punish anyone who refuses to wear one. This is a serious case of religious coercion and abuse of the army's punitive authority, aim[ed] at deterring secular soldiers from refusing to wear one in the future.[5]

Such secular views represented by *Haaretz* are less influential in a society that is heading for a large majority of religious and ultra-Orthodox members in the near future, based on population trends and the much higher birth rate of such communities. The editorial is a whistle in the dark on the part of a secular community, already isolated and cowered by religious propaganda and aggressive practices that have become the default in Israeli-Jewish society. The process of *Hadatha* (religionization) is seemingly unstoppable, strongly supported by most of the Jewish political elite, even by those who are secular. The fightback by *Haaretz* is noteworthy but hopeless.

Thus, the role of the army in society has indeed changed but has not lessened in importance. As before, it is a mainstay of the economy, with most of Israel's exports being directly or indirectly security related. The IDF is also a major force on university campuses,[6] a partner and developer of software, hardware, and a wide range of other equipment selling in Israel and abroad and a driver of research and development. The IDF (with its appendages) is the largest employer in Israel as well as the wealthiest organization in the country and the base of the new social and political elites.

Some of the most important changes resulting from the IDF's fighting civilians and facing militias, such as in Lebanon or Gaza and the West Bank, have taken years to be formulated into doctrines and regulations. The most famous of these might be the Hannibal Directive and the Dahiya doctrine, with both dating back to fighting in Lebanon and Gaza.

Even after leaving Lebanon in 2000, the IDF has returned a number of times to hammer Beirut and other parts of South Lebanon. During the last major invasion in 2006, not only were large parts of the Lebanese urban landscape devastated again, but more than 1,100 people were killed and more than 4,400 were wounded. The IDF also incurred heavy losses in comparison to earlier conflicts in Lebanon, and this led to reevaluation in IDF and government circles. The problem was not the enormous damage caused by an unjustified and illegal invasion—it was that heavy Israeli losses were suffered, and Hezbollah had managed to capture IDF soldiers, in some cases alive. A change was called for.

First came the Dahiya doctrine, named after the southern suburb of Beirut with large apartment blocks, mostly flattened by IDF bombs. The doctrine came to light through the WikiLeaks release of a US State Department document. In it, Gadi Eisenkot (who has since served as chief of general staff) outlined the main reasons for the new doctrine; it merits quoting at length:

> Section 2. (SBU) In his first interview in four years, OC Northern Command Maj. Gen. Gadi Eisenkot described a tense situation along the northern Israeli border—and suggested a crushing Israeli response should fighting resume. ... He argued that the Second Lebanon War was allowed to continue for too long; the next war —if it breaks out—"should be decided quickly and powerfully, without winking to world public opinion." Eisenkot made these comments to the Israeli newspaper Yedi'oth Ahronoth on Oct. 3.
>
> Section 6. (S) Eisenkot labeled any Israeli response to resumed conflict the "Dahiya doctrine" in reference to the leveled Dahiya quarter in Beirut during the Second Lebanon War in 2006. He said Israel will use disproportionate force upon any village that fires upon Israel, "causing great damage and destruction." Eisenkot made very clear: this is not a recommendation, but an already approved plan—from the Israeli perspective, these are "not civilian villages, they are military bases." Eisenkot in this statement echoed earlier private statements made by IDF Chief of General Staff Gabi Ashkenazi, who said the next fight in Southern Lebanon would come at a much higher cost for both sides—and that the IDF would not hold back.[7]

While most of the comments were made during interviews that the Israeli public was obviously privy to, it is of interest that the report itself remained secret. The publication of the interviews in

Israel worked to dispel the criticism of IDF conduct during the war, promising the Israeli public a tough stance in future conflicts, as though thus far the IDF had been wearing kid gloves and adhering to the Queensberry Rules. It also served to spread fear beyond Israel's "borders," especially in Palestine itself.

The Palestinians did not have to wait long. During the Christmas festivities of 2008, Israel moved into Gaza with enormous force, causing the kind of damage and destruction described by the doctrine. Almost 1,500 Palestinians were killed, mostly civilians, of which more than 400 were children. The devastation in Gaza was more intense than in any previous IDF assault. Israel hit at the infrastructure—electricity, gas, and water systems, fragile in Gaza at the best of times and already affected by the illegal blockade announced by Israel in 2006. All was damaged beyond repair, plunging the small enclave into a deep humanitarian crisis.

Neither was this to be the last instance of the Dahiya doctrine in practice, and as promised by Eisenkot, each time the IDF was to flex its muscles, the result was more terrifying. In the following two incursions into Gaza, during November 2012 and July/August 2014, the already terrifying level of death and destruction was surpassed: the IDF killed 2,310 Gazans in the summer of 2014 and wounded 10,626. Entire swathes of Gaza neighborhoods disappeared overnight, with more than 120,000 people losing their homes. The infrastructure, partially repaired since the last attacks, lay in ruins.

The massive cruelty of the attack was directed at breaking the resistance organizations in Gaza. Yet they fought back, causing relatively heavy Israeli casualties: sixty-six soldiers dead, and 469 soldiers wounded. In other words, the IDF paid almost as heavy a price in Gaza as it did in Lebanon in 2006 (though their losses were dwarfed in comparison with Palestinian losses). An excellent demonstration of this twisted logic can be gleaned from the 2019 election candidate's video clip for the freshly retired Chief of General Staff General Benny Gantz, head of the new party Hosen Le'Israel (Might for Israel). The first election video promo was a drone shot over the destruction in Gaza during the 2014 brutal attack, which Gantz had commanded; actually, the shot was taken by a Palestinian drone to show the extent of the destruction. Gantz used the clip and superimposed the following text: "Hamas is suffering one blow after another; 6,231 targets were destroyed; 1,364 terrorists killed; 3.5 years of quietness; Only the Strong Win; Gantz—Israel before everything!"

Here was someone responsible for the mass murder of civilians, using his "achievements" as election propaganda, apparently with great success.[8] His polling numbers greatly improved after the broadcast. A week later, he was due to win around thirty-five seats, with Netanyahu predicted second with around thirty seats. A few days later, Gantz and a few other generals fronted the new "party" Kahol Lavan (Blue and White) at its Tel Aviv launch, with Gantz flanked by General Gabi Ashkenazi and General Moshe Ya'alon. All three generals are proud holders of an ethnic-cleansing record in Gaza, a fact they all trumpet.[9] To balance the statistics, their slate also included Israel's first female general for good measure—Orna Barbivai.

All in all, the "party" included eight senior IDF and security services candidates, with Netanyahu only managing three in comparison.[10] That Gantz was so high in the polls had nothing to do with his politics, which no one was clear about, apart from the fact he was very fierce, a message he had until now radiated. His sudden ascent from the political nowhere to mass popularity was due purely to his army career. Even his sister admitted she did not know his political leanings.[11] Achieving the same number of Knesset members as Netanyahu is quite an outstanding result, obviously, but this in the end did not make him the prime minister, despite his murderous credentials.

What is of special interest here is the behavior of the "left" in Israel after the bloodletting in Gaza. Netanyahu was blamed for "not finishing Hamas" and "being too soft" by the leader of the Zionist bloc, Yitzhak Herzog, who then lost the election that followed the attack, despite his jingoistic prattle. The voters chose the one who hits hard, rather than the one who talks harder. Herzog was only reiterating the Dahiya doctrine in political terms: "In regards to security, I am more extreme than Netanyahu," he told the *Jerusalem Post* in July 2015, speaking about the new Iran deal to curb Iranian nuclear development: "Voicing his disappointment over the deal, Herzog noted that a country that 'funds, trains and nurtures terrorist organizations' was both detrimental to Israel and to its future."[12] The result of the election was the further shifting of the Israeli polity to the extreme right, the racialization of the public debate, the dehumanization of society, and the delegitimization of Palestinians as human beings.

While this sounds like political posturing, such gestures have real and devastating results. At one time, the Dahiya doctrine was also considered a political gesture, which was reevaluated in light of its implementation. In the reactions and initiatives of Netanyahu, Gantz, and Herzog, one can detect distinct signs that Israel is an army with a state. The army commanders form the agenda and lay down the rules of engagement and the terminology for politicians to adopt and follow. The circle opened in 1948 was closed in 2006, and ever since, it has been the rule: no leading Israeli politician may voice a position not mandated by the IDF. The Israeli Jewish–only democracy is underwritten by the IDF similar to the way the Turkish state was once underwritten by the armed forces. This political arena, operating within larger, stronger IDF circles in Israel makes the conflict unresolvable. To the IDF, reality appears as a series of targets through the gunsight.

The Hannibal Directive forms and dictates Israeli political decisions. This shadowy article of military faith, kept secret until the Gaza incursion in 2014, was apparently coined and formed in 1986, based on some IDF actions that placed soldiers in dangerous situations, especially the risk of being taken prisoner—dead or alive. Hezbollah had extracted a price when negotiating the return of Israeli soldiers, especially in 1996. Releases of large groups of resistance fighters had been forced on the army by public pressure within Israel; families of captured IDF soldiers demanded their release.

To resolve this dilemma, the Hannibal Directive dictates that captured IDF soldiers or their corpses must be found and returned to Israel, at whatever cost. In practice, it meant killing the captured soldiers, to prevent their becoming used by the enemy as a bargaining chip. This has recently attracted attention during the especially brutal attack on Gaza in 2014. During that operation, Lieutenant Hadar Goldin was reportedly captured by Hamas fighters, and in order to make it impossible for his captors to escape with his body, a huge military operation was initiated, including massive bombing of the whole area around the Rafah crossing, in order to make escape through the tunnels impossible. Almost a hundred Palestinian civilians were killed and hundreds of houses destroyed, as well as the infrastructure of the area. An Israeli probe by the army has found that Goldin was actually killed by IDF soldiers, who implemented the Hannibal Directive to avoid his capture alive by Hamas.[13] A year passed before the stark facts could be revealed, involuntarily and piecemeal. By releasing them in this way the IDF is forcing politicians into accepting IDF policies and actions after the event, without real control or influence.

Under the guise of the code word *security*, many of Israel's crimes were committed, hidden, and later justified. "Security" was used for instigating the Nakba, the great ethnic cleansing operation started in the winter of 1947 that lasted fourteen months. In 1948–49 it was used for hiding worse crimes and supplying distorted reports to the local and international press throughout the war. After the war, it was used to justify refusing refugees the right to return to their homes[14] and to attack Palestinian villages in the West Bank and Gaza before 1956, claiming the attacks had never happened. The same lies were told about the Lavon affair and the Israeli-led terror cells in Cairo in 1954. It was used ever since 1948 in every case when Israel committed war crimes or refused to implement UN resolutions, or both. It will later serve to justify the bombing of Beirut, of Syria and Iraq, the occupation of South Lebanon, further bombings of Beirut and Gaza, and operations in Libya and the Horn of Africa, as well as on the high seas and air piracy.

"Security" is the magical word that in Israel justifies any and all actions. The IDF has formed a political system in its own image so that such decisions can be taken democratically, making them seem more rational and acceptable not just to the population of Israel, but far beyond.

The Infantilization of the IDF Soldier

A fascinating note to end this chapter with is the presentation of IDF soldiers on Israeli media as "our children," "boys," "our boys," and "our sons" by commentators and politicians alike. This follows years of such characterization of Israeli soldiers in Israeli cinema, from *Beaufort* to *Waltz with Bashir*. These films presented the Israeli soldier as young, inexperienced, deeply troubled, confused, sensitive, or even poetic. This presentation would be quite disturbing for anyone who has watched the news on television in the last few decades.

The presentation of the indefatigable IDF as a collection of oversensitive children is certainly not accidental. In the now infamous case of nineteen-year-old medic Elor Azaria murdering a bleeding Palestinian lying on the ground, the medic had no regrets: "I would act exactly the same, because that's how I needed to act."[15] Immediately after the shooting, Azaria was called "the son of us all" by many Israeli politicians and commentators, and the same slogan was used in the many public demonstrations in support of the murderer. Elor Azaria has become a poster boy for the IDF, a nice boy, our child.

The choice of words is very interesting. Israelis know that most of the dead in the attacks on Gaza are children, among them many babies. To describe those who killed them as mere boys and children is to separate them from those other children, the ones without names, the ones one does not recognize. While Palestinian male children are described as young men when they die, Israeli soldiers remain boys when they kill. This infantilization of the soldier is also an infantalization of the public arena, of the debate, of the language itself. It is an Orwellian act of betraying truth and reality. We know from other arenas of mass murder that without dehumanization of the victims, it is not possible to get thousands of soldiers to kill, day after day, without guilt or remorse.

PART III

CONCLUSION: WHITHER ISRAEL?

12

More Divided than Ever

The mythologised sabra signified paradoxically the destruction of the Diasporic Jewish entity it claimed to rescue.

—Ella Habiba Shohat, "A Reluctant Eulogy: Fragments from the Memories of an Arab-Jew"

The aspirations of Israel's leaders, especially Ben-Gurion, were always directed at creating a nation through the IDF, because this was supposedly a democratic army, at least where Jews are concerned. Despite more than seven decades spent on the national formation project, it is clear today that this was not a successful process and that what divides Israelis is much more important than what unites them. It may also be one of the reasons that this "army of the nation" is failing in its military tasks, despite (and maybe because of) the increasing brutality it employs against noncombatants on a daily basis. By examining the history of the nation project's formation, we may get answers to both questions.

As a young state without the burden of an entrenched class society, one might have assumed that Israel might become a modern society unhindered by the historical weight of class politics. However, this has not been the case. The European Jewish society was itself atypical, as pointed out by the early-twentieth-century socialist thinker Dov Ber Borochov. In his first major work "The National Question and the Class Struggle," published in 1905, he argued that economically the Jews constituted an inverted pyramid. Whereas in a typical society, industrial and agricultural workers make up the base of the pyramid, among the Jews, few filled these roles, and the lack of normal productive life made them vulnerable to forces outside their control and led to their wanderings from country to country.[1] According to Borochov, this irregular feature of Jewish society must lead to the majority of Jews emigrating to Palestine—though at the time of his writing, only a tiny minority of Jewish immigrants had gone there. In Palestine, he believed, Jewish society might attain a normative pyramid structure, a conducive base for progressive social transformation along Marxist lines.

But Borochov represented a tiny, insignificant faction within the incoming settlers of the Second Aliya, during 1904–14. Though technically Ben-Gurion also belonged to that faction, he never took their positions seriously, abandoned them almost on arrival, and was hostile to Marxism throughout his life. Zeev Sternhell notes that any other form of progressive socialism, such as Marxism or democratic socialism, was automatically rejected by the Zionist leadership of Palestine because they were not based on the nation or ethnicity.[2] As a result, the type of socialism chosen and practiced by left Zionists was the degraded variety of national socialism, which the Nazis unfortunately immortalized in their party's name. Sternhell preferred the use of the label *nationalist socialism*, where the national was certainly more influential than the other half of the moniker.

This type of socialism appeared in the last years of the nineteenth century in Europe and "preached the organic unity of the nation and the mobilization of all classes of society for the achievement of national objectives."[3] It was not surprising that this tendency was socialist only in name; in practice it eschewed class conflict, as all classes had to "play their part in nation-building." Such ideology suited Zionism perfectly, with its echoes of organic development: "Nationalist socialism was based on the idea of the nation as a cultural, historical, and biological unit, or, figuratively, an extended family."[4] It is clear why Zionism found such an ideology attractive; national(ist) socialism was an exclusivist, organic, and racialized theory and practice, suitable for recruiting large nations for war or colonial settlement.

> All the life-force of the labor movement was directed toward the conquest of the land. Thus, a tribal view of the world grew in Palestine ... All the attention of the labor movement was focused

on the attainment and exercise of power, not in order to bring about a social revolution, nor to realize universal values, but for the sake of a national revolution.[5]

Clearly, this was not realized at the time by most progressives in either the West or the East; with few exceptions, this remains the situation today. With its large and powerful workers' parties, the Kibbutz movement and other socialist window dressing such as the use of socialist terminology at meetings abroad, the Israeli Labour movement was successful in persuading Stalin (in 1947–48) that it was a genuine socialist movement, while presenting Truman with a more accurate version of their aims. That this was a socialism for Jews only, and not even for all Jews, was never properly understood. One is reminded of the slogan of workers in apartheid South Africa: "White Workers of the World—Unite!" The Histadrut, the 1920 federation of Hebrew workers, was exclusively Jewish, as was the Kibbutz movement; why would they be anything else, if their main task was to "conquer the land"?[6]

Another irregularity that we have already touched upon is the unusual make-up of the Jewish population in Israel. Before the 1948 war, most Palestine Jews were European-born Ashkenazis, with the minority being Palestine-born Sephardis, living mainly in Jerusalem, Jaffa, Safad, and Tiberias. The Ashkenazis came as would-be agricultural workers, but most abandoned this calling after arrival and started living in the urban parts of Palestine. The Sephardis eschewed agricultural work altogether.[7] This left a very small group of agricultural workers, albeit quite influential, mainly in the kibbutzim. Moreover, the incoming immigrants from Russia and Poland included very few industrial workers, because most Jewish industrial workers in Eastern Europe were Communists or Bund supporters and avowedly anti-Zionist, unlikely to emigrate to Palestine.

Thus, despite the analysis and prognosis by Borochov, most emigrating Jewish industrial workers ended up in the United States and the Americas, and a small minority (less than 1 percent) went to Palestine. The Jewish pyramid remained an upside-down affair—imbalanced, irregular, and "unhealthy" in the words of Borochov and Herzl. The Histadrut organized mainly Ashkenazis and avoided accepting Arab workers altogether, being part and parcel of the leading Zionist settler-colonial institutions: "With the founding of the Histadrut, socialism became merely a tool of national aims, and the Labor movement unhesitatingly took the path of nationalist socialism."[8]

This nationalistic variety of European socialism seems to have contributed to a rather homogenous ethnic constitution in pre-1948 Palestine, but the society was made up of a large majority of traders and service personnel, with a minority of manual workers. The war and its aftermath brought much hardship—food was in short supply, the economy was in dire straits, foreign currency was badly depleted, and morale was low. Something drastic had to happen, and quickly. Despite the common perception that pre-1948 Palestine attracted only or mainly Ashkenazi Jews, research has shown that Jews from the Arab world came to Palestine in similar proportions to Ashkenazi Jews. It has been argued that they were ignored and discriminated against because of their smaller numbers; 90 percent of all Jews in 1895 were Ashkenazi.[9]

But not all Ashkenazis were the same. The sons and daughters of the prestate generation, those proud warriors for the two campaigns, Conquest of Labor and Conquest of the Land, looked down upon the newcomers from the Displaced Persons camps in Europe. The coded expression was "they went like lambs to the slaughter"; in other words, "they are not like us, the New militarized Jew." The fact that numerically, Jews were more heavily represented in forces fighting the Nazis than any other national or ethnic group was not yet known or understood.

Yitzhak Laor points out that the self-image of the new Sabra was distinctly Aryan:

> As you delve into Hebrew literature of the 1940s to 1970s, you meet more blue-eyed blonds than you could meet anywhere in Israel ... In Yizhar's works, the "others" are ugly Jews of Middle Eastern descent, or Diaspora Jews from Europe. In Shamir's novel that I spoke of earlier, the Other is a Diaspora Jew, a Holocaust survivor, described as a "podgy bald man," and also as a despicable crook.[10]

The survivors filling the abysmal Ma'abarot (transit camps) mingled with the Mizrahis—Jews from the Arab countries—who arrived after them, but most Ashkenazis had some family members who helped them out of the camps and into one of the large conurbations. Not so the Mizrahis, who were neglected and shunned.

Historians and sociologists studying the systemic inequalities endemic to Israeli society have traced them to a rather persistent form of Orientalism: "One important reason for this neglect was that the Zionist movement shared the Orientalist outlook of Europe and the European colonial movements, and considered its project an outpost of European civilization in the barbaric East."[11] Thus, the Arab Jews have been considered, like their brethren in the Middle East and North Africa, uneducated barbarians, inferior to the incoming Ashkenazis, the ones with the power to impose such definitions. Such Jews were more or less invisible for the leaders of Ashkenazi Zionism before 1945 and before the grim realities that emerged about the destruction of European Jewry.

For many decades ignored by Zionist planners as the "wrong human material" for Zionist purposes, the Arab Jews—from Iraq, Syria, Egypt, Yemen, Libya, Tunisia, Algiers, and Morocco—and the Jews of Iran and India had to be taken into account, as the potential of the Displaced Persons camps in Europe was exhausted. Even with many of the refugees who survived the Holocaust, Israel only numbered almost 650,000 Jews in 1948, and most of these were not agricultural or industrial workers. The Arab Jews (as well as the Jews of Iran and the Balkans) had to be considered, despite their "lower quality." Israel had to build two bodies of great importance for its future: its military forces and its labor force.

While the Jews of the Middle East and Egypt had mainly arrived by 1956, the Jews of the Maghreb came mainly in the 1960s; while the first group was settled in urban centers, and hence started to integrate early, the second group, mainly from Morocco, were forced to live in Ma'abarot in the periphery, surrounded by barbed-wire fences, and armed guards made sure they stayed put. Public housing of low quality was built for them in the Negev and the Galilee. In such locations they found it nearly impossible to find employment due to the isolation of the development towns and the conditions there. Many would have left Israel to return to their countries, or to live in France, if they had had that option.

Despite the substantial differences in the conditions facing the two groups, and their different degrees of integration, the Ashkenazi elite —officials, politicians, local and central government, media channels, intellectuals, artists, and academics, as well as functionaries like physicians and lawyers—viewed this large group of incoming Jews as seriously deficient, and treated them with racist disdain.[12] Not surprisingly, such orientalizing racism went hand in hand with a lack of discussion of socialist theory and what Strenhell refers to as the "poverty of intellectual life" within the Zionist Labor movement,[13] as well as a history of democratic lacunae and ossification, and the creation of a "political oligarchy." These warped attitudes leached into the wider polity and became entrenched.[14]

This discriminatory policy was inwardly justified by an implicit classification of the incoming immigrants. The first two definitions for incoming Jews covered the Jews of Europe: *chalutz* (pioneer) and *ole* (pilgrim), while the Arab Jews were considered mere *mehagrim* (immigrants). While the chalutzim were the leaders of the movement, using ideology to direct the activities of the large group of olim— themselves part of the Zionist movement and able to benefit from its ideology and be motivated by it, the Mehagrim were conceived of as a mass of individuals, impervious to Zionist propaganda, and able to serve only as an army of labor and cannon fodder, as far as the Ashkenazi elite and their intellectual trailblazers were concerned.[15]

During the late 1950s and early 1960s the country was covered by large transit camps, without being able to organize productive employment for the newcomers, exacerbating its existing difficulties. The newcomers were either traders or artisans, with few farmers or industrial workers among them, and Israel's plans of building a modern industrial base supported by this workforce had to be left in abeyance. The class structure emerging was again distorted; the power relationship reflected almost entirely discrimination along ethnic lines. The Arab Jews became the new underclass and would remain so for decades. They were not a working class in the European sense—they eschewed industrial and political organization, which was not part of their tradition and experience, and in most cases they were unable to secure well-paid or permanent positions.

Most Arab Jews remained unemployed, at best getting a few days' work every month on useless public works programs with miserable pay; sometimes they were not paid for months.[16] The resulting distress, deep shock, and violence in reaction to the official racism, denial, and lack of plans for resolving their plight was expressed through numerous outbursts concerning this long-term structural feature of Israeli-Jewish society. As a consequence, the Israeli bureaucracy and its compliant social scientists working for the government typified the mizrahis as

> frozen in their ways and deep in mental stupor ... were even lower than the city Arab population and resembled their feral Moslem neighbours. The majority of the Jewish population was illiterate, and like their neighbours—holding bizarre prejudices; their social life was totally frozen: no social movement of any kind.[17]

With such views, it is hardly surprising that the IDF could not be expected to patch the results of a racist social system. The percentage of Mizrahi officers was derisory, even among the lower officer grades in the IDF, and the most they could attain was the rank of non-commissioned officer. This locked them out of the social ladder that the IDF presented to the Ashkenazi Jews.

Indeed, the newcomers were isolated from all aspects of Zionist culture, then completely secular and leftwing. While in Iraq, and to a degree in Egypt and Lebanon, Jews had played an important role in the development of communist and socialist movements, such Jews have typically gone to Europe. The Jews of Morocco, the large single group, were mostly not acquainted with either communist, socialist, or Western liberal traditions, on the whole, and most were religious, all of which separated them from those supposedly looking after them; their Hebrew was biblical, used for religious rituals rather than daily conversation, and their other languages, Arabic, English, and French, were of no practical use and shunned by the authorities and by the elite. Writing about a new documentary series on television called *Ma'abarot,* Ariana Melamed aptly described the people they met in Israel: "Ashkenazi, secular, full of their new Hebrew superiority, extremely hostile to the Diaspora and those who have just left it; mainly wishing for immigrants in their own image."[18] Instead of finding the Land of Promise, they ended up in (for them) a Kafkaesque, labyrinthine culture to which they had no key. The political oligarchy allowed no space for opposition and built an exclusivist pyramid of power.[19] This could only lead to widespread alienation and protest.

Most North African Jews experienced European racism for the first time in their life, and despite the many reports about their ineptness, pessimism, and lack of energy or intellect, they mounted a campaign against the state that had treated them with such contempt. Sami Shalom Chetrit is sanguine about their reaction to the mistreatment they faced:

> One can clearly say that in the first few years the *Mizrahis* were not passive, and that the resistance had already been present in the *Ma'abarot* and the camps, and later in the *Moshavim* (cooperative farms) and the towns and more established city neighbourhoods they found themselves in, though this resistance was not recognised by the ruling establishment and academia as legitimate social struggle, but was dubbed criminal activity.[20]

The single social trait surviving the seven decades of Israel's existence, apart from its treatment of the Palestinians, is the power relationship between Ashkenazis and Mizrahis, typifying and bedeviling all aspects of Israeli society.

While the Mizrahis were crucial for the building of the army and the nation—that elusive entity absent in the 1950s according to Ben-Gurion's analysis—they were slow to organize along class lines, but started instead building associations predicated on ethnic identity, such as Moroccan or Yemenite. Indeed, the recent concept of grouping Mizrahis only emerged from decades of struggle, as the sectors of the oppressed and discriminated-against combined forces. The ethnically centered associations were easy prey for Mapai, which took them over (and in some cases also set them up[21]); these served as machineries for bonding the newcomers into the nation-in-the-making, through the crucible of the IDF; they also helped to subdue any radical efforts against the ruling Mapai.[22]

The Ashkenazi industrial and agricultural workers were as hostile to their "brethren" from the Arab world as they were toward Arab citizens of Israel. And while they were dragooned to join the Histadrut, they were never treated as equals and partners. There was no choice about joining the Histadrut, of course; workers of the period joined the Histadrut not as ideological partners of its mission, but as serfs or company workers joining the workers club in a company town. To enjoy basic rights, such as employment, medical treatment, and education, one had to be a member.

What is socially fascinating is the fact that being members of this powerful organization did not create the least sense of partnership, agency, or workers' solidarity in the Mizrahi workers who joined it. Due to the racism and discrimination so embedded in the state machinery, the Mizrahim espoused deep resentment and hostility toward the nationalist-socialist organs of Zionism, recognizing them as tools of oppression under fake socialist claims and pretensions. This would lead, eventually, to the solid support offered to Begin's rightwing party Herut, and to the great backlash in 1977, when Begin won power and changed Israeli political history.

One of the main losers of the backlash were the kibbutzim. While not losing financially—they were too strong for that—they gradually lost the stranglehold they had over the political arena. And by the mid-1980s they were no longer in power, especially in the IDF, which had started to fall under the influence of the Right and the settlers. After seven decades, the role of the Kibbutz movement was exhausted, and the widespread privatization of the Israeli economy, which has affected the Histadrut, had reached the kibbutz. There was much to gain for the small group of tycoons who controlled the economy, as they purchased the rumps of Histadrut industrial capital and globalized the economy. The kibbutzim turned into sought-after housing for lifestyle seekers wishing for greenery and birdsong, now that nationalist socialism had departed.

This raises the issue of class definitions within Israeli society. In no way can class be understood in traditional European terms when observing Israeli society; there is no history of class divisions that can be compared to European historical structures, as a result of decades of nationalist socialism. Neither the old Sephardi society, that existed before the beginning of Zionist colonization, nor the incoming Ashkenazi immigrants corresponded in any way to classical definitions of class; both consisted of a mixture of traders and artisans, without the classical polarities of capitalists and industrial workers or peasants.

The kibbutz members who definitely described themselves as the working class of Palestine were nothing of the sort. They could not exist without the support of the middle class and rich donors abroad and were a tiny elite of the Ashkenazi population. The less-than-genuine nature of their left tendencies is evidenced by the fact that their main struggle was not against Jewish capitalist employers; instead, they formed the battering ram of the colonial attack on the indigenous Palestinians and their labor, as discussed above. Indeed, the ruling party had at all costs eschewed the use of the term *proletariat*, recognizing that its members do not fit this description.[23]

Even the only group that eventually approximates the traditional position of the sub-working class, the Mizrahim, was nothing like the lumpenproletariat in Europe. What was more important in Israel was ethnic rather than class identity of people, disproving the Zionist claim that all Jews shared a common ethnicity and fate. Israel has resembled an ancient tribal society more than a modern class society. To a degree this is still the case despite the obvious mixing of the "tribes" through the IDF, education, intermarriages, employment, and social contact.

Historically, the incoming Mizrahis were settled according to a coalition formula—so many towns or villages were Mapai territory, so many others the latifundia of the National Religious Party, and so on. Such towns were one-party affairs until relatively recently. Each party supported the interests of its protectorates, so as to guarantee votes at election time. To this explosive ethnic mix were added the Orthodox religious parties. Those of the religious Ashkenazis, such as Degel Hathorah (Thora Flag) vied with the newcomer Mizrahi Shas for influence and power in the process of coalition-making after each election.

Furthermore, the Israeli electoral system of proportional representation has created a strong incentive for ethnic and religious parties to exist and thrive; they have become the crucial coalition partners of the larger parties of the left and the right, as Mapai and Herut presented themselves. Such distinctions between left and right in Israel, questionable as they were even in 1948, have no real value today. Their real identity is related to the historical battle between Mapai and Herut in the 1930s—two white Ashkenazi parties, coming out of and relating to a European historical setting and having no value in the Palestinian or Israeli context in which they operated. That one of the party leaders, Menachem Begin, understood the potential of Sephardi and Mizrahi voters meant that he was able to use the bitterness of such groups, at one time a majority of the Jewish population, to destabilize and eventually defeat the Mapai power base in 1977. This certainly was not a class struggle, but an intertribal conflict with racist and ethnic overtones.

This rather unique feature of the Jewish social structure in Israel has caused much confusion in the West, where parties of the left have traditionally and uncritically supported Mapai and other parties of the Zionist left, which ruled Israel until 1977 and continue to support their stunted and rightwing descendants such as Ha'Avoda (Labor), later rebranded the Zionist camp, as if these were parties of the left. The continued membership of such parties in the Socialist International (Ha'Avoda and Meretz are members) and the strong involvement of key Israeli politicians such as Shimon Peres and Ehud Barak were certainly reasons for this inappropriate partnership between European and other social-democratic parties with the ruling group of a settler-colonial regime of an occupying apartheid state.

The Political Economy of Settlement

Another development in recent decades is the growing power of the settlements, as they became larger and more populous. The population of the settlements constituted a whopping 13 percent of Israel's Jews in December 2015, and it is growing very rapidly.[24] This "community" of interest, residing illegally outside Israel's boundaries, has gradually come to realize that its interests are best represented by setting up its own political organizations, and the result is the new party Habayit Hayehudi (Jewish Home), which has been part of Netanyahu's coalition more or less since its inception.

Thus, to the various ethnic-tribal parties of Israeli Jews was added the party of the settlements' tribe. Like the multiplying settlements, they now have three extremist parties running for the April 2019 elections. One of the new parties, Otzma Yehudit (Jewish Power) is none other than the banned organization Kach (Thus), which has been described by the United States and the European Union as a terrorist organization and banned from elections in the past. This newish tribal grouping is probably the strongest power broker in Israel and is vying for influence with the "Russian" tribal party of Avigdor Lieberman, Yisrael Beitenu (Israel Is Our Home). Both parties find most of their support in the settlements, but while Lieberman's party is secular (to retain the support of the many "Russian" non-Jews), the party led by Naftali Bennett is Jewish-nationalist-religious. He has since created an even more extreme party, Yamin Hadash (new Right).

The racialization of Israeli politics has been completed by the coming together of the Arab-Palestinian political parties, forced by an ever-increasing spate of racist legislation aimed at annulling Arab representation in the Knesset. Recognizing this danger, the various small Arab parties came together to form the Joint List in 2015; they hold some 10 percent of the Knesset seats, while representing 22 percent of Israel's population. While not set up as an ethnic party, it nonetheless operates like one in what is basically a Zionist tribal, racist context. Some of its constituent parties have successfully fielded Jewish non-Zionists, like Offer Kassif, who was elected in September 2019.

With fifteen seats in the current Knesset (May 2020), the Joint List is the main opposition party, if not the only one. Needless to say, it does not enjoy the normative rights and privileges of an official opposition party in a democracy and is fighting for its life against further hostile legislation designed to remove it from the Knesset. In a recent article in the *Guardian*, Amjad Iraqi of the Palestinian legal rights group Adalah exposes Netanyahu's position:

> Netanyahu is right: Israel is not a country built for all its people, a fifth of whom (more than 1.5 million) are Palestinian Arab citizens of the state. This was the case long before the "Jewish nation-state law" was enacted last year, and long before Netanyahu returned as prime minister a decade ago. Since the state's establishment in 1948, more than 65 laws have been used to restrict the rights of Palestinian citizens in all fields of life, with more being passed every year.[25]

While the make-up of the Knesset is no longer static as it was during the 1950s, many of the new parties are just rehashes of old bodies in need of a new face. The Labor Party has had quite a few makeovers during the past century of its existence. It started as Poalei Zion (Workers of Zion) and then became, at different times and as different factions, Achdut Ha'Avoda (Unity of Labor), Mapai (Party of Palestine Workers), Ma'arach (The Alignment), Ha'Avoda (Labor), and recently, Ha'machaneh HaZioni (Zionist Camp). The Herut Party, a descendant of Jabotinsky's Beitar movement in the 1930s, has also gone through several facelifts and emerged as Gahal and Likud.

Other parties have also metamorphosed into new alignments and new identities, but, as Jean-Baptiste Alphonse Karr has put it, "Plus ça change, plus c'est la même chose." The only real changes are those that affected and were caused by real power shifts: the setting up of Mizrahi parties such as Shas; the establishment of the "Russian" parties such as Yisrael Beitenu; and the new party representing the settlements, which is itself a reworking of a past empty shell, Mafdal (National Religious Party), which existed for over ninety years but has dwindled away. New parties appear as new social actors form; this was true in the three cases mentioned.

The other obvious development, not at all unique to Israel, is the shifting of the national narrative and attitudes sharply to the right in the last decades of the twentieth century, becoming neoliberal and globalist. This has meant that the last vestiges of leftwing pretense have been removed from parties such as Ha'Avoda and Meretz, which represented what's left of the Labor Settlement Movement (the colonizing movement of the left in Palestine), which also meant the Ashkenazi population in the kibbutzim and cities. The grouping excludes other social groups, such as Palestinians (obviously), Mizrahis, and black Jews from Ethiopia, not to mention migrant workers or refugees.

This shift has mirrored the larger one in Europe and elsewhere, as the power of the Soviet Union waned and then disappeared altogether. As the United States has replaced France as Israel's military, financial, and political/diplomatic sponsoring empire after 1967, the social narrative has sharply shifted toward unfettered market capitalism, and new groups of business forces have joined the older, established ones. The arrival of the "Russians" since the 1970s has given Israel a crucial technical boost with hundreds of thousands of Soviet engineers and scientists with advanced knowledge, especially in a range of military industries, helping to transform Israel into an information technology giant in the 1990s and beyond.

Again, it was the Ashkenazi Jews who reaped the benefits of this development, making their "tribe" even more powerful than before; this required some adaptations of old narratives. The open racism against Mizrahis, so prevalent in the early years, had to undergo refinement. In order to pretend that parties are representative, most had to add Mizrahi candidates in realistic locations to their electoral lists, but such changes remained mainly cosmetic, with financial, political, cultural, military, scientific, and judicial power remaining securely in the hands of the Ashkenazi elites. The makeup of such elites has changed, though. From constituting the leadership of the kibbutzim and Labor Settlement Movement, it has turned into a much more complex grouping, comprising the new bourgeoisie of the Likud and its rightwing partners, with information technology and defense contractors joining the ranks of billionaires and replacing some of the old players.

This transformation was also structural. The large conglomerate that Ben-Gurion constructed around the Histadrut in the 1920s and 1930s, one which until the 1960s had controlled more than 60 percent of industry and the economy, has been gradually dismantled under Likud governments, to the point that it has even started concentrating on some of its neglected activities, such as representing (Jewish) workers. The Histadrut and the government-controlled utilities were privatized, as Israel turned into a laissez-faire economy, enriching the rich who snapped up existing sectors and established new industrial sectors of their own; the result was enormous growth in social inequality, with Israel in leading positions (sometimes first, when the Occupied Palestinian Territories are factored in) on the World Inequality Database.[26]

This rapid growth of wealth has brought about renewed unrest. As occurred in the United States, where black people saw each incoming wave of immigrants gaining social positions above them, so did the Mizrahis in Israel, not to mention the Palestinians. Despite political fig-leaf Mizrahi candidates, this large social sector is still suffering discrimination, and there is little evidence that this will change soon.

During the last decades of the twentieth century, Israel went through the spasms of removing power from the Histadrut, and one of the related issues was the privatization of the Kibbutz movement. Not only have the kibbutzim become privatized, but another, more important change has taken place; the land they control has also been privatized. This land, stolen from the Palestinians, had been a ward of the state, which freely loaned it to the kibbutzim for decades. With privatization, the land has become the property of the individual kibbutz members, who could use it in any way they wished. Thus, the members who have never paid for their homes or the land on which these were built (because they had been financed by the Zionist movement) became millionaires overnight.

This has justly angered Mizrahis, living mostly in moshavim and development towns (euphemisms for underdeveloped areas). They did not qualify for such largesse and instead live in small, badly built flats and pay monthly rent, without the ability to buy them or get them gifted.

An important radical organization, the Kedma Educational Trust, researched the unequal educational infrastructure for Ashkenazis and Mizrahis. Set up in 1993 by Mizrahi academics and activists, Kedma has established a number of model schools for Mizrahi children and worked on an alternative syllabus for schools, with some noted successes. As a result of its long-term action, some of the shortcomings of the education offered to Mizrahim were corrected, if only partially and grudgingly. But such successes were short-lived, and most such schools were closed by the government or local authorities.

Toward the end of the 1990s, Kedma publicized the deep housing inequalities; it was spurred into action by the glaring injustice of the kibbutz privatization saga. This led to rising anger in Mizrahi circles, not quite resolved by buying off local activists with choice appointments. This issue continues to boil under the surface, never totally extinguished, as the discrimination continues.

This came to a head recently during 2011, when a large movement for social change seemed to have been forming in Israel, influenced by the early, misleading achievements of the so-called Arab Spring. In hindsight, it seems clear that this was not a movement of the disenfranchised, but of newly impoverished and dislocated young Ashkenazi, who like other Western middle-class members, saw their social power shrink and costs soaring. The unaffordability of urban housing drove the protesters of the Tents Movement, with hundreds of thousands setting up tents at urban centers. The organizers, young Ashkenazi students, made some symbolic moves toward poor Mizrahis but failed to get them recruited in meaningful numbers.

This movement fizzled out within three months. It achieved little for its followers, apart from the placing of two of their leaders in the Knesset, the tried and tested method of ending social unrest in Israel. A few months after the protest, Netanyahu (who was trailing badly in the polls) decided to give the public its favorite medicine, an attack on Gaza. The 2012 attack on Gaza put an end to the social protest and boosted Netanyahu to a leading position again. While the Israeli-Jewish tribes are deeply divided on many issues, the "security" issue serves as the most effective social adhesive. The issues raised by the tents protesters have all disappeared into the dustbin of Israeli politics.

The positioning of ultra-Orthodox and nationalist-religious communities in Israel deserves some mention. The number of religious Jews, the great majority of Jews living in Palestine under the Ottomans, shrank to a small minority as secular Zionists have arrived since 1905 and especially after 1918. At the formation of the Jewish State in 1948, religious Jews were a small minority of the 650,000 Jews then living there, with 85 percent being secular; by 2010, only 42 percent of Israelis described themselves as secular. How did this come about?

As pointed out by Sternhell, "The religio-historical element as a focus of national identity has even greater importance in Zionism than in other national movements."[27] Since its formation, Israel has passed a series of legal exemptions for religious Jews, making their social status somewhat exceptional. The origin was the famous status quo agreed to by Ben-Gurion in 1948, based on a letter he wrote to the leaders of the ultra-religious party Agudath Israel on June 19, 1947. Agudath Israel leaders had demanded special legal status; they argued that most orthodox Jewish communities had been liquidated during the Holocaust, and they demanded special measures to keep the status quo in place, one being the release of all Yeshiva (religious seminary) students from military service, a provision that has survived, almost intact, to this day.

When Ben-Gurion asked how many would be affected by this proviso, he was told two to three hundred. Today, the number is closer to half a million. Yeshiva students are financed by the state through special coalition agreements, getting free education, unlike all other students, and they are further subsidized through a generous system of state bursaries based on family size. This has brought about a population explosion: the Yeshivas in Israel hold many more students than Yeshivas in the whole of Europe and Russia held before the Holocaust.

The enormous growth in ultra-Orthodox numbers in Israel does not necessarily represent the turning to God by millions:

> Some research shows that at least 40 percent of the members of ultra-Orthodox (Haredi) communities in Israel maintain the strict discipline of the Jewish Halacha not out of deep religious conviction, but more to comply with social and family pressure. It is widely known that in the world of the Yeshivas and study halls, much undercover secular activity is taking place.[28]

With the special payments guaranteed for Yeshiva students, most of them remain "students" throughout their lives, because the system allows for this fiction; they turn into professional Jews. Many of them are not up to academic rigor and would not be accepted by other academic institutions in Israel. By "studying" at the Yeshiva, they have a simple line to making a living:

> Most *Haredim* make their living out of being *Haredi*. The jobs of kosher controllers, yeshiva teachers, instructors or house masters, *mikve* (ritual bath) operators, undertakers, and many others are reserved for the large army of clerks and officials belonging to the steadily growing religious state bureaucracy.[29]

Thus, Israel has created a large nonproductive, dependent sector, without duties but with many extra privileges. This has been a leading issue in most Israeli elections, with the release from IDF service especially angering voters. The latest manipulator of this issue is Yair

Lapid, the chair and leader of Yesh Athid (There's a Future) Party who has capitalized on the promise to change the exemption policy once he is in government. Once he got into government in March 2013, after his party won nineteen seats in the Knesset, this promise, which brought him great popularity, was quietly abandoned.

While Israelis like to speak about the "demographic time bomb," referring to the fact that Palestinians are now (or soon will be, depending on which demographer one is listening to) the majority between the Jordan River and the sea, the real demographic problem lies in the Jewish sector in Israel. The growing Haredi community is now a real burden on the rest of society, especially the poor Mizrahim and the Palestinian citizens of Israel. The huge subventions to Yeshivot and the many tens of types of spurious and costly fictitious institutions, many of which exist only on paper, are impairing Israel's ability to deal with its serious social inequalities. Indeed, the Haredis themselves are enslaved by such support mechanisms—eschewing any type of birth control, their families are very large, and even with the generous benefits based on the number of children, they do not manage to rise above the poverty line. As many depend on the benefits system as their main or only income, this has become a rod for their own backs, while also punishing the rest of society.

Politically, much recent research has proven that the Haredi and national-religious communities are the most antidemocratic in Israel. This is clearly demonstrated in a recent Pew Research Center report.[30] According to this report, only 40 percent of Israelis are secular, with Jewish religious communities making up 41 percent of all Israelis. When asked about which should take priority, democratic principles or *Halacha* (religious law), 89 percent of all Haredis surveyed preferred Halacha, while 3 percent supported democratic principles. When asked whether the state should give preferential treatment to Jews, 97 percent of Haredis and 96 percent of national-religious said yes.[31] On many other questions the two groups have sharply diverged from the secular population.

This fault line is one of the most damaging for the future—not just of the Zionist entity, but of any chance for a just termination of the settler-colonial project. One can hardly debate democracy with political powers committed to divine rights and promises and whose idea of justice is racist and xenophobic. The same two groups are also very clear about the conflict: 59 percent of Haredi and 71 percent of national-religious are of the opinion that all Arabs (Palestinians) should be expelled, while the same view is held by 36 percent of secular Israelis. The net result is that 48 percent of Israelis support expulsion and only 46 percent are against it. By removing those who replied "don't know," the picture becomes clearer: more than 51 percent of Israelis support expulsion. With the swift natural growth of the Haredi and national-religious communities, this becomes an obvious conclusion: there is never likely to be a majority of Israelis for democracy and just peace. That so many secular Israelis also support expulsion is a mark of the influence of the religious majority and its priorities on the rest of Jewish Israelis. In less than a decade, with the majority of Haredi and national-religious even bigger, one can safely assume that the conflict will no longer be resolvable peacefully. With President Trump in the White House, any such hopes are injudicious.

While the conflict-resolution angle is the most important, such a religious majority in Israel will affect society in many other ways. The financial viability of the Israeli economy depends on a majority of the working population in real jobs, in production and services, as well as in the public sector. With the Haredi population growing at the rate it does—average number of children per family being 7.7—and the net cost to the economy it represents through the various benefits they accrue, both communally and individually, the typically high GDP of the Israel economy can no longer be sustained. This will lead, in the end, to social benefits being reduced or cut altogether. In Israel, a country constantly kept on the edge of war, such a hurtful reduction in social benefits may prove to be too much for the population to accept. The usual solution to social unrest in Israel is a military attack or invasion of Gaza, Lebanon, or somewhere else. While the suffering will be borne by Arab victims, it will also further reduce the economic freedom of such a government, making the social crisis even worse.

But the real problem for Israeli society can be seen in the missing element from the Jewish State, the Palestinian population. I have dealt elsewhere with the war of statistics and the use of statistics as a weapon by Israeli academics and politicians. The table and figure below clearly demonstrate this.

Census results of major faith communities in Israel, 1949–2014 (in thousands)

Year	*Druze*	*Christians*	*Muslims*	*Jews*	*Total*
1949	14.5	34	111.5	1,013.9	1,173.9
1960	23.3	49.6	166.3	1,911.3	2,150.4
1970	35.9	75.5	328.6	2,582	3,022.1
1980	50.7	89.9	498.3	3,282.7	3,921.7
1990	82.6	114.7	677.7	3,946.7	4,821.7
2000	103.8	135.1	970.0	4,955.4	6,369.3
2010	127.5	153.4	1,320.5	5,802.4	7,695.1
2011	129.8	155.1	1,354.3	5,907.5	7,836.6
2012	131.5	158.4	1,387.5	5,999.6	7,984.5
2013	133.4	160.9	1,420.3	6,104.5	8,134.5
2014	135.4	163.5	1,453.8	6,219.2	8,296.9

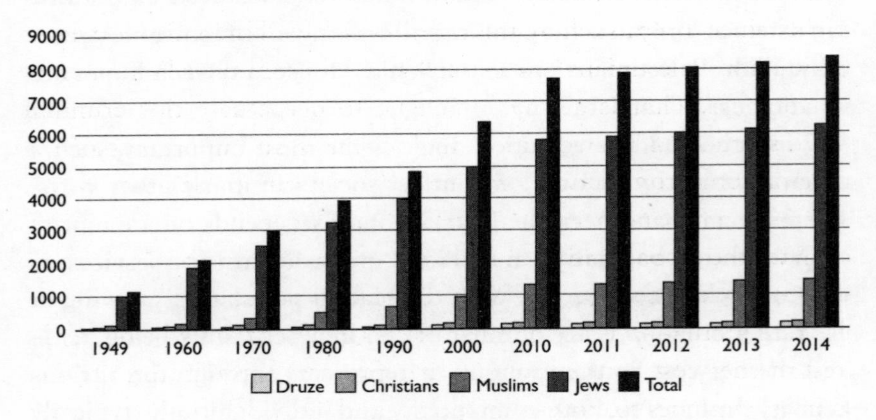

Israeli social groups

Source: *Statistical Abstract of Israel*, 2015, cbs.gov.il, accessed January 5, 2017.

First of all, the Palestinian population is split into three religious communities, while the Jewish population is not, though in reality it is. The Haredis do not even consider Reform Jews to be Jewish, for example, not to mention the secular Jews (who are worse than non-Jews in their opinion). Once we add the three Palestinian communities up, we receive the number 1,752,700, which together with the invisible Palestinians in Gaza and the West Bank, makes more than 6,200,000 Palestinians in Palestine. The number of Jews is also bumped up by including some 400,000 Orthodox Christians from the former Soviet Union, who, while being openly Christian (some even actively antisemitic) are still counted as Jews in order to bump up the figures.

Another way of looking at the figures is an overall analysis: Between 1949 and 2014, the number of Jews rose sixfold, artificially boosted by two large immigration waves—the Arab Jews in the 1950s and the Russian Jews starting in 1970. During the same period, the number of Palestinians under Israeli rule went up by a factor of 13, though they suffered deprivation and forced ethnic cleansing. This simple formulation is a clear picture of the "problem" faced by Israel.

Whichever set of figures one chooses, it is clear that due to the somewhat higher rate of natural growth of the Palestinian society, there will be more Palestinians than Israeli Jews within the historical Palestine, now completely under Israeli control. Most of this population has been living without any form of human or political rights since 1967, such as the right for self-determination. At least six million Palestinians are also living away from their land, mainly as refugees. That Israel has managed to perpetuate this criminal state since 1948 is evidence of the fickleness and unfairness of the international community, but one certainly cannot depend on this continuing for another seven decades.

With the global spread of the Boycott, Divestment, and Sanctions (BDS) movement, and with increasing opposition to Israeli apartheid, this situation is no longer static. As Israel prepares to annex the rest of the West Bank, probably in stages, the opposition to its war crimes continues to coalesce; Israel plans to annex the land but is certainly unwilling to annex the people and to give them full citizenship rights; doing so would annul its claim to being a Jewish State and is not acceptable to any Zionist party (in some ways, especially on the left). That is why they are flogging the dead horse of the two-state solution, one which Israel itself made impossible.

While the Palestinians in the Occupied Territories lack any semblance of human or political rights, including the rights to travel, work, property, and medical treatment, the Palestinian citizens of Israel are losing the limited rights they once held. The right to vote and stand for election is clearly being undermined to the point it can no longer be considered real. Palestinian citizens of Israel in many parts, like central Israel, the Negev (Naqab), and the Wadi Arra Triangle are having their homes permanently destroyed, unthinkable had they been Jews. Normally the excuse for home demolitions is the lack of a building license; in fact, a license cannot be acquired, as the authorities have never agreed to a local development plan by Arab municipalities, hence no building can legally take place.

Thus, the fastest growing community in Israel is not allowed to build on its own land, not to mention Palestinians in the Occupied Territories. The clearly racist motive of the authorities is to make life impossible for Palestinians, as a way of making them leave the country. Thousands of Palestinians have done so every single year, giving up the struggle to live as non-Jews in Israel. Many thousands have left the West Bank, as mentioned earlier, for the same reasons. This war on the Palestinians has been intensified recently by Netanyahu, seeking both to harm the Palestinians and to appeal to the most extreme sectors within Jewish Israel. When writing about the overnight destruction of eleven homes in one village alone, Kalansua, the editor of *Haaretz* notes:

> Although there was no urgent need to tear down the structures, Prime Minister Benjamin Netanyahu wishes to aggressively advance "enforcing building laws in the Arab communities."
>
> He and Public Security Minister Gilad Erdan know perfectly well that such acts will not solve the problem, but only exacerbate it. There are more than 50,000 houses that were built without a permit in Arab and Druze communities in Israel. Applying the law "blindly" means erasing entire neighborhoods and leaving half a million people without a roof over their heads.[32]

Lest we forget, he is speaking here of a quarter of the Palestinian citizens of the Israeli state.

Hence, while Israel was never democratic before the 1967 war —when its Palestinian population lived under a brutal military government of occupation until 1966—since 1967 the Israeli illegal rule was extended to all Palestinians left in historical Palestine. This circle cannot be squared without ending the Zionist project and the Jewish settler-colonial state with its racist apartheid regime. While this conundrum is badly understood outside Israel/Palestine, the BDS movement is acting to remedy this lacuna, and international public awareness of the illegality of the Israeli regime in Palestine is growing. This will become a crucial factor in enabling a peaceful and just solution for Palestine and all the people living in it, as well as the refugees living elsewhere; the next decade may be the only period when it can still be peacefully resolved.

It is clear that even today, real class solidarity is not possible in Israel, not even among its Jewish population, due to the great dividing lines drawn along national, tribal, ethnic, and cultural lines, not to mention religious differences. This also means that using the traditional labels of left and right, as if one were in 1970s Europe, is worse than meaningless. As most manual workers in Israel come from groups that are ideologically proscribed as lying "outside of the consensus" or at best on its periphery—foreign migrant workers, Palestinians, and some Mizrahi and Ethiopian Jews—solidarity between these groups is currently out of the question. Class in Israel is a much less meaningful identification than the other ones mentioned, because identity in Israel is tribal or ethnic. This point is crucial to understand, especially for Marxists (such as the current author) who habitually use this distinction as overwriting or overriding all others. If this insight is important elsewhere, it is much more crucial when looking at Israeli society and its complex networks of identity and belonging.

13

Is Israel a Democracy?

The nationalist ideology of the Jewish Labor movement was to conquer as much land as possible.

—Zeev Sternhell, *The Founding Myths of Israel*

In Israel and abroad, the standard position of Zionist apologists is that Israel has been a (Western) democracy ever since it declared statehood in May 1948, unlike the Arab states surrounding it. The propaganda insists that Israel remains "the only democracy in the Middle East." Interestingly, in its Proclamation of Independence, signed on May 14, 1948, Israel declared itself a Jewish State; the word "democracy" or "democratic" is never mentioned in the official translation of the Proclamation into English (1,075 words). Israel's propagandists claim that the concept of democracy is inherent in the text, partially quoted below:

> ACCORDINGLY WE, MEMBERS OF THE PEOPLE'S COUNCIL, REPRESENTATIVES OF THE JEWISH COMMUNITY OF ERETZ-ISRAEL AND OF THE ZIONIST MOVEMENT, ARE HERE ASSEMBLED ON THE DAY OF THE TERMINATION OF THE BRITISH MANDATE OVER ERETZ-ISRAEL AND, BY VIRTUE OF OUR NATURAL AND HISTORIC RIGHT AND ON THE BASIS OF THE RESOLUTION OF THE UNITED NATIONS GENERAL ASSEMBLY, HEREBY DECLARE THE ESTABLISHMENT OF A JEWISH STATE IN ERETZ-ISRAEL, TO BE KNOWN AS THE STATE OF ISRAEL.

> The State of Israel will be open for Jewish immigration and for the Ingathering of the Exiles; it will foster the development of the country for the benefit of all its inhabitants; it will be based on freedom, justice and peace as envisaged by the prophets of Israel; it will ensure complete equality of social and political rights to all its inhabitants irrespective of religion, race or sex; it will guarantee freedom of religion, conscience, language, education and culture; it will safeguard the Holy Places of all religions; and it will be faithful to the principles of the Charter of the United Nations.[1]

The long preamble in the Proclamation, only part of which is quoted above, so heavily qualifies the nature of this state that the claim of equalities that follows is rather meaningless. From the moment of its inception, Israel was intended as a state for Jews only, promising to treat others well, while not making them full citizens in any real sense of the word. As non-Jews, they would be lesser beings, existing in sufferance in the Jewish State.

The sentence on equality for all has never been honored in the seven decades since, as we shall see below. Indeed, from the outset, it became clear that inequality before the law is the main characteristic of the new state.[2] By carefully avoiding the word "democracy," the Proclamation gets around the depleted status of Palestinian Arabs in Israel. In other words, Israel was to be a herrenvolk democracy from its inception. If South Africa, about which this term was coined, was a democracy for whites only, Israel is a democracy for Jews only.

The most significant tests of the new state have been its relationship and treatment of the indigenous Palestinian population during the 1948 war and the period immediately following it. We have already learned about the many massacres and forced expulsions of the Palestinians before the May date, and later during the war, but it did not end there. At the end of the war, when the cease-fire agreements with Arab states were signed, Israel refused to retreat to the lines that Resolution 181 had drawn for the Jewish State and also refused to even discuss the return of the expelled population to their homes.

With their number exceeding 750,000, this meant that over half of the Palestinians were now refugees, and the UN passed Resolution 194 demanding their immediate return to their homes. The Arab states negotiating with Israel in Rhodes demanded that Israel honor the resolution, but once it became clear that Israel had no intention of doing so, and that no other country would intervene and enforce it, the cease-fire agreements went ahead, with Arab states refusing to sign peace agreements until the refugees were allowed to return. In so doing, Arab states honored the UN resolutions, while Israel has flouted them.

This brazen defiance by Israel and its refusal to carry out Resolution 194 was crucial. As an early test of the UN—itself very young, having been set up in late 1946—it showed that even a small and relatively weak state like Israel could defy the UN General Assembly with impunity. In fact, the Israeli settler-colonial state forcibly displaced most of the Palestinian indigenous population within a few months, the first time such a drastic act took place in modern times, with nothing done to enforce the resolution. Thus, it became clear that UN resolutions were merely placatory. This sorry state of affairs quickly became an established norm.

Israel's refusal to respect UN resolutions also flew in the face of the equality clauses of its own Proclamation of Independence. But this was easily explained: the expelled were not citizens of the new state and would never be allowed to return or to attain such citizenship. They were doomed to remain refugees with no redress or compensation. Most citizens of Israel accepted this racist, inhumane measure as necessary to retain the "Jewish character of Israel as a Jewish State."

With nationality defined as Jewish, rather than Israeli, non-Jews were automatically defined as others, not belonging to the Israeli polis. As such, many illegal measures could be taken against them with impunity; they were a "minority" since most of the Palestinians had been illegally expelled. As the Jewish population of the new state was 650,000 on the eve of its birth, the 750,000 Palestinian refugees were seen as an existential danger to the Jewish State, with their civic rights rejected out of hand. Some 154,000 Palestinians remained within Israel[3]—covering 78 percent of Palestine—making up one-fifth of the population of the new state. To leave them as full citizens with equal rights would be the least one may have expected of the new Jewish state, one that for self-evident reasons never adopted a written constitution; a written constitution could not have denied citizens full rights, unless Israel was prepared to declare itself an apartheid state, like South Africa had been since 1948. That option was shunned by Israel—it was simpler and neater to avoid a written constitution altogether.

Having failed the first test by denying the return of the refugees, Israel followed this with more draconian measures. After the ceasefire agreements were signed, Israel imposed a Military Government on all Palestinian Arabs living within its boundaries. The legislation in force was based on the Emergency Laws passed by the British Mandate authorities in 1936 and 1945 and controlled all aspects of daily life—severe restriction on movement, employment, political activity, including many punitive measures, ranging from curfew and administrative detention to forced exile.[4] The control established by the Military Government regulations gave Israel total power over its disenfranchised Palestinian population, leaving them in a state of siege for eighteen years, until 1966.

When the Mandate authorities introduced these emergency powers, the regulations were strongly resisted by the Zionist leadership:

> The laws were violently opposed by Jewish settlers in Palestine various ways. ... Dr. Dunkelbaum [speaking in a conference of 400 Jewish lawyers opposing the regulations], later to become a Supreme Court Judge, said: "These laws, of course, constitute a danger to (Jewish) settlement on the whole, but we, as lawyers, are particularly concerned with them. The laws contradict the most fundamental principles of law, justice and jurisprudence. They give the administrative and military authorities the power to impose penalties which, even had they been ratified by a legislative body, could only be regarded as anarchical and irregular. Defence Laws abolish the rights of the individual and grant unlimited power to the administration.[5]

This accurate description of the Emergency Regulations did not stop the astute Dr. Dunkelbaum from renewing the same regulations in 1950 and for years later, in his role as an Israeli Supreme Court judge.

Every single civic appointment, from a kindergarten teacher and driver to policeman and judge, was in the hands of the appropriate military governor, with the Shabak (General Security Service) controlling the population through a system of spies, agents, favors and punishments, administrative detention without trial, and denial of services to communities not servile enough. This has led, for example, to the existence of the "unrecognized villages," where many large communities have been denied roads, water, health, transport, electricity, or sewage services. Many such communities still dot the country, deliberately omitted from official maps. Collaborators act as agents of the regime in every village and town, offering costly interventions on behalf of the trapped Arab population in its dealings with the occupation powers. Many of these agents are themselves trapped, becoming collaborators only after having been jailed and tortured.

The IDF was, from its very inception, the main conduit of the state for all its Arab subjects, as most other normative state services were withheld or rendered meaningless. That this was in flagrant contradiction to the one piece of legislation that had any constitutional standing in Israel seems not to have impressed the Jewish population, apart from the few members of the Communist Party (which was not outlawed outright but was so carefully monitored and cajoled that it offered no cause for concern). The rest of the Jewish population willingly accepted the severe limitations imposed on the Arab population as necessary and justified. This denial of civil rights to indigenous citizens by the state created a virtual apartheid system, where people were defined according to their ethno-national origin, but more importantly, their colonial status. Colonizer Jews had full rights, while "Arabs" lived under military occupation, Emergency Regulations, and laws that applied only to them.

This pattern still applies to all thinking and actions in Israel and the Occupied Palestinian Territories. From that poisoned font of the Emergency Regulations flows all Israel's juridical systems applied to non-Jews to this very day. The Emergency Regulations, now in their seventieth year, have been annually renewed by the Knesset with the votes of the Israeli Zionist-Jewish parties, extending the system of double legislation into the second millennium, in defiance of any civic rights and normative legislation.

It was not long before another crucial bit of legislation, also of a constitutional standing, was passed by the Knesset. On July 5, 1950, the Law of Return was passed. This law gave every Jew in the world, wherever he or she lived, an inalienable right to full citizenship. This legislation, which put into jurisprudence the Zionist wish of "ingathering the exiles," was a further formation of the nature of the new state as a herrenvolk democracy.

By default, this law made the historically incorrect assumption that all Jews in the world belonged to one "Jewish" nationality. In this way, Israel has become the first and only state that denied its nationality, where its citizens may not hold Israeli nationality. This separation of ethnicity, language, religion, citizenship, and nationality allows the convoluted legal maneuvers that discriminate under the law.

Following quickly on the heels of the Law of Return, another law was passed that allowed the confiscation of absentee landlord property. This crucial legislation, itself illegal under the Geneva Convention and international law and contrary to the UN resolutions of 1949 and 1950, legitimated the illegal confiscation of most of the land area of Palestine, as well as all real estate of Palestinians banned from return. By confiscating their land and property, their fate was sealed; new immigrants, mainly of Arab origin, were placed in their homes and new settlements were built on their land.

Herzl had not gone that far. He assumed that Jews of the West belonged to the same ethnicity and religion—these were categories understood at the time. He did not consider Jews everywhere as being of the same nation or ethnicity. This ideology has been challenged, both at the time and since, especially recently by the innovative work of Israeli historian Shlomo Sand.[6] Sand maps the complex development of the concept of the *Judentum*—an amalgam of Jewishness and Judaism used in Germany to denote the totality of Jewish existence[7] —not as a nation, but as a group with distinctive common characteristics. The concept of Jews as a people is a very modern one. It was first projected by antisemites who bundled together Jewish European communities, perceiving them to be part of a plot to take over the world. When Herzl developed political Zionism as a cultural and political reaction to antisemitic pressures on the Jews of Europe and Russia, toward the end of the nineteenth century, he borrowed this racist typification from the enemies of *Judentum*. Only later did he include the Jews of Russia in his Zionist program, once he realized that they had reacted more positively and with greater commitment to his call for action than had his own "community" in central and western Europe. About the Jews of the Mashriq and Maghrib, Herzl knew next to nothing and did not include them in his projections, perceiving them as alien to European traditions and ways of life he mistakenly conceived as essentially Jewish. The Zionist leaders of Israel in the period following the establishment of the state shared Herzl's biases against Arab Jews, as examined in an earlier chapter, but had no choice but to reconsider; like Herzl, they found their call to Western Jews (their preferred communities) to emigrate to Israel totally unsuccessful, so they were forced to consider "lesser" groups in order to build up the fledgling state and bolster its defenses. The most pressing need certainly was to build the IDF.

If Zionism has been successful in controlling Palestine over the last seven decades, it was mainly due to the success of the IDF as a nation-building apparatus. This massive project of social engineering has indeed produced the Israeli/Hebrew nation. This entity is united mainly through the perceived common "enemy" and the process of subjugating Palestine and its people. To put this differently: it is the very act of oppressing the Palestinians that has produced and shaped Israeli identity. The flip side of this fact is the simple definition of Israeli identity as designated negatively, equaling the non-Palestinian.

Nations do not spring, fully fledged and armed, from Zeus's head. All are invented and constructed, as argued by Anderson, Sand, Renan, and others, and their traditions and histories are also the subject of invention and forgetfulness, as Hobsbawm and Renan have reminded us.[8] Most nations take more time to invent and fashion themselves than Israel had at its disposal in 1948. The imagining process started in 1897, but it could build on the long history of Judentum. Despite this background, the need to cobble a nation out of the discrete and distinct communities that cohabited in 1948 Palestine was most urgent.

What exactly could hold together ultra-Orthodox European Jews, the old Sephardi population of Palestine Jews, Zionist Jews of Eastern Europe, socialists of the Kibbutz movement, survivors of the death camps without any choice other than to migrate to Israel, and the Arab Jews that started to arrive from the Mashriq (Iraq) and the Maghrib (Tunisia, Morocco, and Algeria)?[9] This was not only a question of social, political, and educational manipulation, but also an urgent legal issue. For very different reasons, the leaders of Israeli

Zionism needed to settle the question of "Who is a Jew?" if Jews were to be treated as a nation, a single national group across all borders and boundaries. This was a crucial definition to quickly formulate, as Jewish identity was bestowing the right to full citizenship of Israel, without any further qualifications or restrictions.

The time to define Israeli nationalities came in 1952, through the passage of the Nationality Law. Intriguingly, the Nationality Law of the State of Israel is no such thing: it does not define Israeli nationality. (In fact, Israeli nationality does not exist, according to Israeli law.) The law defines the rights to citizenship of Israel or the rights of residency in Israel (and later in 1970, the Occupied Palestinian Territories). The main nationality identified in the Nationality Law of 1952 is Jewish, not Israeli. This was almost an obvious need—to define an Israeli nationality, the Knesset would have to pass legislation based on the actual population of Israel and hence could not define Israel as a Jewish State with extraterritorial rights for Jews, if Arabs were also to be Israeli nationals. The solution was to define not Israeli nationality but rather Jewish nationality as the main and majority nationality in Israel and to reserve Arab, Druze, or Bedouin nationality for the Palestinians.

The Jews became citizens of Israel; to a lesser extent, so too did the Palestinians living in Israel. The Palestinian citizens of Israel are, therefore, not living there by right, as do Jews, but by dint of residence, which may be revoked. While citizenship is difficult to annul, a process never applied to any Israeli Jew, residency can be annulled at any time. An example of this are the Palestinians of East Jerusalem, annexed to Israel in June 1967, immediately after the war. They became residents rather than citizens of the state, and hence their right to reside in Jerusalem (and nowhere in Israel) can be (and is in fact) easily annulled. In thousands of cases, a Palestinian leaving Jerusalem to study abroad, for instance, is denied re-entry into the country under the pretense that they have forfeited their residency rights by leaving the city.

Thus, the Nationality Law creates a dual system of legal apartheid —full citizenship for Jews and a reduced one for Palestinians. Though the definitions are treated as national identities, they are nothing of the kind. They combine religion, ethnicity, language, and culture to isolate and separate Palestinians into various identities (none of which is Palestinian). Meanwhile it treats Jews as a singular, unified national group, even though the common denominator between them is religious rather than ethnic or national. Thus, for example, Palestinians are divided into Arabs, Bedouins, Druze, Armenians, Circassians, and many other categories, while Jews of all ethnicities and nationalities become a unitary imagined community, the nation of Jews, a nation that has never existed anywhere prior to 1948.[10]

Hence, by a stroke of the juridical pen, Israel has become officially racist, denying not only the traditional equal rights normative within democratic societies but contradicting its own proclamation. While the Military Government has actively and practically controlled Palestinians in a stage of siege under military occupation since 1948, the Nationality Law has converted this division into juridical racism, a legally binding and enforced apartheid.

In 1953, the Land Acquisition Law was passed, confiscating the land of 349 Arab villages with a total area of 1.2 million dunams (around 468 square miles, or more than 1,500 square kilometres—most of the arable land in Palestine) for the exclusive use of Jewish Israelis. In the same year, the Knesset empowered the Jewish National Fund to purchase land and manage "state land" for the exclusive use of Jewish Israelis. In 1960, a law was passed stipulating that "the ownership of 'Israel lands'—namely the 93 percent of land under the control of the state, the Jewish National Fund and the Development Authority—cannot be transferred in any manner."[11]

An apartheid state requires extreme measures to retain its control. Israelis were not allowed to find out about the Lavon affair because military censorship had banned any mention of it on Israeli media outlets. The elite had been shaken severely by the affair, and the "activists"—those in Mapai who wished for military conflict with the Arab states—wanted Ben-Gurion to immediately replace Lavon and Sharett, since he had held both roles before retiring in 1953. Eventually, secret machinations, hidden from public scrutiny, engineered the return of Ben-Gurion on January 21, 1955, to the role of defense minister. One month later he initiated the Gaza attack, coded Black Arrow. He continued to undermine Sharett as prime minister and replaced him altogether in November 1955. Now Israel had two hawks heading its civic and military institutions: Ben-Gurion as prime minister and defense minister and Moshe Dayan as Chief of General Staff. Dayan had been champing at the bit for some time, arguing for a comprehensive military conflict with Egypt:

> This was not the first time that Dayan has expressed his desire for an all-out clash with Egypt—on the assumption that only in this way could Israel bring an end to the *fedayeen* attacks. The Chief of Staff first proposed mounting a major military operation against Egypt at the end of 1953, after the Straits of Tiran were closed to all shipping to and from Israel. In September 1955 Cairo announced that it was extending the blockade to flying over the straits—forcing El Al, Israel's national carrier, to suspend its flights to Africa—and on October 23 Ben-Gurion told Dayan to have the IDF ready to capture the straits. The army began drawing up plans and even sent a reconnaissance unit deep into Egyptian territory to establish routes for troop movement. Four days after the conversation between Ben-Gurion and Dayan, Nasser disclosed that he had signed a major arms' deal with Czechoslovakia.[12]

When Nasser moved to protect Egypt by rearming, IDF preparations for the 1956 Suez War were well advanced, but the problems were political, not military. Ben-Gurion was forced to report to Dayan in December 1955 that the plan they both hatched to capture the straits was voted down by the cabinet.[13] The cabinet was well aware that both its military leaders were prone to adventurism and demonstrative violence and took the logical way out of unnecessary and dangerous conflict. The assault initially was voted down by the cabinet. Ben-Gurion and Dayan knew that another way would have to be found in order to attack Egypt, and Shimon Peres was tasked with establishing the casus belli for this attack.[14]

Ben-Gurion and Dayan had a set of principles for military intervention, which would indeed worry any cabinet:

> The search for a well-calibrated retaliation policy has been going on since 1950, when the first infiltrations began. ... Jerusalem's retaliation policy from the outset was that the only way to deter the Arabs was to act according to the following three dictums:

1. Arabs understand and appreciate only the language of force.
2. If attacked, Israel must not remain passive; restraint will be interpreted as either fear or weakness and invite additional violence.
3. A Quid pro quo is not enough; Israel must exact twice the price—or more—for every attack.

Only in this way would the Arabs be moved to reconsider the advisability of every assault they planned against Israel.[15]

Schiff, a military correspondent for *Haaretz*, was obviously not critical of this orientalist, imperially inclined credo of the Israeli state when writing in 1974. In this he is typical of the Israeli elite who were in the know about the "irregularities" of Israel's policies but supported them with uncritical silence. Like all his contemporaries (excluding Uri Avnery, editor of the weekly *Ha'olam Haze*), Schiff yielded to the military censor and avoided telling the Israeli public about the Lavon affair. Such are clearly the signs of an autocratic, undemocratic, or even antidemocratic society in action. Schiff and his colleagues willingly accepted such limitations on their duty to inform the public; this they did because they lived in an apartheid state. Arabs were unwanted, unequal citizens, and because of their "alien" nationality, always suspect; nothing could be discussed openly and truthfully within Israel. Instead, the nod-and-wink system replaced free and unbiased journalism, and all Zionist papers and their journalists accepted the need to censor. Thus, one of the mainstays of any democratic polity, press and media freedom, was abandoned in order to bolster a racist, unequal society. This silent assent to undemocratic means has been one of the most important vehicles of social control in Israel, not only of the Arab minority but the entire social structure. By dubbing any issue as one involving "security," the state and the elites have literally removed it from the public sphere, from debate and discussion, and from any form of democratic scrutiny. This tendency, which came about after the establishment of the state, has never changed. That the Arab minority lived under Military Government until 1966 helped to mark them as the "enemy within."

The massacre of Kafr Qassem in which innocent and unarmed civilians were killed in a few minutes and without provocation was another example of how Israel viewed its Arab citizens. But for the intense pressure of the Communist Party, the story would have remained unknown. Under the racist military regime of occupation, Israeli Arabs were not considered (nor did they consider themselves to be) full citizens of the Jewish State.

An interesting but almost unknown feature of Israeli society is that the IDF controls, directly or indirectly, 80 percent of the land area in Israel. Moti Basok writes:

> 39% of Israel's state land area, covering a total of 8.7 million dunams (not including the West Bank) are controlled by the IDF; additionally, the IDF has enforced restrictions on civilian usage over an additional 8.8 million dunams, which are 40% of the Israeli land area—this emerges from the annual Report of the State Comptroller, the judge (retired) Micha Lindenstraus, on bodies and entities in the Israel Security system. This means that the IDF controls the large majority of Israeli state land.[16]

The report also notes that there are no legal agreements covering land use and control between the IDF and the Israel Land Authority, which has apportioned such use to the IDF.[17] The full report is even more worrying than the press reports.[18]

Another of the fundamental failings of Israel as a would-be democracy is its tribal social structure; since its inception, the party controlling the government has treated power as a right to which it is entitled, allowing it to spread the dividends of power exclusively among its supporters. Israel is supposed to have picked up some of the trappings of the British Civil Service, from whom it took power in May 1948. However, the only traceable heritage it followed was the undemocratic and draconian heritage of the Mandate legislation—the Emergency Regulations, implemented to crush Palestinian nationalism. This means that the "Arab Sector" (as it is euphemistically called in Israel) is the poorest part of society and has remained so for seven decades. It is also the least educated part of the population. This factoid is carefully husbanded by Education Ministry calculations, which afford Arab education half the government funding given to Jewish student education.[19] Adalah, the influential Arab legal centre, identifies an even deeper inequality in its report, quoting Israeli official government statistics:

> Arab citizens of Israel comprise approximately 20% of the population in Israel, and 25% of the country's school students. Throughout their schooling, from elementary to high school, Arab and Jewish students primarily attend separate schools. The education system in Israel is a centralized system administered through the Ministry of Education. Among the main responsibilities of the Ministry of Education are the development of curricula, the supervision of teachers, and the construction of school buildings. Local municipalities are primarily responsible for the maintenance of school buildings and their equipment. For the academic year 2000-2001, the government invested a total of NIS 534 per student on average for Palestinians, compared with NIS 1,779 per Jewish student.[20]

It seems a Jew is worth 3.2 Arabs. The educational anti-Arab bias does not end there; Nurith Peled-Elhanan, a Hebrew University professor of education, notes how Israel's education system defines and describes its Arab citizens:

> The Arab with a camel, in an Ali Baba dress. They describe then as vile and deviant and criminal, people who don't want to pay taxes, people who live off the state, people who don't want to develop ... The only representation is as refugees, primitive farmers, and terrorists.[21]

Is it a wonder that Jewish Israelis are habitually racist?

Similar points have been made about health, community support, roads, communication—on every aspect of public life in the "Arab Sector" in Israel. Obviously, any society where social investment and spending is based on ethnic, racial, or national identifiers is a racist society, not just in principle, but in daily practice. That this structural and legal discrimination has been part of Israel since its inception is even a deeper indictment of its civil society and its continued denial of this racist reality. No real democracy could allow such inbuilt inequality.

In the last few years a legal blitz has disenfranchised Palestinian citizens of Israel, further isolating and victimizing them. It is clear that in Israel jurisprudence and legislation were used as an extension of military occupation—a tool of normalizing and "making kosher" a wide series of internationally condemned military and political moves that Israel has taken ever since 1948. Israeli democracy is an oxymoron; it is not even a democracy for its Jewish population, let alone its Palestinian citizenry. Indeed, Israeli sociologists and political geographers have examined Israel in order to find an accurate definition that properly describes its peculiar nature. Oren Yiftachel has called Israel an ethnocracy[22] while Sami Smooha defined it as an ethnic democracy.[23] Other historians looking at similar examples, such as apartheid South Africa, have used a German term of the Nazi period and have defined regimes based on racialized marginalization as herrenvolk democracy,[24] which offers a more nuanced and less atypical definition. After all, Israel and Israeli academics have continuously argued for the exclusivity and exceptionalism of Israel, presenting Israel as unique in world history. Instead, one may examine the normative definition to check their fit for this case.

This elitist form of government is typically employed by the majority group as a way to maintain control and power within the system, and it typically coincides with the false pretense of egalitarianism. There is a prevailing view that as people of the majority gain freedom and liberty and as egalitarian principles are advanced, the minority is repressed and prevented from being involved in the government.[25]

For people in the West who are either unconcerned about or dismissive of Arabs in general, and of the fate of the Palestinians in particular, defining Israel as a democracy may seem unproblematic. For others, such self-delusion may be a travesty. While Israel was not a democracy even before the Netanyahu government took over in 2009, it is clear that the few threads that still connected its social structure to that of normative democracies were removed in the last decade, making way for a proto-fascist apartheid state.

Such understanding is now confined to the narrow elite represented by the daily *Haaretz*—increasingly worried about the ground disappearing beneath their small edifice. In a recent editorial, the now normalized Israeli disavowal of reality was avoided; in discussing the treatment of the Palestinian population under military occupation in the gray zone between Israel and the Palestinian Authority, controlled totally by Israel and considered its territory in defiance of international law, the editorial is clearer than ever before regarding Israeli hatred of the term apartheid:

Over the years, the High Court of Justice has refused to touch the hot potato of the settlements' legality. Rather, it has ruled that as long as settlers are living in the territories, their needs have to be met like the needs of any person living in a certain place. But what does that have to do with taking private Palestinian lands on which to build or expand the settlements?

The Justice Ministry's reliance on a controversial arrangement, used in Cyprus, is nothing more than a superfluous, damaging and manipulative action. The best solution for Israel is the two-state solution. The other possibility—granting full civil rights and equality before the law for Jews and Palestinians—would mean the end of the Jewish state. The third possibility is called apartheid.[26]

The small group of lefties at *Haaretz* is no longer part of the Israeli consensus and holds a marginal position within Israel. *Haaretz* is voicing right-wing positions more than at any time in its history. The clarity of definition is forced on the editors by the drastic moves of the rest of the Israeli polis toward racialized fascism. The tone of the editorial is of course guarded—it presents the situation as if not yet solidified, as if a proper argument may shake the Israeli public into reversing its slide toward the hated apartheid regime title. However, it is clear from the hostile comments by some *Haaretz* readers that its fragile social base is much reduced as the Netanyahu regime turns more and more of the social structure into a proper herrenvolk democracy.

Since this editorial was published, much has taken place in Israel, pushing society even further to the right, while leaving *Haaretz* and some other insignificant sectors in a lonely corner. One cannot enumerate all the laws that the Netanyahu government has passed denying basic civic freedoms, but one must deal with the main one—the Nation-State Law, or to give the full name (evincing its narrative), "Basic Law on Israel as the Nation-State of the Jewish People."

This law—the apotheosis of Netanyahu's destructive politics—is seen by even ardent Zionists as the end of any pretense of democracy in Israel. The law enforces legal apartheid, where before it was applied by custom and practice. One of the giveaway clauses of the law is the support it gives to "Jewish-only" communities, making it impossible for Arab Palestinians to live in such communities—taking its cue from the South African Group Areas Act, excluding black people from "White areas."

Mordechai Kremnitzer, a renowned human rights lawyer, puts it aptly:

> How far is it between this constitutional provision and the call of racist rabbis not to sell or rent housing to Arabs (and who were, despite these calls, not indicted for racist incitement)? How far is it from another slogan that might have been appropriate before Israel's establishment—"Hebrew labor"? Which meant, don't hire Arabs for these jobs unless it's for jobs not suitable for Jews.[27]

By mentioning the prestate campaign against Arab workers and describing a slogan as one that "might have been appropriate" at that time, Kremnitzer clearly points out the continuity of Israeli and Zionist apartheid. He clarifies what this means for Palestinian citizens of Israel:

> For Arab citizens to come to terms with the law (they couldn't possibly agree with it), they would have to come to terms with being absent, or inferior; according to the attorney general, they would have to accept having no collective rights as a national minority. Nobody could come to terms with things like that involving such humiliation, and no decent person should lend a hand to a "law" that requires things a law must not require. This is not a deed worthy of being called legislation; this is aberrant use of the arbitrary force of the majority to deliberately harm the minority by enacting a Basic Law. If this is allowed to pass, we won't have anywhere to hide our shame.[28]

The law also guarantees that in cases where the legal code and the Jewish Halacha differ, the Jewish Halacha shall prevail, which simply makes Israel a Judaic Republic, sister to the Islamic Republic:

> The fact that the state is defining itself as religious—a definition that was deliberately omitted from the Declaration of Independence—provides more encouragement for the caustic processes of religious coercion and the destructiveness inherent in the bill. There's no

> doubt that this will be used to justify demands to intensify the influence of religion on the state ... The problem with the nation-state bill is not only what's in it, but also what's not. The bill's authors made sure to remove any commitment to the principles of the Declaration of Independence, most prominently the principle of equal rights for all its citizens.[29]

It is clear that the law is not pointing only to the future, ominous enough in itself, but legally regulates the corrosive processes which have transformed the Israeli society into an apartheid state—it puts in legal parlance the realities on the streets, normalizing and advocating it as correct and just. Like similar processes in Nazi Germany, the future can only be worse. Once a group of people, in this case the Palestinian citizens, is marked unwanted and lacks a legal base and justification (in their own country), the road is open to more terrifying steps. The Palestinians have been marked as *homo sacer,* and harming them is now within the spirit of the law.

This painful and fast transition of Israeli society is yet to be noticed by most of the commentators of Western media outlets. Such media, especially in Anglophone countries, are strongly influenced by Zionist gatekeepers who are careful to occlude signs of this process and present the adage "the only democracy in the Middle East" as a magic formulation that may defeat reality. Such influential figures as Thomas Friedman in the *New York Times* or Jonathan Freedland in the *Guardian* are guaranteeing that their own interpretation of the political map is accepted as standard. Until such time that the biased gatekeepers lose their stranglehold on what can be published about Israel, one suspects that such unthinking inanities as the "only democracy in the Middle East" will continue to rule dominant media, and critics of Israel and its corrosive racism will be dubbed antisemitic.

Nevertheless, the prevalence and popularity of alternative media and online sources and forums are slowly undermining the fictional certainties of the Israel Lobby in the West, introducing a more apposite understanding of the facts than allowed on mainstream media. This may become an important lever to change the situation in Palestine.

Afterword

Can There Be Another Israel?

This book is certainly not an easy read. I have argued in this work that there is a certain predisposition innate to the realities of settler-colonial states, which structures their raison d'être so that the whole society is geared toward the continuation of the project. What emerged in Europe as a liberation movement of oppressed Jews became in Palestine a settler-colonial project; what started as popular socialism turned quickly into national(ist) socialism, to use Zeev Sternhell's phrase.[1] The dreams and projections of Herzl, a nineteenth-century rightwing liberal,[2] will turn into a hard-baked apartheid based on deep racism toward the colonial Other, the indigenous "native," whose territory the project requires, while trying to expel the population in its entirety.

For those of us who would dearly love to see a just solution to the injustices in Palestine, thinking about the end game is part of our makeup. It may be useful to start by examining the claim by many liberal Zionists, who argue that Zionism had other options that it has, unfortunately, abandoned. The largest such grouping are the leftwing Zionists who, like the late Amos Oz, argue that if Israel could go back to the Green Line of June 4, 1967, everything would be fine—a Palestinian "state" could be set up alongside Israel, the "occupation" would end, injustices of the past five decades would evaporate, and peace shall reign, at least in some form. I shall call this grouping the *two staters*. They have a following in many Western countries among those who know very little about the history of Zionism and would like the conflict to vanish, the same way that hunger and disease should.

Some, who are more acquainted with the history of the settler-colonial project in Palestine, argue that the point of no return took place in 1948. Until then, they believe, Zionism was conciliatory, but somehow the events of 1948 and the great success of the IDF in expelling the Palestinian population and defeating the Arab armies converted a basically decent Jewish liberation movement into a colonial modus operandi. I tend to think about this group as the Romantics. They have a deep need to believe in the progressive nature of historical Zionism and see the future in terms of some cloudy vision of a return to a state that never existed.

Both such positions are not based on historical knowledge and understanding, but on exclusively Zionist wishful thinking, or more accurately, the need to convince others of a view they themselves would wish to take. They do not accept the stark realities of the settler-colonial project—by speaking instead of the dreams and visions of liberating Jewish life in Europe, or the mystical vistas of messianic Zionism, always unconnected to the realities in Palestine, to the persistence of the non-Jewish Other. This solipsistic conception of Zionism is specific to leftwing and liberal Zionists and their many supporters, while the Zionist right, from Jabotinsky onward, were more open and honest about the plans of clearing the land of non-Jews, as was Herzl in his *Diaries*. With the right taking over control for most of the period since 1977, such a rosy, befuddled vision is difficult to hold on to or justify, despite the great efforts of its cultural celebrants. Menachem Begin, Yitzhak Shamir, Ehud Olmert, Ariel Sharon, and now Benjamin Netanyahu do not pretend that they wish to live in peace and harmony with the indigenous population of Palestine, and they have done all they could to continue the dispossession started by Labor Zionism, which shared the same aims but adopted a liberal-friendly presentation stratagem—"shooting and crying," as it was called after 1967.

Such attempts deny the very core of the Zionist project and the dream it is predicated on—control of the whole of Palestine without a non-Jewish presence.

Could There Have Been Another Israel?

The basic contradiction in Zionism—the wish to control the whole of Palestine while stating this can be done by a Jewish State—is there because it is impossible without genocide or racialized expulsion. Once a decision was made to control the land through settler-colonialism, then other choices automatically followed—the creation of the IDF and its military–industrial complex; dependence on arms exports and an (evil) empire (there are no other empires); and living by the sword. Once that happens, one cannot wean oneself from it. There is no way to turn swords into ploughshares.

Writing about the sociopolitical drive behind science fiction, Frederic Jameson discussed the difference between utopia and dystopia, noting that in the United States, utopian films are never produced by Hollywood, while cinematic dystopias are ten a penny. He argues that in a society that believes its way of life is the best there is, there is no call or even potential for utopian imaginative creativity, either written or in media forms.[3] One is already living the ultimate utopia.

Similarly, but for different reasons, Israelis are unable to perceive peace, life without a military edge, without the need to kill, expel, maim, and harm large numbers of innocents and even die themselves, in order to defend their way of life, which is a way of fighting life. By scrutinizing the various wars and conflicts since 1948, it becomes clear that Israelis only feel safe in armed conflict, which they feel they can control; the concept of peace emanates a strangely menacing aura for most of them. Peace (a condition they have never experienced) cannot be trusted, for deeply justified reasons—it cannot be achieved without first assenting to some simple conditions: return of the refugees, equality before the law, repeal of racist legislation, and the abolition of for special privileges for Jews—in essence, normalizing the state by de-Zionizing it, making it a state for all its citizens, with total equality. While most Israelis do not necessarily understand themselves as Zionists (most of them have only the faintest idea of what Zionism actually means), they nonetheless are mindful of their privileges and racist enough to oppose equality. Writing in reply to the actress Rotem Sela, who has argued for "a state of all its citizens"—on the face of it, something so obvious one should not need to argue for it, Netanyahu reminded the Palestinian citizens of Israel that they are not real citizens. He uploaded a picture of himself against the backdrop of an Israeli flag, and wrote: "Dear Rotem, an important correction: Israel is not a state of all its citizens. According to the Nation-State Law that we passed, Israel is the nation-state of the Jewish People—and them alone."[4] So Israel may be the only state which does not comprise all its citizens and, conversely, belongs to a group that is mostly noncitizens!

During a toxic election campaign likely to be his last, Netanyahu pulled no punches—he felt that the best way to win was to voice his real beliefs, rather than tone them down—he knows his public and played a major role in navigating them into the racist, extreme-right corner they now occupy. In the last election on September 17, 2019, more generals were elected to the Knesset than ever before, most of them ex-chiefs-of-general-staff.[5] There is no clearer indication of the direction the Israeli polity is taking. Israel is preparing for more wars and more bloodshed.

Thus, the situation Israel has created—endless conflict, danger, destruction, and mayhem—is presented as the Zionist utopia, simply because Israelis cannot imagine a better situation, one in which they will be better off. One suspects that most Germans between 1933 and 1945 felt similarly, as do most Americans nowadays—when one's country seems to be at the zenith of its power, few understand that this state is likely to change, and drastically so.

Israelis cannot imagine living without the apparatus of security, which brings them no security. The IDF has built an immense war machine—thousands of tanks, armored vehicles, fighter jets, missile boats, nuclear submarines, cyber-power networks, and hundreds of nuclear weapons. This enormous might is poised in mid-air, lacking an enemy to strike, with the Arab states removed as potential enemies. The IDF and its related military–industrial complex are the heart of Israeli existence and have reached a phase of inflated growth that is clearly superfluous and unwarranted—plainly overblown, mismatched for the task of policing its settler-colonial holdings. But the idea of changing tack, of experimenting with nonlethal methods of negotiating reality seems too far-fetched, too risky for Israelis to envision. And what cannot be imagined cannot ever materialize. Israeli imagination is now limited to better, more sophisticated ways of seeing all, knowing all, killing all, more quickly and efficiently. That has positive market value, is saleable, while peace is an obvious danger. What will Israel produce under a regime of peace? It is not immediately clear. After all, Israelis spent seven decades building up their specific security utopia and would feel naked without it.

Another difficulty is Israel's rife social divisions, which can only be held at bay by the tried-and-tested method of frequent and periodic attacks on "enemies." Perfected in 1948, it has worked ever since. As pointed out by Zeev Sternhell, Israel has never experimented with either peace or real socialism as solutions for vulnerability and inequality:

> Thus, it is particularly significant that at the end of these long years of dominance a movement that claimed to be socialist, had not created a society that was special in any way. There was no more justice or equality there than in Western Europe, differences in standard of living were just as pronounced, and there was no special attempt to improve the lot of the disadvantaged.[6]

If this was the case under Ben-Gurion's Mapai, it is even plainer under governments since then, which do not claim socialism of any kind, but its reverse. Israel has turned into a society like so many others that advantages only the elites.

While this standard set of limited social aspirations may work in normal conditions—meaning societies not involved in constant military conflict and active settler-colonialism—it certainly does not work in Israel, as the widespread tent protests of the summer of 2011 have clearly demonstrated; Netanyahu terminated the protest by attacking Gaza, delaying the explosion. The particular conditions of Jewish-Israeli life are both stressful and dehumanizing (although certainly not as stressful and dehumanizing as their actions are for the Palestinians)—too exacting for a nation wishing to believe its own projected utopian fictions. To continue acting as a colon, Jewish-Israelis must be united in the struggle for control, well beyond its undefined, elastic borders, changing every time a new map is drawn. By choosing the security mass neurosis, Israel has chosen high-cost, low-quality life—a life under threat from its own machinery of conflict dissemination.

The conviction and insistence that Judaism is a nation rather than a religion or culture, and one that must control its own national territory, be the majority in it, supress other groups within and without it, and insist on undemocratic and racist superiority, is a road to apartheid, a road now travelled for over seven decades. While in the past this stark reality was easier to ignore, in the face of so many privileges, the developments since 2000 make denying realities more complex; the wars fought are increasingly difficult to justify and even to "win." As the range of costs—human, political, financial, cultural, and social, not to mention emotional—are mounting, more seems to be at stake, and the denial now requires the Jewish population to agree to an openly racist, exclusivist, undemocratic apartheid polity. While this still functions—most Jews have gone along with this despite the consequences—it is no longer possible to claim, as was the case in the past, that Israel is a just, democratic, and rational Jewish State. Israel was never such a state, but now even the claim has become impossible. Some Israelis now understand that their society is becoming an international pariah, despite the vast political support it still enjoys from most governments—in most people's mind it is becoming as isolated as apartheid South Africa once was. Typically, this understanding is shared by a small segment of the elites, mainly academics and professionals, creative media and artists. What such people may do with this realization is unclear; one has not been able to notice an effort to stem the tide of racist legislation that continues to further undermine the few rights still enjoyed, in theory, by 1.8 million Palestinian citizens of Israel, not to mention more than four millions living without rights of any kind, in the Occupied Palestinian Territories.

IDF Direct Involvement in Politics

Throughout this volume, the underlying assumption was that the IDF does not pose a direct threat to the Israeli political elite, and in that sense Israel is similar to states that have experienced military coups, from Chile to Greece and from Argentina to Pakistan. While this is indeed true, due to the almost total control of the political agenda by the military, there were some important and telling examples of the

IDF dictating the political agenda directly and aggressively, which need to be considered.

The most emblematic of such exceptional political moments was probably during late 2000, as Prime Minister Ehud Barak, himself an ex-Chief of General Staff of the IDF, was conducting his ill-intended and deceitful "peace talks" with Yasser Arafat. In hindsight, Barak was certainly the weakest, most indecisive, and detrimental prime minister Israel has ever had. Facing the enfeebled Arafat, he could have reached an agreement that, while far from just, would have served Israel in bringing an end to the armed conflict, though at a terrible cost to the Palestinians. Such an agreement would have the support of the United States and President Bill Clinton, the initiator of the negotiations; Western countries, and most if not all Arab states. But Barak was unable to walk the extra yard.

The reason was not the intransigence of Arafat; rather, Barak was terrified of rear action by the Israeli right. Its leader, Ariel Sharon, was at a low ebb, being edged out by the former Prime Minister Benjamin Netanyahu, who was trying to eclipse Sharon by overtaking him on the right. Most Likud Party members were seriously tempted, supporting Netanyahu against Sharon, as a secret poll commissioned by Sharon had ascertained.[7] To survive, Sharon had to prove to the Likud voters that he is at least as aggressive as Netanyahu and as capable of trouncing the limping negotiations between Barak and Arafat, removing the "dangers of peace." Barak's fears had a basis in reality—the memory of Rabin's murder less than a decade earlier was fresh in the public's mind, as was Netanyahu's incendiary acts of bringing about the assassination. While Barak could easily defeat Arafat at the negotiation table, his real enemies were Sharon and Netanyahu, vying for the crown of the wildest Israeli politician.

While Barak was watching his back, making sure he could not be seen as giving an inch to Arafat, Sharon took the initiative and made his move, one that would change the course of the conflict for decades to come. He announced that he would make a visit to the holy Moslem site of Haram al-Sharif (or as Israelis prefer to call it, the Temple Mount). This high Jerusalem plateau bears the two mosques most sacred to Muslims after Mecca and Medina, and the site of Mohammad's ascent on his "night flight." This move by

Sharon, the Israeli military leader most hated by Palestinians, was a serious provocation, endangering the flagging and half-hearted peace talks and further inflaming the armed conflict. Barak ignored strong advice against allowing the Sharon visit; he was reassured by the General Security Service (Shabak) and the IDF Intelligence Branch that the visit posed no serious danger and that there is no risk of rioting.[8] He would later rue the day he had so misjudged the situation, accepting this skewed advice.

This was a trap laid for Barak, not just by Sharon, but by his own IDF General Staff. As the ensuing riots would develop into what became known as the al-Aqsa intifada, Barak would try to mitigate and pacify the situation, but to no avail; the IDF opposed him. Both the Chief of General Staff General Shaul Mofaz and his deputy, Moshe Yaalon, were hard liners of the Sharon mould and in league with him in the struggle to stop the negotiations and oust Barak. They totally ignored Barak's orders to "contain the hostility and lower the flames," promising "a bullet for every [Palestinian] child," further entrenching the challenge created by the Sharon visit and getting as close as Israel got to a coup d'état. Barak was told that "from the chief of staff to the last of the sergeants at the roadblocks, no one is implementing your policy."[9] The peace talks were dead in the water, and so was Barak himself; his government survived for a further two months before he called an election, which he lost by a large margin to the ascendant Sharon.

This episode, a reinforcement of the lessons of the Rabin assassination in 1995, would dictate the rules of the political game in Israel for the next two decades. No leader, not even celebrated ex-chiefs of general staff like Rabin and Barak, could survive an attempt to force a policy that the IDF General Staff considered unacceptable. Not only was the tenure of such politicians doomed—their very existence was doomed if they tried to persist in the face of the IDF threats. With the IDF now under the spell of the settlements' leadership and the extreme right, no political leader could ever come to power unless they excelled in the unlikely task of overtaking Netanyahu on the right. Some, like the doomed leaders of the Labor Party—Yitzhak Herzog and Avi Gabai—have tried this circus act of out-Heroding Herod, to no avail. Israeli voters trust the genuine article, fascism of the right rather than the left.

This shackling (*Hishuk*) of the political leadership in Israel is widely supported and well understood by the public. As popular violence is the habitual tool of the political right (as is political assassination), any threats of such measures, were they to be made by the left, would lack any credibility. The threats of popular violence of the right is credible and effective, as has been proven time and again, and recently in the United States in 2016. The Israeli right is, if anything, even more intimidating and the left in Israel has long disappeared as a political force to reckon with. In these circumstances, the range of political options in Israel has never been clearer or narrower. This should be borne in mind by all who claim that the solution to the conflict will come from within the Israeli-Jewish public itself. Such a claim has never been less credible than it now is.

The Cultural Claims of Extreme Zionism

The recent clarification of Zionist aims and methods, through a series of apartheid legislation, murderous attacks on Gaza, racist limitation of rights within Israel, and total rejection of the Oslo Accords, limp as these were—all mark the terrifying direction of travel. Such developments are neither new nor surprising—Israelis justify them by the Judaic historical tradition, as part of a projected Jewish identity endeavor, premised on the Bible and other historical myths. It will be useful to list the main components of this systematic frontal attack on historical facts.

From the long and complex history of Jewish existence, Zionism distilled some of the most damaging elements, because they serve its military and political agenda. It has denied or abandoned the most revered, universalist fundamentals of Jewish history and culture as redundant, concentrating on the nationalist, racist, populist, and martial elements of the Jewish historical narrative. In a sense, such elements are exactly the ones that contradict and annul this history.

The Chosen People Myth

There is no denying that the chosen people myth originates in the Hebrew scriptures and infuses large sections of the sacred text. Today we know that such texts, purporting to be written in a period starting some fifteen to seventeen centuries before Christ, were composed and edited more than a thousand years later, forming what Shlomo Sand calls a *mythistory*.[10] Sand points out that the creation of a nation, as argued by Renan, is based not only on imaginative invention but on constant rewriting and re-editing of the mythistoy: "To create a new paradigm of time, it was necessary to demolish the 'faulty and harmful' previous one. To begin the construction of a nation, it was necessary to reject those writings that failed to recognise its primary scaffolding."[11] This is another way of expressing what Renan called "forgetfulness":

> Forgetting, I would even say historical error, is an essential factor in the creation of a nation and it is for this reason that the progress of historical studies often poses a threat to nationality. Historical inquiry, in effect, throws light on the violent acts that have taken place at the origin of every political formation, even those that have been the most benevolent in their consequences.[12]

Renan also realizes that nations are modern and "not something eternal. They had their beginnings and they will end."[13] Thus, the "scaffolding" of Jewish history as constructed by Zionism stands out as an extreme piece of national cultural capital, in a Frankenstein-fashion creation of mythistory from the dead bits—and the bad bits—of Jewish historical narratives; it claims that the resulting monster is eternal, beyond history. As most historians of nationalism clarify: "Nations, in this sense of the word, are something fairly new in history."[14] One cannot speak of nations in antiquity, as most people were not literate, and ideas about identity were limited to clan, tribe, and locality, and later, religion; wider-ranging entities such as the Roman empire were culturally and religiously permissive and permeable and were not based on the concept of nation.

Thus, the Zionist construction of the nationalist scaffolding, extending backward and forward in time into the ancient past and the distant future is not just ahistorical but lacks factual and rational foundation. When Netanyahu repeatedly claims that Jerusalem is Israeli forever or when his rival, former General Benny Gantz recently claimed that Israel will never "withdraw from the Golan Heights," they are referring to their specific concept of the Hebrew-Jewish nation.[15] This construct references biblical texts projecting a mythistory of a Jewish people as a nation "chosen by God," its beginnings shrouded in the mists of time and its future being "eternal." The nation thus created becomes itself eternal.

For nations to claim they are special is par for the course and should surprise no one. For a group of people to claim selection by God is more exclusive, designed to further ground their national rights—as God cannot be temporal, so too is the purported nation eternal. For Jews this naive belief in their eternal selection was a suitable salve for the cruel Judeophobic history they had faced in Europe, and it certainly helped their survival. For Israelis, it provides a way of ignoring both reality and history—if one is selected by God, the daily cruelties become a mere detail, needed for achieving higher, eternal objectives. Such an outlook also assuages obvious fears that Israeli power may not last, and a day of reckoning may prove even harsher than the brutality of the IDF—if the existence of the nation is eternal, guaranteed by God, there is nothing to fret about. This is the view even though many Israelis are not religious—as the saying goes, "Israelis do not believe in God, but believe he has given them the land of Israel." One may contend that for a small nation like Israel, not enjoying the obvious benefits of size and stature of states like the United States or Russia, to claim exceptionality requires the use of divine selection, as well as the abusive use of the Holocaust narrative. There is also no point in asking why God, having chosen the Jews, then went on to inflict on them every possible disaster. The sport of biblical quote-swinging may not yet be an Olympic sport, but Israelis are experts at proving the impossible, using *Credo quia absurdum* as their guiding principle. God and the Bible are great for supplying the justification for mass crimes and have functioned well in that role for many centuries, across the religious and national spectrum.

The Tradition of Zealotry

Zealotry during the Second Temple period has been an emerging, particular Jewish tradition; after lying fallow for two millennia, it was rekindled by Zionism into an extreme, world-defying phenomenon. The historical "paradigm" of zealotry in Jewish history has been most harmful, to be sure. Both the fall of Masada and the destruction of the Temple and of Jerusalem by the Romans can be attributed to that propensity for conflict, which Jews have been shunning since then. There are even rabbinical universal exhortations against this predisposition, which religious authorities have strictly prohibited as *'olim bahoma*, spiteful acts of zealotry endangering Jewish communities through the ire of the gentile nations. The zealots in Jerusalem have acted irrationally, believing they can win against the massive legions of the Roman empire and against the wishes of the Jerusalem population, which they have terrorized.

It is important to mention that the historical thinker and religious leader held in highest esteem by contemporary Jewry, Rabbi Yohanan Ben-Zakai, secretly contacted the Romans during the siege of Jerusalem, striking a deal to spare the academic and religious elite, so as to build an alternative cultural center at Yavne, under his leadership. This was a clear move against the extremist zealots, which forced destruction on the citizenry. Ben-Zakai went down in history as the saviour of Jewish identity in the face of a terrible calamity and the total destruction of the temple and Jerusalem.

Undoubtedly, Judaism suffered from extremism at an early phase in its formation as a religion. Freud speaks of this tendency as being part of monotheism itself, which he dates back to Amenhotep IV, who was the first to try to create a monotheistic religion, "the first attempt of the kind, so far as we know, in the history of the world, and along with the belief in a single god religious intolerance was inevitably born, which had previously been alien to the ancient world and remained so long afterwards."[16] It is certainly true that such zealotry also typified early Christianity, which emerged from this period in Judaism as a subset of it, another sect added to the many that existed at the time, but destined to become a new religion. This will be repeated later by early Islam and the great divide between Sunni and Shia followers, as well as during the fifteenth and sixteenth centuries in the religious wars of Europe. While zealotry is a typical part of the formative period of a religion, it may be also a phenomenon of nationalist or religious revival; it is so in the case of Israel, where it seems to be used by extremist Zionists as a way of justifying ethnic cleansing. An example of this tendency in Israel today is a plan for the conquest, ethnic cleansing, and repopulating of Gaza by Israelis, published by the deputy speaker of the Israeli Knesset, Moshe Feiglin: "His detailed plan, which calls for the use of concentration camps, amounts to direct and public incitement to genocide—a punishable crime under the Genocide Convention."[17]

It is clear that the terrifying destruction inflicted by Israel on the Gaza Strip at the time of this plan's publication did not satisfy Knesset member Moshe Feiglin; he was searching for a more final solution, it seems. This he justified by national-religious zealotry of the kind discussed above. Since that time, many more have joined his approach to a "resolution" of the "Gaza Problem," and plans are getting even more extreme. With the combined religionization of Israel and the rightward shift of the political agenda, such zealotry is extremely threatening, with the clear potential to bring a new catastrophe in Palestine.

The Practice of Excommunication of Critics

Excommunication was a specific practice of monotheistic religions, and Jewish communities were no exception. The excommunication of the philosopher Baruch (Benedict) Spinoza is a famous case—he was expelled from the Jewish community of Amsterdam in July 1656 after he published his doubts about the origination myth of the Hebrew Bible and the nature of the divinity. In later life, his written work was also proscribed by the Catholic Church in Holland, so both Jews and Catholics were barred from reading his work or communicating with him.

It is interesting that the best-known case of an even more brutal form of excommunication in modern times—by Zionism—was that of another Dutchman, the poet and author Jacob Israel de Haan, a Haredi Jew who was murdered by the Hagana in June 1924 as he left a Jerusalem synagogue. De Haan was assassinated because of his public anti-Zionism and his political activity—he had been attempting to negotiate a peaceful settlement with Arab leaders.[18] Such views were typical among Haredi Jews in Jerusalem; the Zionist project was perceived as a modern example of 'olim bahoma, something that was religiously proscribed and politically disastrous. They were not wrong. They were acting in the spirit of Ben-Zakai.

Though the Mandate authorities knew the identity of the murderer dispatched by the Hagana, Avraham Tehomi, they were either unable or (more likely) unwilling to bring him to justice. For the Hagana, the murder of Jacob de Haan was intended to issue a stark warning to the Jewish Haredi community—the Hagana would stop at nothing to reach their political aims, and Jewish resistance to Zionism would not be tolerated. While the views of Haredi Jews did not change and de Haan was viewed as a martyr by the community, it became much more muted in its criticism of Zionist aggression. It is indeed ironic that Zionism, which the Haredis then considered "godless," later adopted the Jewish religion as its mainstay in furthering its political aims, through returning to the earlier examples of religious zealotry as newfound icons of heroism. With the rapid religionization of Israeli society, this is only likely to escalate toward a political breakdown.

The Practice of Isolationism and Separateness

European Jews suffering from antisemitism were treated as pariahs by the Christian cultures they lived among. The above-mentioned belief in divine selection, as well as the strictures of the Jewish religion itself (which have become more stringent over time), have also helped to isolate Jewish communities, but fear of antisemitism and pogroms was the main driver of isolation. Arguably, the dependence of such communities on potentates in most cities and countries was a way of protecting them from the crowds; they provided important services to the power-that-be, usually in the form of financial skills. This special protection was not deemed justified by antisemitic crowds and often fuelled their rage. As is well-known, Jews lived very close to the palace or in ghettoes near it. While such isolation allowed Jewish life to thrive and continue despite the many difficulties imposed by society, it also severely limited the options open to Jews; it cut Jews off from the main thrust of European life. There is no doubt that Jewish social character was deeply affected by such isolation and by the fear and loathing they experienced daily. This isolation was only ended by the French Revolution and Napoleon's military expeditions across the continent, which granted Jews rights for the first time in modern Europe. The ghettoes could now open their gates and destroy the walls separating the Jews from society and many did.

Ironically, the changes wrought by granting (limited) rights to Jews during the nineteenth century led to the development of modern, nationalist antisemitism—the middle classes were now competing with bright Jews in the professions and later also in academia and politics, leading to the emergence of Judeophobic political organizations. If before Jews had been hated as the European Other, now they were despised and hounded because they wished to wholeheartedly belong and become citizens with full rights and duties.

The reaction to modern antisemitism split Jewish society in Europe. The majority were strongly behind a drive to achieve full rights and were prepared to pay the price—an abandonment of their insularity and isolation—and shedding some of the more extreme characteristics that have arisen during the centuries of pariah life. For them, fighting antisemitism meant joining the rest of the progressive forces in society and struggling for full civil rights—universal suffrage, free association, security and rights for workers, universal education, free movement, freedom of speech. These were iconic struggles of the modern world, joined and sometimes led by European Jews, who were offered this potential by the age of enlightenment.

A small minority of Jews in Europe became Zionists (typically between 1 and 2 percent). Believing in an ahistorical antisemitism, supposedly innate in the non-Jew, they believed that resisting it was pointless and that the only solution was separation—leaving Europe altogether and building a Jewish-only polity in Palestine, a kind of mega-ghetto in the Middle East. But this time they would have proper sanitation and a powerful army to guarantee the takeover of an Arab country with the support of an imperial power.

This second choice represented (then and now) a return to the isolation and segregation of the Jew, based on a deep conviction that life among the gentiles is forever dangerous. Ironically, Zionism is the closing of the circle—one of life apart, so strongly supported by all the developments described above; biblical expressions describing the Israelites as *Am Levadad Yishkon* (a people apart) have allowed Jews in the ghettoes to suffer isolation with dignity. Zionism has turned this into ideological racism, based on a pathology of fear and loathing. In Palestine, such a pathology was turned against the Palestinians to justify racialized genocide; current trends of Zionism in Israel only exacerbate this development.

Military Particularism Rather than Universalist Humanism

For most of the period since the destruction of Jerusalem by the Romans, Jews lived abroad, mainly in Europe where 95 percent of all Jews existed before the end of the nineteenth century. During this period, barring a short time after leaving Palestine, Jews had shunned most forms of military service, preferring instead civil professions as tradesmen and artisans. Their typical command of a number of languages and familiarity with various regions—through frequently being expelled from various locations—had turned them into specialized facilitators of administrative and political ruling elites. In so doing, Jewish communities wished to secure a safer grounding of their limited status by eschewing military skills altogether, as a gesture of submission and obedience. Civil, financial, and administrative services were the mainstay of Jewish existence for nearly two millennia, assisting the growth of an intellectual Jewish class.

All this changed with the appearance of Zionism at the end of the nineteenth century. The building of agricultural settler-colonial habitations in Palestine, mostly through the disenfranchisement of the local fellahin, gave rise to early antagonism by Palestinians, creating the need to build a military force to defend both extant and future colonies. Such a force only exacerbated the resentment of Palestinians, leading to a further increase in militarizing the colonial conflict with the indigenous people. A vicious circle was set in motion, as was the case in other settler-colonial projects. The existence of large Arab communities surrounding Palestine only helped to further fuel the explosive conflict.

The fast and efficient development of the Zionist military forces (and after 1948, the IDF) is clear evidence both of intention and ability in this field. For a settler-colonial society, there was no choice but to develop a powerful and destructive army—the prospects for existence of the racialized colony depended on it. That this armed force was also used to form the nation, as well as to bolster it financially, made it the most important institution in the young state. The talent for developing military skills, systems, and equipment was a surprising departure from the long civil Jewish history, but it was clearly required by the colonial need to subdue and expel the indigenous Palestinians. In this way, the IDF became the formative agency of Zionism, its most influential social machinery; it has directed the thinking, analysis, options, and actions of the Israeli polity ever since the beginning of the Zionist project and certainly since the end of World War I.

As long as this is the case, and one can see no change on the horizon (or even a recognition of the need for such a change), the Israeli society and state are, inevitably, ready and waiting for another cataclysmic clash with the Palestinians. Arguably, every single decision and action of the state is directed to a future showdown and is shaped to insure Israeli victory in this terrifying future calamitous confrontation. In the meantime, all is done to expel and inconvenience as many Palestinians as humanly possible, with some marked successes. Most Palestinian Christians have been made to leave their country; they are more acceptable to racist Western societies than their Muslim brethren, so they are an easy target for the Israeli state. Israeli Arabs have lost most of their limited rights through a system of apartheid legislation and are probably about to lose their representatives in the Knesset, if the decisions by the electoral commission—the body monitoring elections in Israel—in March 2019 are allowed to stand by the Supreme Court. Plans to expel them to the Occupied Palestinian Territories have been published by Avigdor Lieberman, the former defense minister, and many others on the right. Gaza is in its twelfth year of blockade and isolation, and the pace of settlement building and Palestinian dispossession has never been higher. All this happened without the slightest pressure emerging from the international community, such as it is these days. Israel knows that it can get away with murder and plans to do just that. If and when the future completion of the task started by Zionism with the 1948 Nakba takes place, the blame will not fall on the culprit alone; Israel will be the machinery that will inflict the blow, but the Western nations that have continuously supported it, financed it, and armed and supplied it with the wherewithal enabling it to prevail in its settler-colonial project will be as guilty as Israel.

As in the case of South African apartheid, Israel and militarized Zionism can only be countered and forced to change through a wide-ranging international campaign of solidarity and boycott. The Boycott, Divestment, and Sanctions movement supplies the model for such action, but to date has had only very limited success, although Israel, recognizing its potential to bring real change to Palestine, has branded it an existential threat.

This is an opportune moment to remind the reader that the Middle East, as ever, is changing at present, and so are power relations within it, as well as without. The inauguration of President Trump has upset many a balance of power in the region and added to the dangers of further chaos and wide-ranging disorder already present and exacerbated since the violent suppression of the Arab Spring uprisings. If early on Israel was rather elated about Trump's overt and unsubtle support of its agenda, then recently it has had serious reasons for concern. The United States' main allies in the region have been greatly inconvenienced by the president's Twitter diplomacy. Saudi Arabia was left dangling after the missile attack on its oil processing plants in September 2019 knocked out half the country's production capacity. Despite Israel's eagerness and encouragement, the United States has avoided military action against Iran and concentrated on defensive measures.[19] Indeed, Saudi Arabia and the Gulf States are now, more than ever, inclined toward a peaceful resolution of the tension with Iran.[20] The great vulnerability to low-altitude cruise missiles exposed in the attack, despite the huge sums invested in the latest US systems, leaves no choice for the Saudi regime. Israel has reevaluated its own defenste strategy following the attack, with Netanyahu speaking of "a national emergency."[21] Netanyahu is seriously worried about retaliation after his many attacks on Iranian targets in the last few years: "Israeli intelligence predicts that vengeance could take the form of cruise missiles or drones sent from Syria or western Iraq, where Shi'ite militias are operating with the support of Iran. The latest Iranian attack in Saudi Arabia rang warning bells in Jerusalem, given the show of high operational ability and potential Israeli weakness in defending itself against similar attacks."[22]

As if this was not bad enough, Trump's overnight decision to remove US troops from northern Syria to make way for the Turkish invasion in October 2019 has increased that sense of insecurity. The cavalier abandonment of the Kurdish allies in Syria by the United States, as well as the arguments Trump attached to his decision, have proven the untrustworthy and unpredictable nature of US support for its allies, as well as the frightening repercussions of the US decision-making process while in the hands of the most irresponsible administration in US history. Israel is worried about the long-term outcomes: "The main change is strategic more than anything operational or tactical: The United States is expediting its departure from the Middle East, and Iran's influence is growing, along with its self-confidence."[23] It would be a mistake to underestimate the long-term effects of this abrupt, ill-considered, and destabilizing change in US policy, though a full evaluation cannot be made of the likely results at this stage. What is quite clear is that it will, in all likelihood, not lead to a more stable political environment and an end to conflict.

The recent announcement by Trump that the US "no longer considers the settlements illegal" is another nail in the coffin of the two-state solution and demonstrates the continued deterioration of the little stability left.[24] This instability is further worsened by the bizarre electoral shenanigans in Israel—three inconclusive elections in a one year, and a possible fourth election on the way, as Netanyahu is neither able to form a government nor willing to allow others to do so; the cynical use of the coronavirus crisis in order to undermine the rule of law and human rights in Israel and the OPT is likely to serve as a prompt for further illegal action by Israel against the Palestinians.

I hope that this monograph serves as a clear warning. Such admonitions now abound—unfortunately without evidence that they are being heeded or that Israel's political, military, and financial supporters are seriously reconsidering their misguided and unjustified stance. The explosion that appears to be coming is not necessarily in the interest of the West, yet it may take place if nothing is done to stay the strongest military machine in the Middle East, one that is ready and waiting to inflict decisive blows within the region as it has done in the past. With Israeli society more inclined than ever to inflict lethal damage on the Palestinians and their aspirations for freedom and equality, denial of the potential for regional disaster is no longer an option.

Notes

Introduction

1 Yosefa Loshitzky, *Identity Politics on the Israeli Screen*, Austin: University of Texas Press, 2001, 50.

2 Nurit Peled-Elhanan, *Palestine in Israeli School Books: Ideology and Propaganda in Education*, London: I. B. Tauris, 2012.

3 Yael Zerubavel, "The "Mythological Sabra" and Jewish Past: Trauma, Memory and Contested Identities," *Israel Studies* 7: 2, 2002, 115–44. Quoted in Peled-Elhanan, *Palestine in Israeli School Books*, 6.

4 Marx is quoted by Vladimir Illich (Ulianov) Lenin in "The Question of Peace," *Lenin Collected Works*, Vol. 21, Moscow: Progress Publishers, 1974, 290–4. This is also claimed by Friedrich Engels in *Eine Polnische Problemation*.

5 Maxime Rodinson, *Israel: A Colonial-Settler State?,* New York: Monad Press, 1973. This is a translation of the original article, "Israel, fait colonial," which was written in 1967, just before the June war between Israel and its Arab neighbors and published in a special issue of *Les Temps Modernes*, "*Le Conflit israelo-arabe,*" in June 1967, just after the war had ended. The English version includes a useful introduction by Peter Buch, setting the scene for the original publication, and contextualizing it within the earlier debate and the ones following its publication.

6 Ibid., 92, quoted from René Maunier, "Introduction à l'étude du contact des races," *Sociologie Coloniale* I, Paris: Domat-Montchrestien, 1932, 37; compare to 21.

7 Rodinson, *Israel*, 91. Note that Rodinson relates to Israel with its pre-June 1967 borders, which had occupied 78 percent of Palestine in 1948.

8 An interesting aspect of combining the roles of victim and executioner, to use Raul Hilberg's terms, is the fact that Israel, the only nuclear power in the Middle East, and one using nuclear processes only militarily, is opposing other countries' right to use the technology even for civilian purposes, such as in the case of Iran's attempt to use nuclear energy in the

production of electricity (a practice normalized elsewhere).

9 Patrick Wolfe, *Settler Colonialism and the Transformation of Anthropology: The Politics and Poetics of an Ethnographic Event*, London and New York: Cassell, 1999, 2.

10 I prefer the spelling *antisemitism*. The traditional spelling, anti-Semitism, is tainted by racist assumptions, suggesting a phobia of Semites, which is obviously not the case with historical antisemitism. A more accurate term is Judeophobia. I use the original spelling in quotes.

11 I have dealt with this issue at length in a recent article; see Haim Bresheeth, "The Israel Lobby, Islamophobia and Judeophobia in Contemporary Europe: Myths and Realities," *Journal of Holy Land and Palestine Studies* 17: 2, 2018, 193–220.

12 Kenneth P. Vickery, "'Herrenvolk' Democracy and Egalitarianism in South Africa and the US South," *Comparative Studies in Society and History* 16: 3, 1974, 309–28.

13 Editorial in the *New Statesman*: "Now Is Not the Time to Give up on a Two-State Solution," *New Statesman*, July 12, 2012, quoting from the interview in *Haaretz*.

14 Julian Borger and Peter Beaumont, "Donald Trump Says US Not Committed to Two-State Israel-Palestine Solution," *Guardian*, February 16, 2017.

15 Quoted in Bernard Avishai, *The Tragedy of Zionism*, New York: Farrar Straus Giroux, 1985, 43. From Harry Zohn, ed., *Zionist Writings*, Vol. 1, Herzl Press, New York, 1973, 153.

16 Amnon Raz-Krakotzkin, "Exile within Sovereignty: A Critique of the 'Negation of Exile' in Israeli Culture," *Theoria Uvikoret*, 4, Fall 1993, 32 (Hebrew; my translation). The article has been published in English, but this section has been edited out in the English version.

17 Yizhak Laor, *The Myth of Liberal Zionism*, London: Verso, 2009, xxiii–v.

18 Jeff Halper, *War against the People: Israel, the Palestinians and Global Pacification*, London: Pluto Press, 2015, 37.

19 Yagil Levy, "Who Controls the IDF: Between an 'Over-Subordinate Army' and 'a Military That Has a State,'" in Elisheva Rosman-Stollman and Aharon Kampinky, eds., *Civil–Military Relation in Israel: Essays in Honor of Stuart A. Cohen*, New York: Lexington Books, 2014, 48.

20 Ibid.

21 Dalia Gavrieli-Nuri, *The Normalization of War in Israeli Discourse, 1967–2008*, New York: Lexington Books, 2013, 1.

22 Zack Baeuchamp, "95% of Jewish Israelis Support the Gaza War," *Vox*, July 31, 2014, vox.com.

23 Yosefa Loshitzky, "Israel's Blonde Bombshells and Real Bombs in Gaza," *Electronic Intifada*, January 9, 2009, electronicintifada.net. Similar points are made by Gavrieli-Nuri, *The Normalization of War in Israeli Discourse, 1967–2008*, 42–3.

24 Gavrieli-Nuri, *The Normalization of War in Israeli Discourse, 1967–2008*, 43.

25 Ibid., 37–8.

26 Ibid., 44.

27 Ibid., 119.

28 Hannah Arendt, *Eichmann in Jerusalem: A Report on the Banality of Evil*, New York: Penguin, 1994.

29 Yosefa Loshitzky, "A Tale of Two Feminists: Hannah Arendt Revisited by Margarethe von Trotta," *Camera Obscura* 101, 2019: 120. For Arendt's own comments comparing Israel to Sparta see Arendt, *The Jewish Writings*, New York: Schocken, 2007, 425.

30 Les Levidow, "Beyond Dual Use: Israeli Universities' Role in the Military-Security-Industrial Complex," *BRICUP Newsletter 121*, bricup.org.uk.

31 Ibid.

32 John J. Mearsheimer and Stephen M. Walt, *The Israel Lobby and U.S. Foreign Policy*, New York: Farrar, Straus and Giroux, 2007.

33 See Sigmund Freud, "Moses and Monotheism," in *The Origin of Religion*, London: Penguin Freud Library, 1964 (1990), Penguin Freud Library no. 13, 237–384.

34 Ibid.

35 Yosefa Loshitzky, "Pathologising Memory: From the Holocaust to the Intifada," *Third Text* 20, 2006: 3–4, 328–9.

36 Zeev Schiff, *A History of the Israeli Army: 1874 to the Present*, London: Sidgwick and Jackson, 1987, 99.

1. The Origins of the IDF

1 Cited in Mary Fulbrook, *Piety and Politics: Religion and the Rise of Absolutism in England, Wurttemberg and Prussia*, Cambridge: Cambridge University Press, 1983, 52, 223.

2 Alex Mintz, "The Military–Industrial Complex: American Concepts and Israeli Realities," *The Journal of Conflict Resolution* 29: 4, 1985, 623–39.

3 Baruch Kimmerling, "Militarism in Israeli Society," *Theory and Criticism* 4, Autumn 1993, 125 (Hebrew). As examples of sociologists who either deny the possibility of militarism in Israel, or explain it away, see Moshe Lissak, "The Israeli Defence Forces as an Agent of Socialization and Education: A Research in Role Expansion in a Democratic Society," in M. R. Van Gils, ed., *The Perceived Role of the Military*, Rotterdam: Rotterdam University Press, 1972, 325–40; and Yoram Peri, *Between Battles and Ballots: Israeli Military in Politics*, Cambridge: Cambridge University Press, 1983.

4 See Volker R. Berghahn, *Militarism: The History of an International Debate 1861–1979*, New York: Cambridge University Press, 1981; Berg, Leamington Spa, Warwicks, 1984; and Michael Mann, *States, War and Capitalism: Studies in Political Sociology*, Oxford: Oxford University Press, 1988.

5 Uri Ben-Eliezer, *The Making of Israeli Militarism*, Bloomington: Indiana University Press, 1998, 7.

6 Lorenzo Veracini, "Introducing Settler Colonial Studies," *Settler Colonial Studies* 1: 1, 2011, 2.

7 Gershon Shafir, "Zionism and Colonialism: A Comparative Approach" in Michael N. Barnett, ed., *Israel in Comparative Perspectives; Challenging the Conventional Wisdom*, Albany: State University of New York Press, 1996, 231–5.

8 It is fascinating to note that Israeli soldiers always use Hebrew to speak to Palestinians at checkpoints, even though the language is not taught in Palestinian schools. It is of course thought necessary for the indigenous population under occupation to learn the language of the colon, but never the reverse.

9 Quoted in Leslie Stein, *The Hope Fulfilled: The Rise of Modern Israel*, Westport: Praeger, 2003, 44.

10 Zeev Schiff, *A History of the Israeli Army: 1874 to the Present*, London: Sidgwick and Jackson, 1987, 2.

11 Ibid; Ben-Eliezer, *The Making of Israeli Militarism*, 2. Ironically, this wording reminds us of the Palestinian resistance slogan of the first intifada: "With faith and blood, Palestine shall be freed."

12 Herzl met with von Plehve in St. Petersburg on August 8 and 13, 1903, and exchanged letters with him afterward, discussing Zionism support of Russian Tsarist tenets and its opposition to social democracy; see Theodor Herzl, *The Diaries of Theodor Herzl*, ed. Martin Lowenthal, New York: Dial Press, 1956, 386–93.

13 Acronym of Netzach Israel Lo Yeshaker (Hebrew) or "The eternity (God) of Israel shall not lie," referencing Samuel I, 15:29.

14 Vladimir Jabotinsky, "We and the Arabs," in *The Iron Wall*, first published in Russian under the title *O Zheleznoi Stene* in *Rassvyet*, November 4, 1923. Published in English in *Jewish Herald* (South Africa), November 26, 1937.

15 Quoted in Ben-Eliezer, *The Making of Israeli Militarism*, 3.

16 Ibid.

17 Lenny Brenner, *The Iron Wall: Zionist Revisionism from Jabotinsky to Shamir*, London: Zed Books, 1984, 97.

18 Walter W. Laqueur, *A History of Zionism: From the French Revolution to the Establishment of the State of Israel*, New York: Schocken Books, 2003, 262.

19 Ibid., 338.

20 Ben-Eliezer, "Chapter Nine: Political Praetorianism in Wartime," *The Making of Israeli Militarism*, 149–68.

21 Ibid., 175.

22 Raphael Patai, ed., *The Complete Diaries of Theodor Herzl*, London: The Herzl Press and Thomas Yosseloff, 1960, 88.

23 David Ben-Gurion, Mapai Faction in the Knesset, April 11, 1951, Mapai Archive, quoted in Ben-Eliezer, *The Making of Israeli Militarism*, 195.

24 Yitzhak Conforti, "'The New Jew' in the Zionist Movement: Ideology and Historiography," *Australian Journal for Jewish Studies*, 25, 2011, 89–92.

25 Benedict Anderson, Imagined Communities: Reflections on the Origin and Spread of Nationalism, London: Verso, 1983, 1991.

26 Ben-Eliezer, *The Making of Israeli Militarism*, 183–201. Gunther Rothenburg, *The Anatomy of the Israeli Army*, London: B.T. Batsford, 1979, 57–8.

27 Ben-Eliezer, *The Making of Israeli Militarism*, 198–9.

28 Quoted in Tom Segev, 1967: Israel, the War, and the Year that Transformed the Middle East, New York: Henry Holt, 2007, 155.

29 A phrase used in virtually all Israeli media coverage of the IDF atrocities in July/August of 2014 in Gaza, which led to some 2,140 Palestinians killed, the great majority being civilians, and some seventy Israelis, sixty-seven of whom were soldiers. Around 96 percent of Jewish Israelis have fully supported the action, according to a range of media polls. The latest examples are quoted in "Bloodbath: How the World Media Covered Israel and the Deadliest Day of the Gaza Protest," *Haaretz*, May 15, 2018.

30 Nira Yuval-Davis, "Front and Rear: The Sexual Division of Labor in the Israeli Army," *Feminist Studies* 11: 3, 1985, 651.

31 Quoted in Dafna Izraeli, "Gender in the IDF Military Service," *Theory and Criticism*, 14, Summer 1999, 85 (Hebrew).

32 Yoana Gonen, "Israel's First Gay Major General Is No Cause for Pride," *Haaretz*, May 17, 2018.

33 Gideon Levy, "The South African Model: Make Israel's Druze 'Honorary Jews,'" *Haaretz*, July 29, 2018.

34 See Yaniv Kubovich, "Israeli Army Suspends Druze Officer Who Published a Facebook Post against the Nation-state Law," *Haaretz*, July 31, 2018.

35 Zionist leaders were greatly invested in Spartan ideas. Ben-Gurion was famously proud of his wood cabin in Sde-Boker in the Negev, where visiting foreign dignitaries never failed to mention his Spartan lifestyle, as well as his penchant for dressing in military khaki shirt and shorts. Moshe Sharett, Levi Eshkol, and Golda Meir were almost as Spartan themselves.

36 See Edwin Black, *The Transfer Agreement*, Washington, DC: Dialog Press, 1989, 2001, 2009.

37 Ilan Troen, "Calculating the 'Economic Absorptive Capacity' of Palestine: A Study of the Political Uses of Scientific Research," *Contemporary Jewry* 10, 2 (September 1989): 19–20.

38 See "German Jewish Refugees, 1933–1939," *Holocaust Encyclopedia*, ushmm.org.

39 See Gershon Shafir, "Zionism and Colonialism," 233–4.

40 Nathan Weinstock, *Zionism: False Messiah*, London: Ink Links, 1979, 180–9; also see Tamar Gozansky, *Formation of Capitalism in Palestine*, Haifa: Machberet, 1986, 171–8 (Hebrew).

41 See Ilan Pappe, *The Ethnic Cleansing of Palestine*, Oxford: Oneworld, 2007, 17–21.

42 Ibid., 19.

43 Sam Katz, *Israeli Elite Units Since 1948*, Oxford: Osprey, 1998, 12–24 (republished in 2005 as *Elite 018: Israeli Units since 1948*).

44 A. J. Sherman, *Mandate Days: British Lives in Palestine, 1918–1948*, Baltimore: Johns Hopkins University Press, 1997, 121.
45 Martin van Creveld, *Moshe Dayan*, London: Weidenfeld and Nicholson, 2004, 46–8.
46 Yoram Kaniuk, *Commander of the Exodus*, Tel Aviv: Hakibbutz Hameuhad, 1999, 48–50 (Hebrew).
47 Ben-Eliezer, *The Making of Israeli Militarism*, 34.
48 Ibid.
49 Ibid., 35.
50 See Gozanski, *Formation of Capitalism in Palestine*, 122–33 (Hebrew).
51 Gilbert Achcar, *The Arabs and the Holocaust: The Arab–Israeli War of Narratives*, New York: Metropolitan Books, 2009, 126.
52 Sherman, *Mandate Days,* 131.
53 Achcar, *The Arabs and the Holocaust*, 126.
54 James L. Gelvin, *The Israel–Palestine Conflict: One Hundred Years of War*, Cambridge: Cambridge University Press, 2007, 12.
55 Benny Morris, ed., *Making Israel*, Ann Arbor: University of Michigan Press, 2007, 18.
56 Ben-Eliezer, *The Making of Israeli Militarism*, 43–7.
57 Ibid., 44.
58 See Ibid.; Uri Brener, *Towards an Independent Jewish Army 1939–1945*, Ramat Efal: Yad Tabenkin, 1985 (Hebrew); David Livnee, *The Order: PALMACH Man Memories*, Jerusalem: Mass Publishers, 1977 (Hebrew); Y. Sade, *What Did the PALMACH Renew?*, Tel Aviv: Hapoalim Books, 1951 (Hebrew); Yehuda Bauer, *From Diplomacy to Resistance: A History of Jewish Palestine*, Philadelphia: Jewish Publication Society of America, 1970; Meir Pail, *From the Hagana to IDF*, Tel Aviv: Zmora Bitan Modan, 1979 (Hebrew).
59 Ben-Eliezer, *The Making of Israeli Militarism*, 44.
60 Ibid., 54, 156.
61 Ibid., 143.
62 Ibid., 149–91.
63 Sherman, *Mandate Days,* 209–12; Tom Segev, *One Palestine Complete*, London: Abacus, 2001, 475–8.
64 Churchill's note on July 6, 1945, quoted in Nicholas Bethel, *The Palestine Triangle: The Struggle between the British, the Jews and the Arabs 1935–48*, London: Andre Deutsch, 1979, 201.
65 Ibid.
66 See George M. Fredrickson, "From Exceptionalism to Variability: Recent Developments in Cross-National Comparative History," *Journal of American History* 82: 2, 1995, 587–604.
67 Pappe, *The Ethnic Cleansing of Palestine*, xiii, 40.
68 Laqueur, *A History of Zionism*, 584.
69 Shlomo Sand, *The Invention of the Jewish People*, London: Verso, 2009, 71–5.
70 Ahmad H. Sa'di and Lila Abu-Lughod, eds., *Nakba: Palestine, 1948, and*

the Claims of Memory, New York: Columbia University Press, 2007, 4.

71 Hagar Shezaf, "Burying the Nakba: How Israel Systematically Hides Evidence of 1948 Expulsion of Arabs," *Haaretz*, July 5, 2019; Uri Ben-Eliezer, "Nation-in-Arms: State, Nation, and Militarism in Israel's First Years," *Comparative Studies in Society and History*, 37:2 (April 1995), 281.

72 Gil Hoffman, Khaled Abu Toameh, Omri Nahmias, "Netanyahu Vows to Annex All Settlements, Starting with Jordan Valley," *Jerusalem Post*, September 11, 2019, jpost.com, accessed on September 13, 2019.

73 In Israel, the term used for turning the settlements from military into civil ones is Izruach, Hebrew for *turning civilian* or *civilizing*.

74 See Benjamin Beit-Hallahmi, *The Israel Connection: Whom Israel Arms and Why*, London: I B Tauris, 1987, 130.

75 See Guy Laron, "Israel's Decision to Launch the 1956 Sinai Campaign," *British Journal of Middle Eastern Studies* 42: 2, 2015, 202.

76 Ibid., 204–6.

77 Ibid.

78 David Ben-Gurion, MAPAI Faction in the Knesset, April 11, 1951, Mapai Archive, quoted in Ben-Eliezer, *The Making of Israeli Militarism,* 195.

79 David Ben-Gurion, "Nezach Israel," *Shenaton Hamemshala*, Israel Government Annual Almanach, Tel Aviv, 1954, 37 (Hebrew).

80 David Ben-Gurion, MAPAI Faction in the Knesset, April 11, 1951, Mapai Archive, 195; Ben-Eliezer, *The Making of Israeli Militarism*, 195; Mapai Political Committee, Mapai Archive, July 24, 1952.

81 Ben-Eliezer, *The Making of Israeli Militarism*, 197.

82 Yossi Verter, "Corruption Scandal in Lieberman's Party Could Redraw the Political Map," *Haaretz*, December 26, 2014.

83 Aluf Benn, "The Secret Hands Guiding Israel's Controversial Pipeline," *Haaretz,* January 6, 2015.

84 See "The Israeli Submarine Scandal: What We Know: Netanyahu Is Taking on Water as a Potential Conflict of Interest Scandal Surrounding Submarine Purchase from German Company Grows," *Haaretz*, November 6, 2017.

85 See Nir Hasson, "Is the Bible a True Story?," *Haaretz*, November 1, 2017.

86 Dan Ben-Amotz and Netiva Ben-Yehuda, *Milon 'Olami Le'Ivrit Meduberet* (World Dictionary for Hebrew Slang), Vol. 1, Tel Aviv: Zmora, Bitan Publishers, 1982 (Hebrew).

87 See en.hebrew-academy.org.il.

88 Ben-Amotz and Ben-Yehuda, *Milon 'Olami Le'Ivrit Meduberet*, 9, 34, 46.

89 Ella Habiba Shohat, "A Reluctant Eulogy: Fragments from the Memories of an Arab-Jew," in Nahla Abdo and Ronit Lentin, eds., *Women and the Politics of Military Confrontation*, London: Berghahn Books, 2002, 268.

90 From *Shironet*, a site of army songs in Hebrew, shironet.mako.co.il (my translation).

91 In the interest of historical accuracy, it is important to point out that I was one of these misguided children.

92 Mehul Srivastava, "WhatsApp Voice Calls Used to Inject Israeli Spyware on Phones," *Financial Times*, May 13, 2019.

93 Shlomo Sand, *The Invention of the Land of Israel*, London: Verso, 2012, 259–81.

2. The 1948 War

1 Avot Yeshuron, *Kol ha-shirim*, eds. Benjamin Harshav and Hilit Yeshurun, vol 1. Tel Aviv: Hakibbutz Hameuhad, 1995, 81 (Hebrew edition of the *Collected Poems of Yeshurun*, translations by Dan Miron, 104).

2 Haim Bresheeth, "Reviving the Palestine Narrative on Film: Negotiating the Future through the Past and Present in *Route 181*," in Karima Laachar and Saeed Talajooy, eds., *Resistance in Contemporary Middle Eastern Cultures: Literature, Cinema and Music*, London: Routledge, 2013, 148.

3 Benny Morris, *1948*, New Haven, CT: Yale University Press, 2008, 83.

4 Ibid., 89–93.

5 Gunther E. Rothenberg, *The Anatomy of the Israeli Army*, London: B. T. Batsford, 1979, 39.

6 Yosef Grodzinsky, *In the Shadow of the Holocaust*, Monroe, ME: Common Courage Press, 2004, 4–9.

7 Jewish returnees to Poland, for example, faced a series of pogroms in 1945–48, where hundreds (and possibly thousands) perished. See Anna Bikont, *The Crime and the Silence*, New York: Farrar, Straus and Giroux, 2015 (2004), 151, 254.

8 See Jan Gross, *Fear: Anti-Semitism in Poland after Auschwitz*, New York: Random House, 2006, 81–116.

9 The phrase used was *chomer 'enoshi tov* (good human material); Grodzinsky, *In the Shadow of the Holocaust*, 9.

10 See the table in Rothenberg, *The Anatomy of the Israeli Army*, 58, which provides the number of military equipment items in December 1947, May 1948, and October 1948.

11 Ibid.

12 Ibid., 58, 63.

13 Morris, *1948*, 87.

14 Benny Morris, *The Birth of the Palestinian Refugee Problem, 1947–1949*, Cambridge: Cambridge University Press, 1987, 21.

15 Avi Shlaim, "Britain and the Arab–Israeli War of 1948," *Journal of Palestine Studies* 16: 4, Summer 1987, 52.

16 Morris, *1948*, 95.

17 Simha Flapan, "The Palestinian Exodus of 1948," *Journal of Palestine Studies* 16: 4, Summer 1987, 5.

18 Ilan Pappe, *The Ethnic Cleansing of Palestine*, Oxford: OneWorld, 2006, 60, 92, 94–5.

19 Ibid., 86–125.

20 Morris, *The Birth of the Palestinian Refugee Problem, 1947–1949,* 17.
21 Morris, *1948,* 105.
22 Shlaim, "Britain and the Arab–Israeli War of 1948," 52–8.
23 Morris, *The Birth of the Palestinian Refugee Problem, 1947–1949,* 41.
24 Ibid., 41.
25 Ibid., 42.
26 Ibid., 43.
27 Ibid., 59–60.
28 Ari Shavit, "Survival of the Fittest: An Interview with Benny Morris," *Haaretz Weekly Magazine*, January 8, 2004.
29 Nur Masalha, *The Palestine Nakba: Decolonising History, Narrating the Subaltern, Reclaiming Memory*, London: Zed Books, 2012, 17.
30 Walter Laqueur, *A History of Zionism: From the French Revolution to the Establishment of the State of Israel*, New York: Schocken Books, 1972, 2003, 584.
31 See statement in Walid Khalidi, ed., *From Haven to Conquest: Readings in Zionism and the Palestine Problem until 1948*, Beirut: Institute of Palestine Studies, 1971, 495–7.
32 Shabtai Teveth, *Ben-Gurion and the Palestine Arabs*, Oxford: Oxford University Press, 1985, 174–5.
33 Ibid., 198.
34 Ibid., 198–9.
35 Their main publications on this topic are Masalha, *The Palestine Nakba*; Pappe, *The Ethnic Cleansing of Palestine.*
36 Masalha, *The Palestine Nakba,* 19–47.
37 Simha Flapan, *The Birth of Israel*, New York: Pantheon Books, 1987, 22.
38 Ibid., 23–4.
39 Ibid., 32.
40 See "Text of Plan Gimmel (Plan C) May 1946: Section on Countermeasures," *Journal of Palestine Studies* 18: 1, Autumn 1988, 20–3; "Text of Plan Dalet (Plan D), 10 March 1948: General Section," *Journal of Palestine Studies* 18: 1, 24–37.
41 Walid Khalidi, "Plan Dalet: Master Plan for the Conquest of Palestine," *Journal of Palestine Studies* 18: 1, Autumn 1988, 4–19.
42 See Walid Khalidi, "The Fall of Haifa," *Middle East Forum*, December 1959, and "Plan Dalet," *Middle East Forum*, November 1961.
43 Samuel Segev, *The Iranian Triangle: The Untold Story of Israel's Role in the Iran-contra Affair*, New York: The Free Press, 1988, 5, 9.
44 Khalidi, "Plan Dalet: Master Plan for the Conquest of Palestine," 9.
45 Ibid., 16.
46 Quoted from the English version in *Journal of Palestine Studies* referred to above, 29. The whole English text of *Plan Dalet* is also available online at jewishvirtuallibrary.org.
47 See Nur Masalha, *Expulsion of the Palestinians: The Concept of Transfer in Zionist Political Thought, 1882–1948*, Washington DC, Institute for Palestine Studies, 1992.

48 Pappe, *The Ethnic Cleansing of Palestine,* 17–22.
49 Masalha, *The Palestine Nakba*; Nur Masalha, *The Politics of Denial: Israel and the Palestinian Refugee Problem*, London: Pluto Press, 2003.
50 Pappe, *The Ethnic Cleansing of Palestine,* 5–7, 37–8, 51–2, 61–4.
51 Nahum Barnea, "Dir Yassin: We Have Returned to You," *Davar*, April 9, 1982. Zvi Ankori commanded the Gadna unit that collected the bodies and buried them. They claimed to have counted 254 dead, and an unspecified number of mutilated bodies. (Ankori later became a historian.) This count was later questioned by Uri Milstein, who claims that around 110 people were killed. A similar number is given by the Red Cross representative, Jacques de Reynier, in Khalidi, *From Haven to Conquest,* 764.
52 This episode was described candidly in Rabin's book but excised by the military censor in 1989, and the book was printed without it. One of the many references to this episode is in Pappe, *The Ethnic Cleansing of Palestine,* 169, and the story is told fully by D. Kurzman, *Soldier of Peace*, 140–1.
53 Yitzhak Laor, *Narratives with No Natives*, Tel Aviv: Hakibbutz Hameuhad, 1995, 125 (Hebrew, my translation).
54 Yossi Amitai, *Brotherhood of Nations under Scrutiny: Mapam 1948–1954*, Tel Aviv: Cherikover, 1988, 37 (Hebrew, my translation).
55 See Ofer Aderet, "What Is Netanyahu Hiding in the Archive from 1948? His Decision to Keep It Closed for Another 20 Years Is Taking Secrecy to Absurd Lengths," *Haaretz*, October 5, 2018.
56 See Jonathan Ofir, "'Barbarism by an Educated and Cultured People'—Dawayima Massacre Was Worse than Deir Yassin," *Mondoweiss*, February 7, 2016, mondoweiss.net.
57 Ibid.
58 Morris, *1948,* 405–7.
59 Quoted in Masalha, *Politics of Denial*, 1.
60 Morris, *The Birth of the Palestinian Refugee Problem, 1947–1949,* 248.
61 Jean Paul Filiu, *Gaza: A History*, Oxford: Oxford University Press, 2014, 92.
62 Flapan, *The Birth of Israel*, 36–7.
63 Filiu, *Gaza*, 92.
64 Quoted in Khalidi, "Plan Dalet: Master Plan for the Conquest of Palestine," 19, from Zerubavel Gil'ad, ed., *Sefer HaPALMACH*, Vol. 2, Tel Aviv: HaKibbutz Hameuhad, 1953, 186 (Hebrew).
65 Shlaim, "Britain and the Arab–Israeli War of 1948," 65.
66 Ibid., 67.
67 Ibid., 64–8.
68 In Israel, the term used for turning the settlements from military into civil ones, is *Izruach*, Hebrew for *making into civilian* or *civilizing*.
69 Yitzhak Rabin, *The Rabin Memoirs,* Los Angeles: University of California Press, 1996, 383: "Allon and I held a consultation. I agreed that it was essential to drive the inhabitants out. We took them on foot toward the Bet Horon road, assuming that the Arab Legion would be obliged to look

after them, thereby shouldering logistic difficulties which would burden its fighting capacity, making things easier for us."

70 Morris, *The Birth of the Palestinian Refugee Problem, 1947–1949*, 292–3.
71 Uri Ben-Eliezer, *The Making of Israeli Militarism*, Bloomington: Indiana University Press, 1998, 215.
72 *Davar*, October 26, 1955, quoted in ibid.
73 Ben-Eliezer, *The Making of Israel Militarism*., 197.
74 Shimon Peres, at the Histadruth Executive, July 19, 1955, Labour Archives in Tel Aviv, quoted in Ibid., 214.
75 Ibid., 207–22.
76 Shlaim, "Britain and the Arab–Israeli War of 1948," 66–7.
77 Ibid.; see also Zeev Tzahor, "The 1949 Air Clash between the Israeli Air Force and the RAF," *Journal of Contemporary History* 28: 1, January 2003, 75–101.
78 Joseph Massad, "Resisting the Nakba," *Al-Ahram Weekly Online*, no. 897, May 15–21, 2008, quoted in Masalha, *The Palestine Nakba*, 43.
79 Mentioned and quoted in Teveth, *Ben-Gurion and the Palestine Arabs*, 188–9. The above translation is the more accurate version published by the Institute of Palestine Studies in Beirut and republished by Jewish Voice for Peace, jewishvoiceforpeace.org.

3. The 1956 War

1 Yisrael Ber, *Bema'agal Be'ayot HaBitahon*, Tel Aviv: Am Oved, 1957, 23 (Hebrew). Bar argues in his book mainly against the supremacy of Britain and the United States and against what he sees as their uncritical and cynical support of the Arab regimes. In 1961, after earlier suspicions were confirmed, he was arrested and tried in secret as a Soviet agent in Israel. He died in prison in 1966. His exact identity was never determined.
2 Ibid., 22.
3 The two authors chose the interesting pseudonym for A. Israeli for the putative author of this book, *Shalom, Shalom Veen Shalom*, published by the authors in Jerusalem in 1961 [1999] (Hebrew).
4 Ibid., 7.
5 Avi Shlaim, "Britain and the Arab–Israeli War of 1948," *Journal of Palestine Studies* 16: 4, Summer 1987, 73.
6 Ibid.
7 Ella Shohat, *Forbidden Memories*, Tel Aviv: Keshet HaMizrah, 2001 (Hebrew); see the chapter "Notes on the Post-Colonial," 274–97.
8 Quoted in Israeli, *Shalom, Shalom Veen Shalom*, 36, from the Knesset minutes.
9 Benjamin Beit-Hallahmi, *The Israeli Connection: Whom Israel Arms and Why*, London: I B Tauris, 1988, 8.
10 See Selwin Lloyd, *Suez 1956: A Personal Account*, London: Jonathan Cape, 1978, 10.

11 William Roger Louis, *Effervescent Adventures with Britannia*, London: I.B. Taurus, 252.
12 Scott Lucas, *Divided We Stand: Britain, the US and the Suez Crisis*, London: Hodder and Stoughton, 1991, 109.
13 Lloyd, *Suez 1956;* about the discussions that brought about the end of the Suez base, see 21–35; see also Erskine Childers, *The Road to Suez: A Study of Western–Arab Relations*, London: MacGibbon & Kee, 1962, 130–1.
14 Lucas, *Divided We Stand*, 82–4; Scott Lucas, "Redefining the Suez 'Collusion'," *Middle Eastern Studies*, 26:1 (Jan. 1990), 102.
15 Childers, *The Road to Suez,* 126–30; Lucas, *Divided We Stand,* 62–3, 101, 103.
16 To read about the range of ideas for a coup d'état against Nasser, see Lucas, *Divided We Stand* 62–3, 86, 101–3, 115.
17 Childers, *The Road to Suez,* 28; Selwyn Lloyd, *Suez 1956*, 28–9.
18 Ibid., 111, quoted from Hansard, July 29, 1954.
19 Ibid., 119–20. This is a translated account of Nasser, written in 1954.
20 Ibid., 101, quoted from Musa Alami, "The Lesson of Palestine," *Middle East Journal* 3: 4, 1949, 373–405.
21 Ibid., 103.
22 See Benny Morris, *Israel's Border Wars 1949–1956: Arab Infiltration, Israeli Retaliation, and the Countdown to the Suez War*, Oxford: Clarendon Press, 1993, 116–21.
23 Ibid., 117, from Weitz letter to Sharett, May 27, 1949.
24 Morris, *Israel's Border Wars 1949–1956,* 121–7.
25 David Tal, "The 1956 Sinai War: A Watershed in the History of Arab–Israeli Conflict," in Simon C. Smith, ed., *Reassessing Suez 1956*, London: Ashgate, 2008, 139.
26 See Morris, *Israel's Border Wars 1949–1956*, 108–15.
27 See ibid., 144 n106, where Morris admits: "The expulsion of infiltrators in late 1948 and the first half of 1949 rested on shaky legal foundations. Israeli law was slow to catch up with the state's political–military needs. Confusion reigned. There was a vague feeling that the expulsions were covered by the British mandatory laws adopted by the Knesset in early 1949." In other words, soldiers and officers did what they thought was expected of them, whether it was legal or not. It was not.
28 See the section "Expelling Infiltrators" in ibid., 143–72.
29 Ibid., 137.
30 Ibid., 143–5.
31 Ibid., 136.
32 Tal, "The 1956 Sinai War," 137–8.
33 Morris, *Israel's Border Wars*, 109–10.
34 Peter V. Pry, *Israel's Nuclear Arsenal*, Boulder, CO: Westview Press, 1984, 6.
35 Ibid.
36 Ibid., 46–8.
37 Morris, *Israel's Border Wars 1949–1956*, 178–9.

38 Zeev Schiff and Ehud Ya'ari, *Israel's Lebanon War*, London: George Allen & Unwin, 1984, 13–14.
39 Ibid., 14.
40 See Childers, *The Road to Suez,* 131.
41 Tal, "The 1956 Sinai War," 141–2.
42 Haggai Eshed, *Who Gave the Order: The Lavon Affair*, Tel Aviv: Idanim, 1979, 29 (Hebrew).
43 Issar Harel, *When Man Rose against Man: A Re-evaluation of the "Lavon Affair,"* Jerusalem: Keter, 1982, 126 (Hebrew).
44 Morris, *Israel's Border Wars 1949–1956*, 181.
45 Moshe Sharett, *Yoman Ishi* [personal diary] iv, 1021, entry for May 26, 1955; quoted in ibid., 182.
46 I prefer this transliteration of the name to the one used below by Israeli sources; it is closer to the Arabic source.
47 Reported in *Davar* daily, October 20, 1953. The English version appears in Livia Rokach, *Israel's Sacred Terrorism: A Study Based on Moshe Sharett's Personal Diary and Other Documents,* New York: AAUG, 1985, Appendix 1, 61–2.
48 Avi Shlaim, *The Iron Wall: Israel and the Arab World*, New York: W.W. Norton, 2001, 95–9; and Morris, *Israel's Border Wars 1949–1956*, 243–62.
49 Morris, *Israel's Border Wars 1949–1956,* Chapters 6, 7, and 8, 173–262.
50 Lucas, *Divided We Stand,* 50–3.
51 Morris, *Israel's Border Wars 1949–1956,* 246.
52 Ibid., 245. As Morris observes in a footnote, Sharett would not find out that he was lied to by Lavon until he replaced Ben-Gurion as prime minister, in 1954. He never acted on this information; Ibid., 245, n84.
53 Both groups faced official shunning of their languages, Yiddish and Arabic, as reported elsewhere in this volume. In the case of Yiddish, a special, unique tax was forced on newsprint in this language to deter readers of the popular Yiddish daily *Lezte Nies* (Latest News), venomously hated by Ben-Gurion. He insisted that Yiddish was not a "proper" language and never mentioned it by name, calling it *jargon* instead.
54 Morris, *Israel's Border Wars 1949–1956,* 243.
55 Ibid., 256–7.
56 Ibid., 327.
57 Ibid., 359.
58 Ibid., 326.
59 Lucas, *Divided We Stand,* 43–5.
60 Ibid.
61 Morris, *Israel's Border Wars 1949–1956,* 331.
62 Childers, *The Road to Suez,* 133.
63 Morris, *Israel's Border Wars 1949–1956,* 289.
64 Ibid.
65 Lucas, *Divided We Stand*, 51.

66 Telegram from US Cairo Embassy to State Department, October 29, 1956, quoted in David Tal, "The 1956 Sinai War," 146.
67 Lucas, *Divided We Stand*, 169, 185.
68 Ibid., 173.
69 Ibid., 174.
70 Roy Fullick and Geoffrey Powell, *Suez: The Double War*, London: Hamish Hamilton, 1979, 89.
71 Merry Bromberger and Serge Bromberger, *Secrets of Suez*, London: Pan Books, 1957, 11–12. See also Fullick and Powell, *Suez*, 89.
72 Fullick and Powell, *Suez*, 91.
73 Paul Johnson, *The Suez War*, MacGibbon and Kee, London, 1957, xi.
74 Ibid.
75 Fullick and Powell, *Suez*, 96.
76 Ibid.
77 Ibid., 139–41.
78 Ibid., 159.
79 Ofer Aderet, "General's Final Confession Links 1956 Massacre to Israel's Secret Plan to Expel Arabs," *Haaretz*, October 13, 2018.
80 Ibid.
81 Koby Niv, "From Talk to Action. From Regularisation to Extermination," *Haaretz*, November 28, 2018 (Hebrew).
82 Yotam Berger, "How Israel Gave Up the Idea of Annexing Sinai and Gaza," *Haaretz*, December 9, 2016.
83 Ibid.
84 Ibid.
85 Noa Landau, "Netanyahu Tells UN Chief: Golan Will Remain Israel's Forever," *Haaretz*, February 16, 2018.

4. The 1967 War

1 William E. Burrows and Robert Windrem, *Critical Mass: The Dangerous Race for Superweapons in a Fragmenting World*, New York: Simon and Schuster, 1994, 280; Avner Cohen, *The Worst Kept Secret*, New York: Columbia University Press, 2010, 72–3.
2 David Ben-Gurion has apparently given up the recapture of Gaza in December 1948, as reported in Tom Segev, *1949: The First Israelis*, New York: Simon and Schuster, 1998 [2008], 3, and had lost the cabinet vote and as a result adjusted his stance; he is quoted arguing that "he preferred a democratic Jewish state, even if it did not include all the land" (19). Hardly a democrat at the best of times, Ben-Gurion, who had no qualms about maintaining the Military Government for Israeli Arabs after the war, could clearly see that Israel could not possibly be a democracy with many more Arabs under its control and would not be Jewish by any measure.

3 Arguably, the weakness of the IDF as a defense force was clearly demonstrated in 1973, when it failed to defend the country against the Egyptian–Syrian surprise attack, losing both ground and morale when forced into defensive action, even as it had the "strategic depth" so desired by Ben-Gurion and Dayan.
4 Tony Judt, *Reappraisals: Reflection on the Forgotten Twentieth Century*, London: William Heinemann, 2008, 271–2.
5 See Jason Thompson, *A History of Egypt: From Earliest Times to the Present*, London: Haus Books, 2008, 301.
6 Ibid.
7 Ibid., 302.
8 Ibid., 311.
9 Ibid., 304.
10 Officially, Egypt remained in the UAR until 1971, when the union was finally dissolved.
11 Thompson, *A History of Egypt*, 305–7.
12 Ibid., 310–11.
13 Ibid., 310.
14 Tom Segev, *1967: Israel, the War, and the Year That Transformed the Middle East*, New York: Henry Holt, 2007, 36–8.
15 Ibid., 126–34.
16 Ibid., 150–1.
17 Ibid., 153.
18 Ibid., 155.
19 Gavriel Ben-Dor, "Israel of 1967: Its International Status and Self-Image before the Six-Day War," in Asher Susser, ed., *Six Days—Thirty Years: New Perspectives on the Six Day War*, Tel Aviv: Am Oved Publishing, 1999, 38 (Hebrew).
20 Father of Yair Lapid, an Israeli broadcaster and leader of a center-right party in the Knesset.
21 Segev, 1967, 162, quoting Yosef Lapid, *Ma'ariv*, May 12, 1967, 11.
22 Segev, *1967*, 163.
23 Ibid., 175.
24 See Ilan Pappe, "Revisiting 1967: The False Paradigm of Peace, Partition and Parity," *Settler Colonial Studies* 3: 3–4, 2013, 342–3.
25 Ibid., 342.
26 Ibid., 343.
27 Quoted in Segev, *1967*, 192.
28 Ibid., 231 (footnote).
29 M. Mayzel, *The Battle over the Golan: June 1967*, Tel Aviv: Ma'arachot, 2001, 93 ff. (Hebrew).
30 Herman Eilts, "The Six-Day War through Egyptian Eyes," in Susser, ed., *Six Days—Thirty Years*, 62–70, 81–3 (Hebrew).
31 Shim'on Shamir, "The Origin of the Escalation in May 1967: The Argument of 'Israeli Threat,'" in Asher Susser, (1999), 62 (Hebrew).
32 Ibid., 66–7.

33 Segev, *1967*, 227.
34 Ibid., 228–9.
35 See Isabella Ginor and Gideon Remez, *Foxbats over Dimona*, New Haven, CT: Yale University Press, 2007, 122.
36 Ibid., 191–205.
37 Segev, *1967*, 238–9.
38 Ginor and Remez, *Foxbats over Dimona*, 142.
39 Avner Cohen, *The Worst-Kept Secret: Israel's Bargain with the Bomb*, New York: Columbia University Press, 2010, 84–5; Seymour M. Hersh, *The Samson Option: Israel, America and the Bomb*, London: Faber and Faber, 1991, 263–8.
40 Warner D. Farr, *The Third Temple's Holy of Holies: Israel's Nuclear Weapons*, Counterproliferation Papers, Future Warfare Series, September 1999, fas.org.
41 See Nic Von Wielligh and Lydia Von Wielligh-Steyn, *The Bomb: South Africa Nuclear Weapons Programme*, Pretoria: Litera, 2015.
42 David Kimche and Dan Bawly, *The Sandstorm: The Arab–Israeli War of June 1967: Prelude and Aftermath*, London: Secker and Warburg, 1968, 179–81.
43 Hal Kosut, ed., *Israel and the Arabs: The June 1967 War*, New York: Facts on File, 1968, 72.
44 Ibid., 74.
45 Kimche and Bawly, *Sandstorm,* 179.
46 Asher Susser, "Jordan and the Six Day War," in Susser, ed., *Six Days—Thirty Years,*110–11 (Hebrew).
47 Kosut, *Israel and the Arabs,* 74.
48 Ibid.
49 Ibid, 75.
50 Ibid, 79.
51 Susser, "Jordan and the Six Day War," 110–11.
52 Avi Raz, *The Bride and the Dowry: Israel, Jordan and the Palestinians in the Aftermath of the June 1967 War*, New Haven, CT: Yale University Press, 2012, 3.
53 Ibid., 25.
54 Ibid., 5.
55 Idit Zertal, *Nation and Death*, Tel Aviv: Dvir, 2001, 170.
56 Ibid.
57 Segev, *1967*, 429–31.
58 Ibid., 430.
59 Giorgio Agamben, *Homo Sacer*: *Sovereign Power and Bare Life*, Palo Alto, CA: Stanford University Press, 1998.
60 A bizarre phenomenon, lasting years, and lending its name to a chapter in Tom Segev's volume, describing the cultural transformation of Israel following the war: "Victory Albums," in Segev, *1967*, 437–54.
61 Moshe Dayan, April 29, 1956, quoted in Chemi Shalev, "Moshe Dayan's Enduring Gaza Eulogy: This Is the Fate of Our Generation," *Haaretz*, July 20, 2014.

62 A recent report on this plan is available online: Rod Such, "Why the Occupation Is No Accident," *Electronic Intifada*, September 18, 2017, electronicintifada.net, accessed on August 17, 2018.
63 Pappe, "Revisiting 1967: The False Paradigm of Peace, Partition and Parity," *Settler Colonial Studies*, 3: 3–4 (2013), 343.
64 Ibid., 345.
65 Ibid., 347.
66 Ibid., 345, 347.

5. The 1973 War

1 Avi Raz, *The Bride and the Dowry: Israel, Jordan and the Palestinians in the Aftermath of the June 1967 War*, New Haven, CT: Yale University Press, 2012, 25–51.
2 Amnon Kapeliuk, *Lo Mehdal: The Politics Which Led to War*, Tel Aviv: Amikam Publishing, 1975, 16 (Hebrew).
3 A speech at the Command and Staff College of IDF, quoted in ibid., 18 (emphasis in original).
4 Benyamin Korn, "Golda Meir Was No J-Streeter," Jewish News Syndicate, September 4, 2015.
5 Kapeliuk, *Lo Mehdal*, 21.
6 Isabella Ginor and Gideon Remez, *The Soviet–Israeli War 1967–1973*, London: Hurst, 2017, 13.
7 Avraham Sela, "The 1973 Arab War Coalition: Aims, Coherence and Gain-Distribution," in P. R. Kumaraswamy, ed., *Revisiting the Yom Kippur War*, London: Routledge, 2013, 41.
8 Ginor and Remez, *The Soviet–Israeli War 1967–1973*, 28–32.
9 Ibid., 55.
10 Kapeliuk, *Lo Mehdal*, 117–8.
11 Ibid., 111–32.
12 Quoted in ibid., 125.
13 Kapeliuk, *Lo Mehdal*, 125.
14 Raz, *The Bride and the Dowry*, 19.
15 Kapeliuk, *Lo Mehdal*, 14.
16 Ibid., 31.
17 Eithan Haber and Zeev Schiff, *Yom Kippur War Lexicon*, Tel Aviv: Yedi'ot Ahronot, 1993, 46, 303, 339, 370.
18 Ginor and Remez, *The Soviet–Israeli War 1967–1973*, 40.
19 Isabella Ginor and Gideon Remez, *Foxbats over Dimona: The Soviets' Nuclear Gamble in the Six-Day War*, New Haven, CT: Yale University Press, 2007, 121–36.
20 Ginor and Remez, *The Soviet–Israeli War 1967–1973*, 10–11.
21 Ronen Bergman and Gideon Meltzer, *The Yom Kippur War: Failures and Lessons*, Tel Aviv: Yedi'ot Ahronot/Hemed, 2004, 175–87 (Hebrew).

22 Zeira, *The Yom Kippur War*, 52–3.
23 Zeira, *The Yom Kippur War*, 53.
24 Zeira, *The Yom Kippur War*, 57–60.
25 Ginor and Remez, *The Soviet–Israeli War 1967–1973*, 281.
26 Zeira, *The Yom Kippur War*, 165–73.
27 Kapelink, *Lo Mehdal*, 125–30, 34.
28 Ibid., 24, 32.
29 Ibid., 24–5.
30 Ibid.
31 Hashavia, *Milhemet Yom Ha'Kippurim*, Tel Aviv: Zamara, Bitan, Modan, 1974, 51.
32 Interview with Moshe Dayan, *Time*, July 30, 1973, quoted in Kapeliuk, *Lo Mehdal*, 34.
33 Ibid., 52.
34 Ibid.
35 Hashavia, *Milhemet Yom Ha'Kippurim*, 25, 27.
36 Ginor and Remez, *The Soviet–Israeli War 1967–1973*, 330–1.
37 Zeira, *The Yom Kippur War*.
38 Ibid., 210–11, 46, 58–9.
39 Kapeliuk, *Lo Mehdal*, 73.
40 Zeira, *The Yom Kippur War*, 89.
41 Ginor and Remez, *The Soviet–Israeli War 1967–1973*, 342.
42 Hashavia, *Milhemet Yom Ha'Kippurim*, 229.
43 Ibid., 252.
44 Riad Ashkar, "The Syrian and Egyptian Campaigns," *Journal of Palestine Studies* 3: 2, Winter 1974, 33.
45 Edmund Ghareeb, "The US Supply to Israel during the War," *Journal of Palestine Studies* 3: 2, Winter 1974, 118–9.
46 Ibid., 114.
47 Ginor and Remez, *The Soviet–Israeli War 1967–1973*, 350.
48 Ibid, 352–4.
49 Mati Peled, in *Ma'ariv*, October 19, 1973, quoted in Elias Shoufani, "Israeli Reactions to the War," in Nasser Hassan Arruri, ed., *Middle East Crucible: Studies on the Arab–Israeli War of October 1973*, Wilmette, IL: Medina University Press, 1975, 49.
50 Kapeliuk, *Lo Mehdal*, 85–102.
51 Edward Said, "Arabs and Jews," *Journal of Palestine Studies* 3: 2, Winter 1974, 7; Khalil Nakhleh, "The Political Effects of the October War on Israeli Society," in Nasser Hassan Arruri, ed., *Middle East Crucible: Studies on the Arab–Israeli War of October 1973*, Wilmette, IL: Medina University Press, 1975, 171.
52 Shoufani, "Israeli Reactions to the War," 46.
53 Kapeliuk, *Lo Mehdal*, 210.
54 Nakhleh, "The Political Effects of the October War on Israeli Society," 167.
55 Kapeliuk, *Lo Mehdal*, 149.
56 Ghareeb, "The US Supply to Israel during the War," 114.

57 Ginor and Remez, *The Soviet–Israeli War 1967–1973*, 360.
58 Ibid, 359.
59 Ibid. 360.

6. Israel's Longest War

1 Claudia Wright, "The Turn of the Screw: The Lebanon War and American Policy," *Journal of Palestine Studies* 11: 4, Summer/Fall 1982, 16.
2 Ibid., 17.
3 Zeev Schiff and Ehud Ya'ari, *Israel's Lebanon War*, London: George Allen & Unwin, 1984, 245.
4 Sheila Ryan, "Israel Invasion of Lebanon: Background to the Crisis," *Journal of Palestine Studies* 12: 1, Summer/Autumn 1982, 24, 26–7.
5 Ibid., 28.
6 Schiff and Ya'ari, *Israel's Lebanon War,* 29.
7 Ibid., 12, 28.
8 Ibid., 11–7.
9 Ibid., 28–37.
10 Ibid., 220.
11 Ibid., 38–9.
12 Ibid., 39.
13 Ibid., 59.
14 Shimon Shiffer, *Snowball: The Story behind the Lebanon War*, Tel Aviv: Idanim–Yediot Ahronot, 1984, 17 (Hebrew).
15 Schiff and Ya'ari, *Israel's Lebanon War,* 52.
16 Ibid., 58.
17 Ibid., 42–3.
18 Ibid., 43.
19 Ibid., 43, 32–51.
20 Ibid., 103.
21 Patrick Tyler, *Fortress Israel: The Inside Story of the Military Elite Who Run the Country and Why They Can't Make Peace*, London: Portobello Books, 2012, 304.
22 Schiff and Ya'ari, *Israel's Lebanon War,* 102.
23 Yezid Sayigh, "Israel's Military Performance in Lebanon, June 1982," *Journal of Palestine Studies* 13: 1, Fall 1983, 45.
24 Yezid Sayigh, "Palestinian Military Performance in the 1982 War," *Journal of Palestine Studies* 12: 4, Summer 1983, 6.
25 Ibid., 17.
26 Sayigh, "Israel's Military Performance in Lebanon, June 1982," 46.
27 Ibid., 51.
28 Ibid., 55.
29 Yezid Sayigh, "Palestinian Military Performance in the 1982 War," 8.
30 Ibid., 7.
31 Schiff and Ya'ari, *Israel's Lebanon War*, 109.

32 Dov Yermiya, *My War Diary: Israel in Lebanon*, London: Pluto Press, 1983, 10.
33 Ibid., 11.
34 Ibid., 48.
35 Jean Genet, "Four Hours in Shatila," *Journal of Palestine Studies* 12: 3, Spring 1983, 4–5.
36 Ibid., 6.
37 Schiff and Ya'ari, *Israel's Lebanon War*, 152.
38 Ibid., 193.
39 Shiffer, *Snowball*, 160 (my translation).
40 Ibid., 162 (my translation).
41 Ibid., 103–4.
42 Schiff and Ya'ari, *Israel's Lebanon War*, 215.
43 Ibid.
44 Ibid., 215–6.
45 Ibid., 213–4.
46 Shiffer, *Snowball*, 108 (my translation).
47 Schiff and Ya'ari, *Israel's Lebanon War*, 246–9.
48 Ibid, 246.
49 Ibid., 245.
50 Ibid., 246.
51 Ibid., 271.
52 Ibid., 280.
53 Ibid., 301.
54 Ibid., 283.
55 Eqbal Ahmad, "The Public Relations of Ethnocide," *Journal of Palestine Studies* 12: 3, Spring 1983, 32.
56 Barak was no stranger to Lebanon; as an officer in Sayeret Matkal, Israel's "elite" death squad, he led an operation in Beirut in 1973 to assassinate leading PLO officials, disguised as a brunette woman.
57 See "Edward Said Joins Stone-Throwers at Israeli Border," *UPI*, July 4, 2000, upi.com.
58 As examples, see Helena Cobban, "The PLO and the "Intifada"," Middle East Journal 44: 2 (Spring, 1990): 207.
59 Tariq Kafala, "Intifada: Then and Now," BBC News, December 8, 2000, bbc.co.uk, accessed December 4, 2017.
60 Edward Said, "Palestinians in the Aftermath of Beirut," *Journal of Palestine Studies* 12: 2, Winter 1983, 4.
61 Ibid., 5.
62 Ibid., 7.
63 Jacobo Timerman, *The Longest War*, London: Pan Books, 1982, 82.
64 Ibid., 83.
65 Ibid.
66 Mordechai Kremnitzer, "Jewish Nation-State Law Makes Discrimination in Israel Constitutional," *Haaretz*, July 20, 2018.
67 Ibid.

7. Lebanon, 2006

1 There are varying English spellings of the Lebanese militia; I chose the spelling used by Arab commentators. In quotes, the spelling in the original source has been retained.
2 Gilbert Achcar, with Michel Warschawski, *The 33-Day War: Israel's War on Hezbollah in Lebanon and Its Aftermath*, London: Saqi Books, 2007, 32.
3 Ibid., 33.
4 Ibid., 16.
5 Ibid., 33–44.
6 Ibid., 37.
7 Eyal Weizman, "The Art of War: Deleuze, Guattari, Debord and the Israeli Defence Force," *Mute*, August 3, 2006, metamute.org; and Amir Rapaport, "The IDF and the Lessons of the Second Lebanon War," Mideast Security and Policy Studies No. 85, Ramat Gan: Bar-Ilan University, The Begin-Sadat Center for Strategic Studies, 2010, 4–7.
8 Weizman, "The Art of War." Weizman interviewed Aviv Kokhavi (now serving as the new Chief of General Staff on September 24 at an Israeli military base near Tel Aviv. Translation from Hebrew by the author; video documentation by Nadav Harel and Zohar Kaniel.
9 Ibid. Kokhavi is the new IDF Chief of Staff.
10 Rapaport, "The IDF and the Lessons of the Second Lebanon War," 4.
11 Quoted in Weizman, "The Art of War," from Sune Segal, "What Lies Beneath: Excerpts from an Invasion," *Palestine Monitor*, November 2002.
12 Weizman, "The Art of War," interview with Eyal Weizman.
13 Rapaport, "The IDF and the Lessons of the Second Lebanon War," 9.
14 Ibid., 5.
15 Rami Zurayk, *War Diary: Lebanon 2006*, Charlottesville, VA: Just World Publications, 2011, 16.
16 Ibid., 6.
17 See Eyal Weizman, *Hollow Land: Israel's Architecture of Occupation*, London: Verso, 2007, 214.
18 Achcar, *The 33-Day War*, 46–7.
19 Ibid., 47.
20 Ibid., 48–50.
21 Rapaport, "The IDF and the Lessons of the Second Lebanon War," 6.
22 Achcar, *The 33-Day War*, 54.
23 Ibid., 86.
24 Rapaport, "The IDF and the Lessons of the Second Lebanon War," 1.
25 Ibid., 13.
26 Ibid.
27 Achcar, *The 33-Day War*, 90.
28 See Weizman, *Hollow Land,* especially 74–5, about Sharon's fighting in 1973 as part of the Likud political campaign.
29 Achcar, *The 33–Day War*, 86.
30 Ibid., 87.

31 Ibid., 89.
32 Moshe Arens, "Let the Devil Take Tomorrow," *Haaretz*, August 13, 2006.
33 Achcar, *The 33-Day War*, 58.
34 Ibid., 61.
35 Ibid., 62.
36 Ibid.
37 Ibid., 68–73.
38 Zurayk, *War Diary*, 23.
39 Ibid., 55.
40 Achcar, *The 33-Day War*, 94.
41 Uzi Rubin, *The Rocket Campaign against Israel during the 2006 Lebanon War*, Mideast Security and Policy Studies no. 71, Ramat Gan: Bar-Ilan, The Begin-Sadat Center for Strategic Studies, 2007, 3.
42 Ibid.
43 Ibid., 9.
44 Ibid., 14.
45 Ibid.
46 Achcar, *The 33-Day War*, 96–9.
47 Uzi Benziman, "Pulling the Wool over Our Eyes," *Haaretz*, August 30, 2006.
48 Achcar, *The 33-Day War*, 100.
49 Nahum Barnea, "Run, Ehud, Run," *Yedi'ot Ahronot*, August 10, 2006, quoted in Amos Harel and Avi Issacharoff, *34 Days: Israel, Hezbollah, and the War in Lebanon*, London: Palgrave Macmillan, 2008, 199.
50 Harel and Issacharoff, *34 Days*, 208–9.
51 Ibid., 189.
52 Ibid., 228–9.
53 Ibid., 243.
54 Ibid., 245.
55 Ibid., 245–6.
56 Achcar, *The 33-Day War*, 107, quoted from Charles Krauthammer, "Israel's Lost Moment," *Washington Post*, August 4, 2006.

8. The Armed Settlements Project

1 James North, "Israel Provoked the Six-Day War in 1967, and It Was Not Fighting for Survival," *Mondoweiss*, June 2, 2017, mondoweiss.net. The *New York Times* article referred to in the quote was written by I. Fisher and titled "Jerusalem Day Celebration Underscores Israeli–Palestinian Divide," published on May 24, 2017.
2 Patrick Tyler, *Fortress Israel: The Inside Story of the Military Elite Who Run the Country—and Why They Can't Make Peace*, New York: Farrar, Straus and Giroux, 2012, 134–90.
3 Ibid., 183.
4 Nur Masalha, *Imperial Israel and the Palestinians: The Politics of Expansion*, London: Pluto Press, 2000, 7; Protocol of the Jewish Agency

Executive meeting on June 7, 1938, in Jerusalem, confidential, Central Zionist Archive 28: 51.

5 Masalha, *Imperial Israel and the Palestinians*, 7.

6 Benny Morris, *Israel's Border Wars, 1949–1956: Arab Infiltration, Israeli Retaliation, and the Countdown to the Suez War,* Oxford: Clarendon Press, 1993, 11.

7 For example, in the summer of 2016, with temperatures above 38°C, Israel cut the water supply to most Palestinians in the West Bank: Amira Hass, "Israel Admits Cutting West Bank Water Supply, but Blames Palestinian Authority," *Haaretz*, June 21, 2016. On the limiting of electricity in Gaza to less than four hours a day, see for example Amos Harel, "As Gaza Races toward Collapse, Israel's Strategy Hampered by Division," *Haaretz*, August 11, 2017.

8 In little more than a decade, from January 1, 2008 to October 24, 2019, there were 5,559 fatalities caused by the IDF and settlers. See "Data on Casualties," OCHA, United Nations Office for the Coordination of Humanitarian Affairs, ochaopt.org, accessed on September 20, 2019. During the Second Intifada, the Palestinian casualties caused by the IDF amounted to 4,791. See "Palestinian Casualties of War," Wikipedia, wikipedia.org, accessed on September 20, 2019. The total runs to tens of thousands, as the second source demonstrates.

9 Neve Gordon, *Israel's Occupation*, Berkeley: University of California Press, 2008, 199–201.

10 See Ilan Pappe, "Revisiting 1967: The False Paradigm of Peace, Partition and Parity," *Settler Colonial Studies* 3: 3–4, 2013, 342–3.

11 Ibid., 347.

12 Gordon, *Israel's Occupation*, 6, 200.

13 Pappe, "Revisiting 1967," 347.

14 Gordon, *Israel's Occupation*, 27–9.

15 Ibid., 31.

16 Ibid., 24, 29.

17 Ibid., 74.

18 Ibid.

19 Ibid., 72.

20 Ibid., 187.

21 Ibid., 127.

22 Ibid., 130–1.

23 Ibid., 178.

24 Ibid., 141.

25 Ibid., 131; also see 118: "The settler population was utilised to police the local population 'just like plain clothes security personnel.'"

26 Chemi Shalev, "For US Jewry, Kahanist Caper Casts Netanyahu as Prince of Darkness and Trump on Steroids," *Haaretz*, February 24, 2019.

27 Noa Landau and Yotam Berger, "Netanyahu Says Israel Will Annex Jordan Valley if Reelected," Haaretz, September 10, 2019, haaretz.com, accessed September 20, 2019.

28 David Hearst, "Could Arab Staying Power Ultimately Defeat Zionism?," *Guardian*, August 5, 2011.
29 Yotam Berger, "Netanyahu Vows Never to Remove Israeli Settlements from West Bank: 'We're Here to Stay, Forever,'" *Haaretz*, August 29, 2017.
30 See Jeff Halper, "A Strategy within a Non-Strategy: *Sumud*, Resistance, Attrition, and Advocacy," *Journal of Palestine Studies* 35: 3, Spring 2006, 46.
31 David Halbfinger, "Netanyahu Vows to Start Annexing West Bank, in Bid to Rally Right," *New York Times*, April 6, 2019.

9. The First Intifada and the Oslo Accords

1 Aaron Klieman, *Israel's Global Reach: Arms Sales as Diplomacy*, Oxford: Brassey's, 1985, 38, 36–46.
2 Jeff Halper, *War against the People: Israel, the Palestinians and Global Pacification*, London: Pluto Press, 2015, 67.
3 Ahmad H. Sa'di, "Incorporation without Integration: Palestinian Citizens in Israel's Labour Market," *History of Human Sciences* 29: 3, August 1995, 429–51.
4 Kenneth W. Stein, "The Intifada and the Uprising of 1936–1939: A Comparison of the Palestinian Arab Communities," in Robert O. Freedman, ed., *The Intifada: Its Impact on Israel, the Arab World, and the Superpowers*, Miami: Florida International University Press, 1991, 6.
5 Ibid., 3.
6 See "Colonel Says Rabin Ordered Breaking of Palestinians' Bones," *Los Angeles Times*, June 22, 1990. Actually, Rabin repeated the call on Israeli television.
7 Ibid.
8 To give an example: The author of this volume was at the time chair of a London film school. Funds were collected and a number of VHS cameras were purchased and brought to Palestine by him and two of his film students, who then went to Gaza and the West Bank to offer the cameras to local activists and train them in using them effectively, during 1988.
9 One such example available to view is Rabin's "Break their Bones" Policy, youtube.com.
10 See "Israel Faces Tough Dilemma over Bodies of Palestinian Terrorists," *Haaretz*, November 2, 2015, haaretz.com, accessed on August 1, 2017.
11 Ibid.
12 Wendy Pearlman, *Violence, Nonviolence, and the Palestinian National Movement*, Cambridge: Cambridge University Press, 2011, 115.
13 Robert O. Freedman, ed., "Introduction," in *The Intifada: Its Impact on Israel, the Arab World, and the Superpowers*, Miami: Florida International University Press, 1991, xv.
14 See Helena Cobban, "The PLO and the Intifada," in ibid., 79.
15 Rami Nasrallah, "The First and Second Palestinian *Intifadas*," in David

Newman and Joel Peters, eds., *Routledge Handbook on the Israeli–Palestinian Conflict,* London: Routledge, 2013, 61.

16 Arthur Neslen, *In Your Eyes a Sandstorm: Ways of Being Palestinian*, Los Angeles: University of California Press, 2011, 122.

17 David Seddon, ed., "Intifada," in *A Political and Economic Dictionary of the Middle East*, London: Taylor & Francis, 2004, 229.

18 Freedman, *The Intifada*, 80.

19 Ibid., 88, quoted from UNLU Leaflet no. 6.

20 Thomas L. Friedman, "Baker, in a Middle East Blueprint, Asks Israel to Reach Out to Arabs," *New York Times*, May 23, 1989.

21 Thomas L. Friedman, "Israel, Ignoring Bush, Presses for Loan Guarantees," *New York Times*, September 7, 1991.

22 Edward Said, "The Morning After," *London Review of Books*, October 21, 1993.

23 Ibid.

24 Ibid.

25 Edward Said, *The End of the Peace Process: Oslo and After*, New York: Vintage Books, 2000 [2001], 5.

26 Ibid.

27 Jack Khoury, "Gaza Cancer Patients: Israel's Refusal to Let Us in for Treatment Is a 'Death Sentence,'" *Haaretz*, January 7, 2017.

28 Said, *End of the Peace Process*, 18.

29 Ibid., 19.

30 Said, "The Morning After."

31 See Adam Whitnall, "Vladimir Putin Says Russia Will Fight for the Right of Palestinians to Their Own State," *Independent*, March 29, 2015.

32 Zvi Barel, "Kushner Reportedly Told Abbas: Stopping Settlement Construction Impossible, It Would Topple Netanyahu," *Haaretz*, August 26, 2017.

33 Barbara Plett-Usher, "Jerusalem Embassy: Why Trump's Move Was Not about Peace," BBC News, May 15, 2018, bbc.com.

10. Israel's Military–Industrial Complex

1 Stewart Reiser, *The Israeli Arms Industry: Foreign Policy, Arms Transfers, and Military Doctrine of a Small State*, New York: Holmes & Meier, 1989, 1–4.

2 Ibid., 7.

3 Bishara Bahbah with Linda Butler, *Israel and Latin America: The Military Connection*, London: Macmillan/Institute for Palestine Studies, 1986, 27.

4 Reiser, *The Israeli Arms Industry*, 9.

5 Ibid., 10, 12–13; "The New Black," *Israeli Air Force Journal* 223 (June 1, 2015) (Hebrew).

6 Ibid., 27–8.

7 Ibid., 29–30.

8 Robert Jackson, *Israeli Air Force Story*, London: Tandem, 1970, 60–2.
9 Reiser, *The Israeli Arms Industry*, 30–1.
10 See David Lennon, "Sharon Takes Control of Israeli Arms Sales for Political Benefits," *Financial Times*, December 18, 1981. Though the figure includes much of the army, it is still staggering. Quoted in Steve Goldfield, *Garrison State*, London: Palestine Focus Publications and Eaford, 1985, 6.
11 Reiser, *The Israeli Arms Industry*, 41–4.
12 Goldfield, *Garrison State*, 11.
13 Ibid., 15.
14 See Reiser, *The Israeli Arms Industry*, 44–5, on the delay of supplying the Phantom F-4 jets to Israel, when President Johnson was advised against selling this advanced fighter jet to Israel. In the end, the IDF did get the Phantoms, after a short delay.
15 Samuel Segev, *The Iranian Triangle: The Untold Story of Israel's Role in the Iran-Contra Affair*, New York: The Free Press and McMillan, 1988, 154–5.
16 Ibid.
17 Benjamin Beit-Hallahmi, *The Israeli Connection: Whom Israel Arms and Why*, London: I B Tauris, 1987, xii.
18 Jeff Halper, *War against the People: Israel, the Palestinians and Global Pacification*, London: Pluto Press, 2015, 38 (emphasis in original).
19 Ibid.
20 Patrick Wolfe, *Traces of History: Elementary Structures of Race*, London: Verso, 2016, 227–38.
21 Reiser, *The Israeli Arms Industry*, 44.
22 Ibid., 50–1.
23 Shir Hever, *The Military–Industrial–Complex of Israel*, youtube.com.
24 See SIPRI Arms Industry Database, 2018, www.sipri.org.
25 See SIPRI, "Military Expenditure by Country as Percentage of Gross Domestic Product," sipri.org.
26 Ibid.
27 See Avner Cohen, *The Worst-Kept Secret: Israel's Bargain with the Bomb*, New York: Columbia University Press, 2010, 84; see also report by Sebastian Robin, "Israel Might Have as Many as 300 Nuclear Weapons. And Some Are in the 'Ocean,' " *National Interest*, July 28, 2018.
28 See Michael Bachner, "Netanyahu and the Submarine Scandal: Everything You Need to Know," *Times of Israel*, March 21, 2019, timesofisrael.com, accessed September 23, 2019.
29 J. M. Sharpe, "US Foreign Aid to Israel," Congressional Research Service, 7-5700, RL33222, April 11, 2014, Washington, www.crs.gov.
30 See "List of Countries by Military Expenditure per Capita," Wikipedia.
31 Beit-Hallahmi, *The Israeli Connection*, 238.
32 Ibid., 61–2.
33 Ibid., 53.
34 Ibid., 41.
35 Ibid., 108–74.

36 Ibid., 131–5.

37 Ibid., 111, 133, 135, and Sasha Polakow-Suranky, *The Unspoken Alliance: Israel's Secret Relationship with Apartheid South Africa*, New York: Vintage Books, 2011, 43.

38 See Charles L. Sulzberger, "Strange Nonalliance," *New York Times*, April 30, 1971, 39.

39 Beit-Hallahmi, *The Israeli Connection*, 161.

40 Ibid.

41 Giorgio Gomel, "Netanyahu's Embrace of Ethno-Nationalists Endangers Jews in Europe," *Haaretz*, February 19, 2019.

42 See Beit-Hallahmi, *The Israeli Connection*, 98–101.

43 Ibid., 76–107.

44 Ibid., 231.

45 Ibid., 232.

46 See Segev, *The Iranian Triangle*, 283–315.

47 See (unattributed) "Son of Israel's Ex-counterterrorism Chief Claims Bush to Blame for 1988 Death," *Haaretz*, June 1, 2014.

48 See "India Largest Purchaser of Israel Arms in 2017," *Middle East Monitor*, May 4, 2018.

49 See Reuters, "Congressional Report: Israel Arms Sales to China Concern US," *Haaretz*, June 15, 2004.

50 See Beit-Hallahmi, *The Israeli Connection*, 175–207.

51 Ibid., 181.

52 Stephen M. Meyer, ed., *The Dynamics of Nuclear Proliferation*, Washington, DC: Lexington Books, 1984, 55; Aaron S. Klieman, *Israel's Global Reach: Arms Sales as Diplomacy*, Washington, DC: Lexington Books, Pergamon-Brassey's International Defense Publishers, 1985, 149; Beit-Hallahmi, *The Israeli Connection*, 208.

53 Lizzy Collingham, *The Hungry Empire: How Britain's Quest for Food Shaped the Modern World*, London: Vintage, 2017, xviii–xix.

54 "Israel to Supply Missile Defence Systems to India for $777M," *India Times*, October 24, 2018.

55 See Stockholm International Peace Research Institute, *2012 Yearbook*, Oxford: Oxford University Press, Table 6.3, 270.

56 See, for example Gilli Cohen, "India Buys $525m Worth of Missiles from Israel, Rejecting Rival U.S. Offer," *Haaretz*, October 25, 2014; and "Israel to Supply […], *India Times*.

57 See (unattributed) "India Largest Purchaser of Israel Arms in 2017," *Middle East Monitor*, May 4, 2018.

11. The IDF Today

1 Uri Ben-Eliezer, *Old Conflict, New War: Israel's Politics toward the Palestinians*, London: Palgrave Macmillan, 2012, 54.

2 Ibid., 54

3 Ibid., 54–5.
4 Ibid., 55.
5 See editorial, "The IDF Is an Army, Not a Synagogue," *Haaretz*, June 4, 2018, haaretz.com, accessed on September 22, 2019.
6 A new development is the building of an IDF camp on the campus of the Hebrew University in Jerusalem, to house the new students from Military Intelligence who will be taught armed and in uniform. That camp will be off-reach for all other staff and students, securitized by a fence, electronics, and biometrics; see Yaniv Kubovich, "Armed Forces Guarding Students: Israeli Army to Open Base in a University," *Haaretz*, March 27, 2019.
7 The Report is a telegram quoted in full on "IDF Commanders Speak out in Press Interviews" and placed on the Public Library of US Diplomacy, wikileaks.org/plusd/cables/08TELAVIV2329_a.html, is dated October 15, 2008, and marked SECRET.
8 See Oren Ziv, "Palestinian Company Says Ex-IDF Chief Stole Footage of Gaza Destruction for Campaign Video," +972, January 27, 2019, 972mag.com.
9 Alison Kaplan Sommer, "Israel's Blue and White Boys Club Wants Your Vote: The Political Landscape of 2019 Is Looking Very Military, Very Male and Really Puts the 'General' in 'General Election,'" *Haaretz*, February 22, 2019.
10 Jonathan Lis, "Fewer Women, More Generals: What Israel's Next Knesset Is Expected to Look Like," *Haaretz*, February 25, 2019.
11 Nir Gontarz, "The Enigma of Benny Gantz: His Own Sister Doesn't Know His Political Views," *Haaretz*, February 25, 2019.
12 Yitzhak Herzog as quoted in JPost.com, Staff, "Amid News of Iran Deal, Opposition Leader Herzog to Fly to US," *Jerusalem Post*, July 14, 2015.
13 This is an interesting case of censorship by Israel's only broadsheet, *Haaretz*. In the Hebrew article by Amos Harel, he writes profusely about the Hannibal Directive and bases his article on the IDF recordings from the battlefield, recently released by the IDF spokesperson, which included the famous Hannibal Directive being enacted as an order by the senior commander. This article was not translated into English, and an English article on the same date by the same correspondent makes no mention of the incident whatsoever. The Hebrew version is available on http://www.haaretz.co.il/news/politics/1.2678461 and includes a verbatim transcript of the recording, as well as some video recordings of this and another similar incident, both recorded by the helmet camera of the senior commander. The English version is on http://www.haaretz.com/news/diplomacy-defense/.premium-1.664908, and is a more analytical article about the position of Hamas, a year later. *Haaretz* uses this procedure often, either by avoiding the translation altogether or by replacing it with another article, as in this case.
14 Moshe Sharett, quoted in Tom Segev, *1949: The First Israelis*, London and New York: The Free Press, 1986, 28n.

15 See Samuel Osborne, "Israeli Soldier Filmed Killing Wounded Palestinian Attacker Says He 'Has No Regrets,' " *Independent*, August 29, 2018.

12. More Divided than Ever

1 David B. Green, "1917: A Great Zionist Mind Dies Young," *Haaretz*, December 17, 2012.
2 Zeev Sternhell, *The Founding Myths of Israel: Nationalism, Socialism, and the Making of the Jewish State*, Princeton, NJ: Princeton University Press, 1998, 13–9.
3 Ibid., 7.
4 Ibid., 8.
5 Ibid., 18.
6 Ibid., 6.
7 James Finn, *Byeways in Palestine*, Boston: Adamant Media, 2002 (reprint of the London 1868 original).
8 Sternhell, *The Founding Myths of Israel*, 19.
9 Gershon Shafir and Yoav Peled, *Being Israeli: The Dynamics of Multiple Citizenship*, Cambridge: Cambridge University Press, 2002, 74–5.
10 Yitzhak Laor, *The Myths of Liberal Zionism*, London: Verso, 2009, xxii–xxxiii.
11 Shafir and Peled, *Being Israeli*, 75.
12 Ibid., 76–7.
13 Sternhell, *The Founding Myths of Israel*, 264–70.
14 Ibid., 270–6.
15 Shafir and Peled, *Being Israeli*.
16 Sami Shalom Chetrit, *ha-Ma'avaḳ ha-mizraḥi be-Yiśrael: ben dikui le-shiḥrur, ben hizdahut le-alṭernaṭivah, 1948–2003*, Tel Aviv: Sifriat Ofakim, 2004, 81–96 (Hebrew). I chose this version because the English translation of the text is inaccurate.
17 Ibid., 64.
18 Ariana Melamed, "Ma'abarot: The Praxis of Bureaucratic Evil," *Haaretz*, February 21, 2019 (Hebrew, my translation). *Haaretz* sometimes refrains from translating what are considered "damaging" articles into its English online edition.
19 Sternhell, *The Founding Myths of Israel*, 277.
20 Chetrit, *ha-Ma'avaḳ ha-mizraḥi be-Yiśrael*, 93.
21 Ibid., 84–5.
22 Ben-Gurion, MAPAI Faction in the Knesset, April 11, 1951, *Mapai Archive*. Quoted in Uri Ben-Eliezer, *Old Conflict, New War: Israel's Politics toward the Palestinians*, London: Palgrave Macmillan, 2012, 195.
23 Sternhell, *The Founding Myths of Israel*, 8, 23.
24 The detailed report, "Population Statistics for Israeli Settlements in the West Bank, is available at en.wikipedia.org/wiki/Population_statistics_for_Israeli_settlements_in_the_West_Bank. It is based on Central Bureau

of Statistics, *Localities in Israel, 2015* (Hebrew). Since then, the pages were removed by the Israeli government and are only available in print. This is part of the new policy of removing sources reported elsewhere in this volume.

25 Amjad Iraqi, "Netanyahu Is Right: Israel Is a Nation with No Interest in Equality," *Guardian*, March 13, 2019.

26 World Inequality Database, "Top 10% National Income Share," wid.world, accessed July 4, 2019.

27 Sternhell, *The Founding Myths of Israel*, 57.

28 See Ofra Yeshua-Lyth, *Politically Incorrect: Why a Jewish State Is a Bad Idea*, Bloxham: Skyscraper Publications, 2016, 75. Yesuah-Lyth is referring to a number of researchers, and mainly to Menachem Friedman of Bar Ilan University.

29 Ibid., 75.

30 See *Israel's Religiously Divided Society*, Pew Research Center, March 8, 2016, pewforum.org.

31 Ibid.

32 See editorial, "Brutality in Kalansua: Applying the Law 'Blindly' Means Erasing Entire Neighbourhoods and Leaving Half a Million People Without a Roof over Their Heads," *Haaretz*, January 13, 2017.

13. Is Israel a Democracy?

1 "Declaration of Establishment of State of Israel," Ministry of Foreign Affairs, mfa.gov. The first paragraph appears in all capital letters in the original.

2 Zeev Sternhell, Th*e Founding Myths of Israel*, Princeton, NJ: Princeton University Press, 1998, 291.

3 See "Palestinian population grew 8 fold since 1948," Ma'an News Agency, May 12, 2011, maannews.com, accessed September 21, 2019; Palestinian population up 9 times since 1948," report in *Xinhua*, May 13, 2019.

4 Kenneth W. Stein, "The Intifada and the 1936–1939 Uprising: A comparison of the Palestine Arab Communities," The Carter Center, March 1990.

5 Sabri Jiryis, *The Arabs in Israel*, Beirut: The Institute for Palestine Studies, 1969, 3.

6 Shlomo Sand, *The Invention of the Jewish People*, London: Verso, 2009; and Shlomo Sand, *The Invention of the Land of Israel: From Holy Land to Homeland*, London: Verso, 2012.

7 See the German definition of *Judentum*: "Unter Judentum (von griechisch ἰουδαϊσμός ioudaismos, hebräisch יהדות Jahadut) versteht man einerseits die Religion, die Traditionen und Lebensweise, die Philosophie und meist auch die Kulturen der Juden (Judaismus) und andererseits die Gesamtheit der Juden. Letztere wird auch Judenheit genannt." See German Wikipedia, de.wikipedia.org.

8 Benedict Anderson, *Imagined Communities: Reflections on the Origin and Spread of Nationalism*, London: Verso, 1983 (1991, revised and

extended). Eric Hobsbawm and T. Ranger, eds., *The Invention of Tradition*, New York: Cambridge University Press, 1983; Sand, *The Invention of the Jewish People*; Ernst Renan, "Qu'est-ce qu'une nation?" Conférence faite en Sorbonne, le 11 Mars 1882, Paris: Calmann Lévy, 1882.

9 Sand, *The Invention of the Jewish People*, 21.

10 Ibid, 19–22.

11 See Roland Nikles,"The Adalah Database of 50 Discriminatory Laws in Israel," *Mondoweiss*, June 14, 2015, mondoweiss.net.

12 Zeev Schiff, *History of the Israeli Army: 1874–1974*, London: Macmillan, 1985 (1974), 84.

13 Ibid, 84–5.

14 Ibid., 85.

15 Ibid., 79.

16 Moti Basok, "The Army State: IDF Controls 80% of Israeli Land," *Marker*, March 30, 2011 (Hebrew), themarker.com/realestate/1.611893.

17 Ibid.

18 See the full report text in Hebrew: mevaker.gov.il/sites/DigitalLibrary/Pages/Reports/779-2.aspx.

19 See "Second Class: Discrimination against Palestinian Children in Israel's Schools," September 30, 2001, special report by Human Rights Watch, hrw.org. See also Adalah, "The Inequality Report: The Palestinian Arab Minority in Israel," March 2011, adalah.org.

20 Adalah, "A Snapshot of the Arab Education System in Israel." In the report, the figures in the excerpt are taken from the following Israel source: Central Bureau of Statistics (CBS), New Survey–Investment in Education 2000/1, Press Release 3, August 2004 (Hebrew).

21 Harriet Sherwood, "Academic Claims Israeli School Textbooks Contain Bias," *Guardian*, August 7, 2011, theguardian.com, accessed January 5, 2019. Also see Nurith Peled-Elhanan, *Palestine in Israeli Schoolbooks: Ideology and Propaganda in Education*, I B Tauris, London, 2011.

22 Oren Yiftachel, *Ethnocracy: Land and Identity Politics in Israel/Palestine*, Philadelphia University Press Philadelphia, 2006.

23 Sami Smooha, "Ethnic Democracy: Israel as an Archetype," *Israel Studies* 2: 2, 1997, 198–241.

24 Kenneth P. Vickery, "'Herrenvolk' Democracy and Egalitarianism in South Africa and the U.S. South," *Comparative Studies in Society and History* 16: 13, June 1974, 309–28.

25 See "Herrenvolk Democracy" on Wikipedia, en.wikipedia.org/wiki/.

26 See "Whitewashing Apartheid" [editorial], *Haaretz*, November 18, 2016, and also Chaim Levinson and Sharon Pulwer, "Israeli Officials Mull Granting Settlers 'Protected Population' Status in the West Bank," *Haaretz*, November 17, 2016.

27 Mordechai Kremnitzer, "Jewish Nation-State Law Makes Discrimination in Israel Constitutional," *Haaretz*, July 20, 2018.

28 Ibid.

29 See editorial, "Don't Soften It, Bury It," *Haaretz,* July 16, 2018.

Afterword

1 See Z. Sternhell, *The Founding Myth of Israel: Nationalism, Socialism, and the Making of the Jewish State*, Princeton, NJ: Princeton University Press, 1998, 3–13.
2 See H. Bresheeth, "Zionism and Liberalism: Complementary or Contradictory," *Open Democracy*, October 31, 2014, opendemocracy.net.
3 Fredric Jameson, *Archaeologies of the Future: The Desire Called Utopia and Other Science Fictions*, London: Verso, 2005; "The Politics of Utopia," *New Left Review*, January/February 2004.
4 Jonathan Lis, "'Israel Is the Nation-State of Jews Alone': Netanyahu Responds to TV Star Who Said Arabs Are Equal Citizens," *Haaretz*, March 11, 2019.
5 Alison Kaplan Sommer, "Israel Election Results: Fewer Women and LGBT People—but Lots of Ex-Generals—in New Knesset," *Haaretz*, September 24, 2019, haaretz.com, accessed September 29, 2019.
6 Sternhell, *The Founding Myth of Israel*, 5.
7 Yossi Verter, "Sharon Will Come Down from the Tree—And Will Take Second Place," *Haaretz*, September 29, 2000, quoted in Idit Zertal and Ehud Ya'ari, *Lords of the Land: The War over Israel's Settlements in the Occupied Territories, 1967–2007*, New York: Nation Books, 2007, 405.
8 Zertal and Ya'ari, *Lords of the Land*, 409–11.
9 Ibid., 413.
10 Shlomo Sand, *The Invention of the Jewish People*, London: Verso, 2009, 71–5.
11 Ibid., 73.
12 Ernst Renan, "What Is a Nation," in Homi Bhabha, ed., *Nation and Narration*, London: Routledge, 1990, 11.
13 Ibid., 20.
14 Ibid., 9.
15 Noa Shpigel, "Gantz Vows 'Never to Withdraw from the Golan Heights,'" *Haaretz*, March 4, 2019.
16 Sigmund Freud, "Moses and Monotheism," in *The Origins of Religion*, London: Penguin, Penguin Freud Library No. 13, 258.
17 See Ali Abunimah, "'Concentrate' and 'Exterminate': Israel Parliament Deputy Speaker's Gaza Genocide Plan," *The Electronic Intifada*, August 3, 2014, electronicintifada.net/blogs.
18 See Eric Shaap, *Mijn lied mijn leed mijn hartstocht: Het leven van Jacob Israël de Haan (1881–1924)*, Amsterdam: Brave New Books, 2018.
19 Benoit Faucon, Gordon Lubold, and Summer Said, "Saudis Heighten Security Defenses after Attacks on Oil Industry," *Wall Street Journal*, October 17, 2019, wsj.com, accessed on October 18, 2019.
20 Imad K. Harb, "Saudi Arabia and Iran May Finally Be Ready for Rapprochement," *Al Jazeera*, October 16, 2019, aljazeera.com, accessed on October 18, 2019.

21 Amos Harel, "Iran and a Dash of Politics: Behind Israel's Scramble to Boost Missile Defenses," *Haaretz*, October 8, 2019, haaretz.com, accessed October 18, 2019.

22 Ibid.

23 Amos Harel, "Trump's About-Face in Syria Forces Israel to Rethink Its Middle East Strategy," *Haaretz*, October 17, 2019, haaretz.com, accessed on October 17, 2019.

24 Michael Moran, "Recognizing Israeli Settlements Marks the Final Collapse of Pax Americana," *Foreign Policy*, November 22, 2019.

Index